MATLAB®

MATLAB®

2nd Edition

by John Paul Mueller and Jim Sizemore

MATLAB® For Dummies®, 2nd Edition

Published by: **John Wiley & Sons, Inc.,** 111 River Street, Hoboken, NJ 07030-5774, www.wiley.com

For general information on our other products and services, please contact our Customer Care Department within the U.S. at 877-762-2974, outside the U.S. at 317-572-3993, or fax 317-572-4002. For technical support, please visit https://hub.wiley.com/community/support/dummies.

Wiley publishes in a variety of print and electronic formats and by print-on-demand. Some material included with standard print versions of this book may not be included in e-books or in print-on-demand. If this book refers to media such as a CD or DVD that is not included in the version you purchased, you may download this material at http://booksupport.wiley.com. For more information about Wiley products, visit www.wiley.com.

Library of Congress Control Number: 2021938618

ISBN: 978-1-119-79688-6

ISBN 978-1-119-79689-3 (ebk); ISBN 978-1-119-79690-9 (ebk)

Manufactured in the United States of America

SKY10027280_110921

Contents at a Glance

Table of Contents

Introduction

MATLAB is an amazing product that helps you perform math-related tasks of all sorts using the same techniques that you'd use if you were performing the task manually (using pencil and paper, slide rule, or abacus if necessary, but more commonly using a calculator). However, MATLAB makes it possible to perform these tasks at a speed that only a computer can provide. In addition, using MATLAB reduces errors, streamlines many tasks, and makes you more efficient.

More important, MATLAB makes sharing your efforts with others incredibly easy. You can use Live Scripts to create report-like output that management can understand, or to develop apps for coworkers to employ when performing their tasks.

MATLAB is also a big product with numerous tools and features that you might never have used in the past. For example, instead of simply working with numbers, you have the ability to plot them in a variety of ways that help you communicate the significance of your data to other people. To get the most from MATLAB, you really need a book like *MATLAB For Dummies*, 2nd Edition.

About This Book

The main purpose of *MATLAB For Dummies*, 2nd Edition is to reduce the learning curve that comes with using a product that offers as much as MATLAB does. When you first start MATLAB, you might become instantly overwhelmed by everything you see. This book helps you get past that stage and become productive quickly so that you can get back to performing amazing feats of math wizardry.

In addition, this book introduces you to techniques that you might not know about or even consider because you haven't been exposed to them before. For example, MATLAB provides a rich plotting environment that helps you not only communicate better but also present numeric information in a manner that helps others see your perspective.

Using scripts and functions will also reduce the amount of work you have to do. This book shows you how to create custom code, which you can use to customize the environment to meet your specific needs. This edition introduces you to Live

Scripts and Live Functions, which enable you to combine code and output into a single report-like version that everyone can use, even if they don't necessarily understand the math. Using classes helps you package your code to make it easier to reuse and understand. If you want to create a form of your code that is accessible to coworkers and people who may not want to know *why* something works, just that it does, you can also discover apps and toolboxes.

After you've successfully installed MATLAB on whatever computer platform you're using, you start with the basics and work your way up. By the time you finish working through the examples in this book, you'll be able to perform a range of simple tasks in MATLAB that includes writing scripts, writing functions, creating plots, and performing advanced equation solving. No, you won't be an expert, but you will be able to use MATLAB to meet specific needs in the job environment.

To make absorbing the concepts even easier, this book uses the following conventions:

- Text that you're meant to type just as it appears in the book is **bold**. The exception is when you're working through a step list: Because each step is bold, the text to type is not bold.
- When you see words in *italics* as part of a typing sequence, you need to replace that value with something that works for you. For example, if you see "Type ***Your Name*** and press Enter," you need to replace *Your Name* with your actual name.
- Web addresses and programming code appear in `monofont`. If you're reading a digital version of this book on a device connected to the Internet, note that you can click the web address to visit that website, like this: `https://www.dummies.com`.
- When you need to type command sequences, you see them separated by a special arrow, like this: File ➪ New File. In this case, you go to the File menu first and then select the New File entry on that menu. The result is that you see a new file created.

Foolish Assumptions

You might find it difficult to believe that we've assumed anything about you — after all, we haven't even met you yet! Although most assumptions are indeed foolish, we made these assumptions to provide a starting point for the book.

Being familiar with the operating system platform you want to use is important because the book doesn't provide any guidance in this regard. (Chapter 2 does provide MATLAB installation instructions.) You really do need to know how to install applications, use applications, and generally work with your chosen platform before you begin working with this book.

This book isn't a math primer. Yes, you see lots of examples of complex math, but the emphasis is on helping you use MATLAB to perform math tasks rather than learn math theory. Chapter 1 helps you understand precisely what you need to know from a math perspective in order to use this book successfully.

This book also assumes that you can access items on the Internet. Sprinkled throughout are numerous references to online material that aren't mandatory but can enhance your learning experience. However, these added sources are useful only if you actually find and use them.

Icons Used in This Book

As you read this book, you see icons in the margins that indicate material of interest (or not, as the case may be). This section briefly describes each icon in this book.

TIP

Tips are nice because they help you save time or perform some task without a lot of extra work. The tips in this book are time-saving techniques or pointers to resources that you should try in order to get the maximum benefit from MATLAB.

WARNING

We don't want to sound like angry parents or some kind of maniacs, but you should avoid doing anything that's marked with a Warning icon. Otherwise, you might find that your application fails to work as expected, you get incorrect answers from seemingly bulletproof equations, or (in the worst-case scenario) you lose data.

TECHNICAL STUFF

Whenever you see this icon, think advanced tip or technique. You might find these tidbits of useful information just too boring for words, or they could contain the solution you need to get a program running. Skip these bits of information whenever you like.

REMEMBER

If you don't get anything else out of a particular chapter or section, remember the material marked by this icon. This text usually contains an essential process or a bit of information that you must know to work with MATLAB successfully.

Beyond the Book

This book isn't the end of your MATLAB experience — it's really just the beginning. We provide online content to make this book more flexible and better able to meet your needs. That way, as we receive email from you, we can address questions and tell you how updates to either MATLAB or its associated add-ons affect book content. In fact, you gain access to all these cool additions:

- **Cheat sheet:** You remember using crib notes in school to make a better mark on a test, don't you? You do? Well, a cheat sheet is sort of like that. It provides you with some special notes about tasks that you can do with MATLAB that not every other person knows. You can find the Cheat Sheet for this book at `www.dummies.com` and entering **MATLAB For Dummies, 2nd Edition** in the Search field. Click Cheat Sheets in the row of options under the book title.
- **Errata:** You can find errata by going to `www.dummies.com/go/matlabfd2e`. Scroll under the image of the book cover to find the Errata link, if there is one. In addition to errata, check out the blog posts with answers to reader questions and demonstrations of useful book-related techniques at `http://blog.johnmuellerbooks.com/`.
- **Companion files:** Hey! Who really wants to type all the code in the book and reconstruct all those plots by hand? Most readers would prefer to spend their time actually working with MATLAB and seeing the interesting things it can do, rather than typing. Fortunately for you, the examples used in the book are available for download. You can find these files by going to `www.dummies.com/go/matlabfd2e`. Scroll under the image of the book cover to find the Downloads link. Alternatively, you can obtain the source code at John Mueller Books Writing with Style (`http://www.johnmuellerbooks.com/source-code/`). Just locate the book's name and click the Download button.

Where to Go from Here

It's time to start your MATLAB adventure! If you're completely new to MATLAB, you should start with Chapter 1 and progress through the book at a pace that allows you to absorb as much of the material as possible.

If you're a novice who's in an absolute rush to get going with MATLAB as quickly as possible, you could skip to Chapter 2 with the understanding that you may find some topics a bit confusing later. Skipping to Chapter 3 is possible if you already have MATLAB installed, but be sure to at least skim Chapter 2 so that you know what assumptions we made writing this book.

Readers who have some exposure to MATLAB can save reading time by moving directly to Chapter 5. You can always go back to earlier chapters as necessary when you have questions. However, it's important that you understand how each technique works before moving to the next one. Every technique, coding example, and procedure has important lessons for you, and you could miss vital content if you start skipping too much information.

1

Getting Started With MATLAB

IN THIS PART . . .

Considering what MATLAB can do for you

Getting MATLAB installed and ready for use

Working with the MATLAB interface

Interacting with the MATLAB files

IN THIS CHAPTER

» **Understanding how MATLAB fits in as a tool for performing math tasks**

» **Seeing where MATLAB is used today**

» **Discovering how to get the most from MATLAB**

» **Overcoming the MATLAB learning curve**

Chapter 1

Introducing MATLAB and Its Many Uses

Math is the basis of all our science and even some of our art. In fact, math itself can be an art form — consider the beauty of fractals (a visual presentation of a specialized equation). However, math is also abstract and can be quite difficult and complex to understand and to use for practical purposes. MATLAB makes performing math-related tasks easier. You use MATLAB to perform math-related tasks such as

- Numerical computation
- Visualization
- Scripting
- Application development
- Machine learning
- Deep learning
- Signal processing
- Other tasks allowed by its various toolboxes (see `https://www.mathworks.com/help/thingspeak/matlab-toolbox-access.html` for details)

This chapter introduces you to MATLAB, an application that performs a variety of math tasks. It helps you understand the role that MATLAB can play in reducing the overall complexity of math and in explaining math-related information to others more easily. You also discover that many organizations and major developers use MATLAB to perform real-world tasks in a manner that improves accuracy, efficiency, and consistency. (A partial list of such tasks appears at `https://stackshare.io/matlab`.) Of course, knowing how you can translate these benefits of MATLAB to your own workplace is important.

REMEMBER

Because MATLAB can do so much, it does have a learning curve. This chapter also discusses what you can do to reduce the learning curve so that you become productive much faster. The less time you spend learning about MATLAB, the more time you spend applying math to your particular specialty, and the better the results you achieve. Getting things done quickly and accurately is the overall goal of MATLAB.

Putting MATLAB in Its Place

MATLAB is all about math. Yes, it's a powerful tool and yes, it includes its own language to make the execution of math-related tasks faster, easier, and more consistent. However, when you get right down to it, the focus of MATLAB is the math. For example, you could type $2 + 2$ as an equation, and MATLAB would dutifully report the sum of 4 as output. Of course, no one would buy an application to compute $2 + 2$ — you could easily do that with a calculator. So you need to understand just what MATLAB can do. The following sections help you put MATLAB into perspective so that you can better understand how you can use it to perform complex math tasks.

Understanding how MATLAB relates to a Turing machine

Today's computers are mostly Turing machines, named after the British mathematician Alan Turing (1912–1954). The main emphasis of a Turing machine is performing tasks step by step. A single processor performs one step at a time. It may work on multiple tasks, but only a single step of a specific task is performed at any given time. Knowing about the Turing machine orientation of computers is important because MATLAB follows precisely the same strategy. It, too, performs tasks one step at a time in a procedural fashion. In fact, you can download an application that simulates a Turing machine using MATLAB at `https://www.mathworks.com/matlabcentral/fileexchange/23006-turing-machine-emulator`. The code is surprisingly short. (Note that the actual design of a computer relies on principles defined by John von Neumann; see https://cacm.acm.org/magazines/2020/1/241712-von-neumann-thought-turings-universal-machine-was-simple-and-neat/fulltext for details.)

UNDERSTANDING HOW COMPUTERS WORK

Many older programmers are geeks who punched cards before TVs had transistors. One advantage of punching cards is getting to physically touch and feel the computer's instructions and data. This physicality gave programmers a good understanding of what happens when a program runs.

Today, the instructions and data are stored as charges of electrons in tiny pieces of silicon too small to be seen through even the most powerful optical microscope. Today's computers can handle much more information much more quickly than early machines. But the way they use that information is basically the same as early computers.

In those old card decks, programmers wrote one instruction on each card. After completing all the instructions, they put the data cards into a card reader. The computer read a card, and the computer did what the card told it to do: Get some data, get more data, add it together, divide, and so on until all the instructions were executed.

A series of instructions is a program. The following figure shows a basic schematic block diagram of how a computer works.

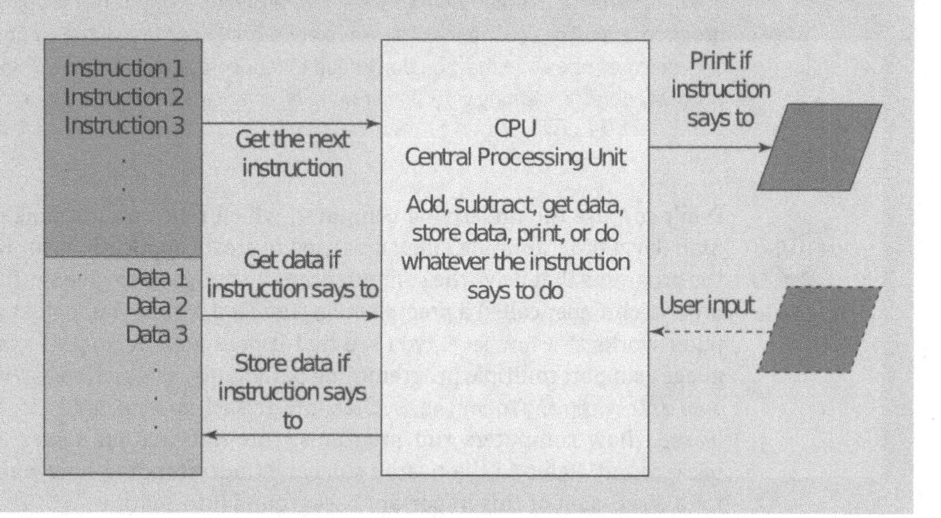

(continued)

(continued)

Unchanged from the old days, when cards were read one at a time, computer instructions continue to be read one at a time at a low level (think machine code). The instruction is executed, and then the computer goes to the next instruction. MATLAB executes programs in this manner as well.

Don't confuse the one-instruction-after-another approach with parallel programming. A parallel program relies on individual processors working in tandem to make application execution faster, but the instructions still execute one at a time on each processor. You can employ parallel programming in MATLAB using the Parallel Computing Toolbox, described at `https://www.mathworks.com/products/parallel-computing.html`.

It's important to realize that the *flow* of a program can change. Computers can make decisions based on a specific criterion (based on one comparison even when an expression contains multiple comparisons), such as whether something is true or false, and take the route indicated for that decision. For example, when the computer has read all the data for a task, the program tells the computer to quit reading data and start doing calculations. One way to map how the computer executes programs is called a *flow chart,* which is similar to a road map with intersections where decisions must be made. MATLAB relies on well-designed flow charts to make it easy to see what the computer will do, when it will do it, and how it will accomplish the required tasks.

The whole concept of a program may seem foreign to many — something that only geeks would ever love — but you've already used the concept of a program before. When using a calculator, you first think of the steps and numbers you want to enter and in what sequence to enter them to solve your problem. A program, including a MATLAB program, is simply a sequence of similar steps stored in a file that the computer reads and executes one at a time. You don't need to fear computer programming — you've probably done something very similar quite often and can easily do it again.

REMEMBER

Don't confuse the underlying computer, which relies on machine code, with the high-level programming languages used to create applications for it. Even though the programs that drive the computer may be designed to give the illusion of some other technique, called a *programming paradigm,* when you look at how the computer works at a low level, you see that it goes step by step. (In fact, some languages support multiple programming paradigms, as described at `https://blog.newrelic.com/engineering/python-programming-styles/`.) If you've never learned how computers run programs, this information serves as meaningful background. Refer to the nearby sidebar "Understanding how computers work" for a discussion of this important background information.

Using MATLAB as more than a calculator

MATLAB is a computer programming language, not merely a calculator. However, you can use it like a calculator, and doing so is a good technique to try ideas that you might use in your program. When you get past the experimentation stage, though, you usually rely on MATLAB to create a program that helps you perform tasks

- » Consistently
- » Easily
- » Quickly

With these three characteristics in mind, the following sections explore the idea of MATLAB's being more than a simple calculator in greater detail. These sections don't tell you everything MATLAB can do, but they do provide you with ideas that you can pursue and use to your own advantage.

Exploring Science, Technology, Engineering, and Mathematics (STEM)

Schools currently have a strong emphasis on Science, Technology, Engineering, and Math (STEM) topics because the world doesn't have enough people who understand these disciplines to get the required work done. Innovation of any sort requires these disciplines, as do many practical trades. MATLAB provides strong support for educational institutions supporting STEM, as described at `https://www.mathworks.com/academia/educators.html`.

Just in case you're wondering about the art you can create using MATLAB, consider the use of fractals in venues like cartoons and movies (`https://www.sciencenewsforstudents.org/article/math-movies-doctor-strange-otherworldly`). Fortunately, you don't even have to write your own fractal code, because other people have created fractal generators for MATLAB (see `https://www.mathworks.com/matlabcentral/fileexchange/78179-the-generator-of-fractal-surfaces-or-images` as an example).

Performing simple tasks

The focus of programming languages today is to make things as simple as possible without loss of functionality. This is the reason that articles like `https://opensource.com/education/15/9/python-in-the-classroom` promote languages like Python, which offer a variety of coding paradigms, yet also provide significant coding power. However, the problem with using a programming language is that you turn into a software developer, impeding your ability to get your research done. When working with MATLAB, you focus on the math, and the interactive environment makes math-specific tasks incredibly easy. In addition, should you need to make your research available to others in a form that allows experimentation, you can design and package your efforts as an application (as described in Chapters 14 and 15) without spending a huge amount of time doing so.

Determining why you need MATLAB

It's important to know *how* to use any application you adopt, but it's equally important to know *when* to use it and what it can actually *do* for your organization. If you don't have a strong reason to use an application, the purchase will eventually sit on the shelf collecting dust. This bit of dust collecting happens far too often in corporations around the world today because people don't have a clear idea of why they even need a particular application. Given that MATLAB can perform so many tasks, you don't want it to just sit on the shelf. The following sections can help you build a case for buying and then using MATLAB in your organization.

Using a product that focuses on math optimization

When you buy a programming language, you get tools that help you create great applications. These tools focus on writing better code, which is nice if you're a developer. However, if you're a researcher, math professional, or someone who is definitely not interested in becoming a developer, you need a tool that helps you write better algorithms, which is where MATLAB excels. As your algorithms increase in complexity, you can use the MATLAB Profiler, as described at `https://www.versionbay.com/articles/2020/matlab-profiler-improving-matlab-code-fibonacci-example/` and in the "Using the MATLAB Profiler to Improve Performance" section of Chapter 8 to improve algorithm performance. The MATLAB Profiler helps you locate precisely where the algorithm bottleneck is so that you can improve your math skills.

Performing what-if analysis quickly

Various kinds of research science require fast answers to what-if analysis. In order to test hypothesis quickly, you need an application that supports quick mock-ups of complex scenarios without a lot of development time. Here are some ways professionals use what-if analysis:

- Medical research professionals use MATLAB to perform visual forms of what-if analysis by relying on the Image Processing Toolbox (`https://www.mathworks.com/products/image.html`)

» You also see MATLAB used for financial analysis with the Risk Modeling and Risk Management Toolbox (`https://www.mathworks.com/help/risk/modeling-risk.html`)

» Electronics designers perform circuit analysis (`https://www.mathworks.com/academia/books/solving-dc-and-ac-circuits-by-example-using-matlab-haskell.html`)

The point is that none of these professionals have time to be a developer if they want to solve problems quickly, which is why they need MATLAB. You can find other examples of professionals who use MATLAB to speed things up in applications like embedded systems, control systems, wireless communication, computer vision, Internet of Things (IoT), testing and measurement, robotics, data analytics, predictive maintenance, power and motor control, and deep learning. Part 5 of the book looks at some key application uses of MATLAB.

Avoiding the complexity of Object-Oriented Programming (OOP)

You may have heard of Object-Oriented Programming (OOP). It's a discipline that helps developers create applications based on real-world models. Every element of an application becomes an object that has specific characteristics and can perform specific tasks. This technology is quite useful to developers because it helps them create extremely complex applications with fewer errors and less coding time.

However, OOP isn't something you need to know in order to work through various types of math problems. Even though you can solve difficult math problems using languages that do support OOP, STEM users can exploit most of MATLAB's power without OOP. The lack of an OOP requirement means that you can get up and running with MATLAB far faster than you could with a conventional modern programming language and without a loss of the functionality that you need to perform math tasks. If you really do want to use OOP with MATLAB, you can find it explained in Chapter 13.

Using the powerful toolbox

MATLAB provides the Symbolic Math toolbox designed to meet the specific needs of STEM users. In contrast to a general programming language, this toolbox provides specific functionality needed to meet certain STEM objectives. Here is just a small sample of the areas that are addressed by the tools you find in the MATLAB Symbolic Math toolbox:

» Calculus

» Equation solving, simplification, and substitution

- Linear algebra
- Visualization
- Variable-precision arithmetic
- Units and dimensional analysis
- Documentation and sharing
- Code generation

Chapters 19 and 20 tell you more about the Symbolic Math toolbox and describe how to perform a variety of tasks with it. You can also see this toolbox described at `https://www.mathworks.com/products/symbolic.html`.

Reducing programming effort with the fourth-generation language

Programming languages are often rated by their generation. For example, a first-generation language works side by side with the hardware. It's the sort of language that programmers used when computers first appeared on the scene. Nothing is wrong with working directly with the hardware, but you need specialized knowledge to do it, and writing such code is time consuming. A first-generation language is so hard to use that even the developers decided to create something better — second-generation languages! (Second-generation languages, such as Macro Assembler [MASM] are somewhat human-readable, must be assembled into executable code before use, and are still specific to a particular processor.)

TECHNICAL STUFF

Most developers today use a combination of third-generation languages such as C, C++, and Java, and fourth-generation languages such as Structured Query Language (SQL) and Python. A few even use fifth-generation languages, such as OPS5 and Mercury. A third-generation language gives the developer the kind of precise control needed to write exceptionally fast applications that can perform a wide array of tasks. Fourth-generation languages make asking for information easier. For the MATLAB user, the promise of fourth-generation languages means being able to work with collections of data, rather than individual bits and bytes, making it easier for you to focus on the task instead of the language. Fifth-generation languages rely on constraints issued to the application and then rely on the application to write the correct code. These languages are used so seldom today that none of them appear on the Tiobe index (see `https://www.tiobe.com/tiobe-index/`), which is a sort of language hall of fame.

REMEMBER

MATLAB employs a fourth-generation language to make your job a lot easier. The language isn't quite human, but it's also a long way away from the machine code that developers formerly wrote to make computers work. Using MATLAB makes you more efficient because the language is specifically designed to meet the needs of STEM users (just as SQL is designed to meet the needs of database administrators and developers who need to access large databases).

Discovering Who Uses MATLAB for Real-World Tasks

An application isn't very useful if you can't perform real-world tasks with it. Many applications are curiosities — they may do something interesting, but they aren't practical. MATLAB is popular among STEM users whose main goal is to productively solve problems in their particular field — not problems unique to computer programming. You can find MATLAB used by the following professionals:

- Scientists
- Engineers
- Mathematicians
- Students
- Teachers
- Professors
- Statisticians
- Control technologists
- Image-processing researchers
- Simulation users

Of course, most people want to hear about actual users who employ the product to do something useful. You can find such a list at `https://www.mathworks.com/company/user_stories/product.html`. Just click the MATLAB entry to see a list of companies that use MATLAB to perform real-world tasks. For example, this list tells you that Bigfoot Biomedical uses MATLAB for diabetes management (see `https://www.mathworks.com/company/user_stories/accelerating-development-of-a-diabetes-management-system-with-model-based-design-qa-with-bigfoot-biomedical.html`). You also find that 3T (an engineering firm) used MATLAB to design robot emergency braking, which you can read about at `https://www.mathworks.com/company/user_stories/3t-develops-robot-emergency-braking-system-with-model-based-design.html`.

Knowing How to Get the Most from MATLAB

At this point, you may have decided that you absolutely can't get by without obtaining your own personal copy of MATLAB. If that's the case, you really do need to know a little more about it in order to get the most value for your money. The following sections provide a brief overview of the skills that are helpful when working with MATLAB. You don't need these skills to perform every task, but they all come in handy for reducing the overall learning curve and making your MATLAB usage experience nicer.

Getting the basic computer skills

Most complex applications require that you have basic computer skills, such as knowing how to use your mouse, work with menu systems, understand what a dialog box is all about, and perform some basic configuration tasks. MATLAB works like other computer programs you own. It has the intuitive and conventional Graphical User Interface (GUI), shown in Figure 1-1, that makes using MATLAB a lot easier than employing pad and pen. If you've learned to use a GUI operating system such as Windows or macOS, and you also know how to use an application such as Word or Excel, you'll be fine.

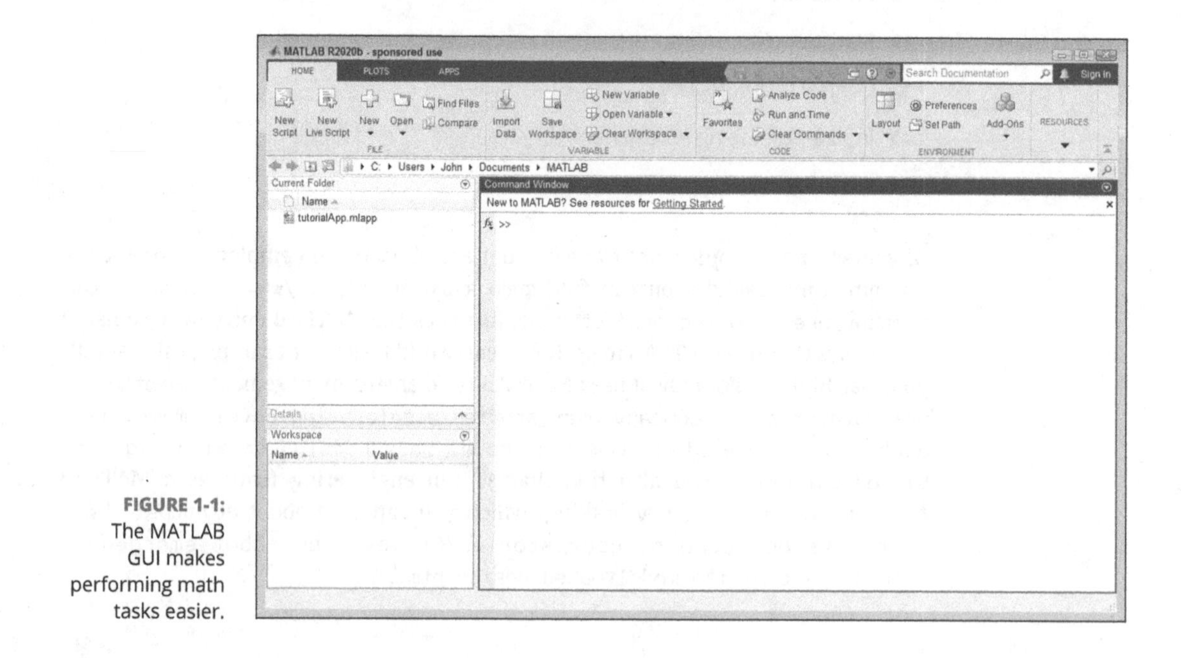

FIGURE 1-1: The MATLAB GUI makes performing math tasks easier.

TIP

This book points out MATLAB peculiarities. In addition, you have access to procedures that you can use to make your tasks easier to perform. The combination of these materials will help you work with MATLAB even if your computer skills aren't as finely honed as they could be. The important thing to remember is that you can't break anything when working with MATLAB. In fact, I encourage trial and error because it's a great learning tool. If you find that an example doesn't quite work as anticipated, close MATLAB, reopen it, and start the example over again. MATLAB and your computer are both more forgiving than others may have led you to believe.

Defining the math requirements

You need to have the right level of knowledge to use MATLAB. Just as using SQL is nearly impossible without any knowledge of database management, using MATLAB is hard without the proper math knowledge. MATLAB's benefits become evident when applied to trigonometry, exponentials, logarithms, and higher math.

TIP

This book assumes that you have a certain level of math knowledge. The math behind the exercises isn't explained to any large degree unless the explanation helps you understand the MATLAB programming language better. However, many sites online cater to math knowledge. For example, you can find a host of tutorials at `https://www.analyzemath.com/` and explanations at Wolfram MathWorld (`https://mathworld.wolfram.com`) in a classroom-like format. These tutorials come complete with exercises that help you understand the math behind the MATLAB examples in this book.

Applying what you know about other procedural languages

One of the more significant problems in understanding how to use any language is the procedure. The point was driven home to one fellow at an early age when his teacher assigned his class the task of writing a procedure for making toast. Every student carefully developed a procedure for making toast, and on the day the papers were turned in, the teacher turned up with a loaf of bread and a toaster. She dutifully followed to the letter the instructions each child provided. All the children failed at the same point. Yes, they forgot to take the bread out of the wrapper. You can imagine what it was like trying to shove a single piece of bread into the toaster when the piece was still in the wrapper along with the rest of the bread.

Programming can be (at times) just like the experiment with the toast. The computer takes you at your word and follows to the letter the instructions you provide. The results may be not what you expected, but the computer always follows the same logical course. Having previous knowledge of a procedural language, such as C, Java, C++, or Python, will help you understand how to write MATLAB procedures as well. You have already developed the skill required to break instructions into small pieces, and you know what to do when a particular piece is missing. Yes, you can use this book without any prior programming experience, but the prior experience will help you get through the chapters faster and with fewer errors.

Understanding how this book will help you

This is a *For Dummies* book, so it takes you by the hand to explore MATLAB and make it as easy to understand as possible. The goal of this book is to help you use MATLAB to perform at least simple feats of mathematical magic. It won't make you a mathematician, and it won't help you become a developer — those are topics for other books. When you finish this book, you will know how to use MATLAB to explore STEM-related topics.

Getting Over the Learning Curve

Even easy programming languages have a learning curve. If nothing else, you need to discover the techniques that developers use to break tasks into small pieces, ensure that all the pieces are actually there, and then place the pieces in a logical order. Creating an orderly flow of steps that the computer can follow can be difficult, but this book leads you through the process a step at a time.

To help you understand MATLAB, this book compares how to accomplish a task in MATLAB with something you're used to using, such as a spreadsheet or calculator. You learn by doing. Try the examples in this book and invent some of your own. Try variations. Experiment. MATLAB's not too tough — you, too, can discover how to use MATLAB!

IN THIS CHAPTER

» **Obtaining and installing your copy of MATLAB**

» **Starting MATLAB and working with the interface**

Chapter 2
Starting Your Copy of MATLAB

Before you can use MATLAB to do anything productive, you need a copy of it installed on your system. Fortunately, you can obtain a free trial version that lasts 30 days from `https://www.mathworks.com/campaigns/products/trials/matlab.html`. If you're diligent, you can easily complete this book in that time and know for certain whether you want to continue using MATLAB as a productivity aid. The point is that you need a good installation, and this book helps you obtain that goal.

TIP

After you have MATLAB installed, it's important to introduce yourself to the interface. This chapter provides you with an overview of the interface, not a detailed look at every feature. However, overviews are really important because working with lower-level interface elements is hard if you don't have the big picture. You may actually want to mark this chapter in some way so that you can refer back to the interface information.

Installing MATLAB

A problem that anyone can encounter is getting a bad product installation or simply not having the right software installed. When you can't use your software properly, the entire application experience is less than it should be. The following sections guide you through the MATLAB installation so that you can have a great experience using it.

Discovering which platforms MATLAB supports

Before you go any further, you need to verify that your system will actually run MATLAB. At a minimum, you need 3.5GB of free hard drive space (5GB to 8GB for a typical installation) and 4GB of RAM (8GB recommended) to use MATLAB effectively. Your system should also have a newer processor with at least four cores. If you install the Parallel Computing Toolbox, you can also make use of GPU processing functionality for display adapters listed at `https://www.mathworks.com/help/parallel-computing/gpu-support-by-release.html`. (It can run on systems with fewer resources, but you won't be happy with the performance.) You also need to know which platforms MATLAB supports. You can use it on these systems:

- Windows
  - Windows 10 (version 1803 or higher)
  - Windows 7 Service Pack 1
  - Windows Server 2019
  - Windows Server 2016
- Mac OS X
  - macOS Big Sur (11)
  - macOS Catalina (10.15)
  - macOS Mojave (10.14)
- Linux
  - Ubuntu 20.04 LTS
  - Ubuntu 18.04 LTS
  - Ubuntu 16.04 LTS
  - Debian 10

- Debian 9
- Red Hat Enterprise Linux 8
- Red Hat Enterprise Linux 7 (minimum 7.5)
- SUSE Linux Enterprise Desktop 12 (minimum SP2)
- SUSE Linux Enterprise Desktop 15
- SUSE Linux Enterprise Server 12 (minimum SP2)
- SUSE Linux Enterprise Server 15

Getting your copy of MATLAB

Before you can work with MATLAB, you need a copy installed on your system. Fortunately, you have a number of methods at your disposal. Here are the three most common ways of getting MATLAB:

» Get the trial version from `https://www.mathworks.com/campaigns/products/trials/matlab.html`.

» Obtain a student version of the product from `https://www.mathworks.com/academia/student_version/`.

» Buy a copy from `http://www.mathworks.com/pricing-licensing/index.html`.

You need to download the copy of MATLAB or the MATLAB installer onto your system after you fill out the required information to get it. In most cases, you install the download on the same system, but there are also methods for downloading and then installing on another system.

Performing the installation

The method you use to install MATLAB depends on the version you obtain and the technique you need to install it. For example, there is one method for installing MATLAB on a separate machine using the file installation key, and there's an entirely different method when you want to download the installer and use an Internet connection. Administrators and users also have different installation procedures, especially when working with multiple machine installations. Use the table at `https://www.mathworks.com/help/install/install-products.html` to determine which installation procedure to use.

TIP

MathWorks provides you with substantial help in performing the installation. Before you contact anyone, be sure to look through the materials on the main installation page at https://www.mathworks.com/help/install/index.html. Take the time to review the material that MathWorks provides before you push the panic button. Doing so will save time and effort.

Activating the product

After you complete the MATLAB installation, you must activate the product. Activation is a verification process. It simply means that MathWorks verifies that you have a valid copy of MATLAB on your system. With a valid copy, you obtain support such as updates to your copy of MATLAB as needed.

REMEMBER

MATLAB automatically asks you about activation after the installation process is complete. You don't need to do anything special. However, in some cases, you might need to perform the task in some other way. The MATLAB Answers page at https://www.mathworks.com/matlabcentral/answers/99457-how-do-i-activate-matlab-or-other-mathworks-products gives you additional details.

USING MATLAB ONLINE

Even though this chapter focuses on using your copy of MATLAB at the desktop, you can also use MATLAB Online if you have the correct license (see https://www.mathworks.com/products/matlab-online.html#license-types for details). MATLAB Online works much like MATLAB desktop, except that you see the interface through a browser, which means that you could potentially use MATLAB on your tablet. To use MATLAB Online, make sure you meet the requirements specified at https://www.mathworks.com/support/requirements/browser-requirements.html. MATLAB Online also comes with some limitations, as described at https://www.mathworks.com/products/matlab-online/limitations.html. In all other respects, if you can perform a task using your desktop copy of MATLAB, you can also perform it using MATLAB Online. Given that the interfaces are essentially the same, this book doesn't mention MATLAB Online again unless there is some special functionality or issue you need to know about. Whether you should use MATLAB Online or not depends on a great many factors, such as your network bandwidth and the capabilities of your machine. To see whether using MATLAB Online will work for you, try running a few examples both locally and by using MATLAB Online to determine the performance and usage advantages of each.

Meeting the MATLAB Interface

Most applications have similar interface functionality. For example, if you click a button, you expect something to happen. The button usually contains text that tells you what will happen when you click it, such as closing a dialog box by clicking OK or Cancel. However, the similarities aren't usually enough to tell you everything you need to know about the interface. The following sections offer an overview of the MATLAB interface so that you can work through the chapters that follow with greater ease. These sections don't tell you everything about the interface, but you do get enough information to feel comfortable using MATLAB.

Starting MATLAB for the first time

When you start MATLAB for the first time (after you activate it), you see a display containing a series of blank windows. It's not all that interesting just yet because you haven't done anything with MATLAB. However, each of the windows has a special purpose, so it's important to know which window to use when you want to perform a task.

You can arrange the windows in any order needed. Figure 2-1 shows the window arrangement used throughout the book, which may not precisely match your display. The "Changing the MATLAB layout" section, later in this chapter, tells you how to rearrange the windows so that you can see them the way that works best for you. Here's a summary of the window functionality:

- **Home tab:** The Home tab of the *Ribbon interface* (a bar that provides access to various MATLAB features, such as creating a new script or importing data) is where you find most of the icons you use to create and use MATLAB formulas. It's the tab you use most often. Also on the interface is a Plots tab (for creating graphic presentations of your formulas) and an Apps tab (for designing your own apps, obtaining new applications to use with MATLAB, installing apps, and packaging apps for use by others). MATLAB calls the Ribbon interface the *Toolstrip,* so that's the name you see throughout the book.
- **Quick Access toolbar:** The Quick Access toolbar (QAT) provides access to the MATLAB features that you use most often. Finding icons on the QAT is often faster and easier than looking them up on the Toolstrip.
- **Command Window:** You type formulas and commands in this window. After you type the formula or command and press Enter, MATLAB determines what it should do with the information you typed. You see the Command Window used later in this chapter.

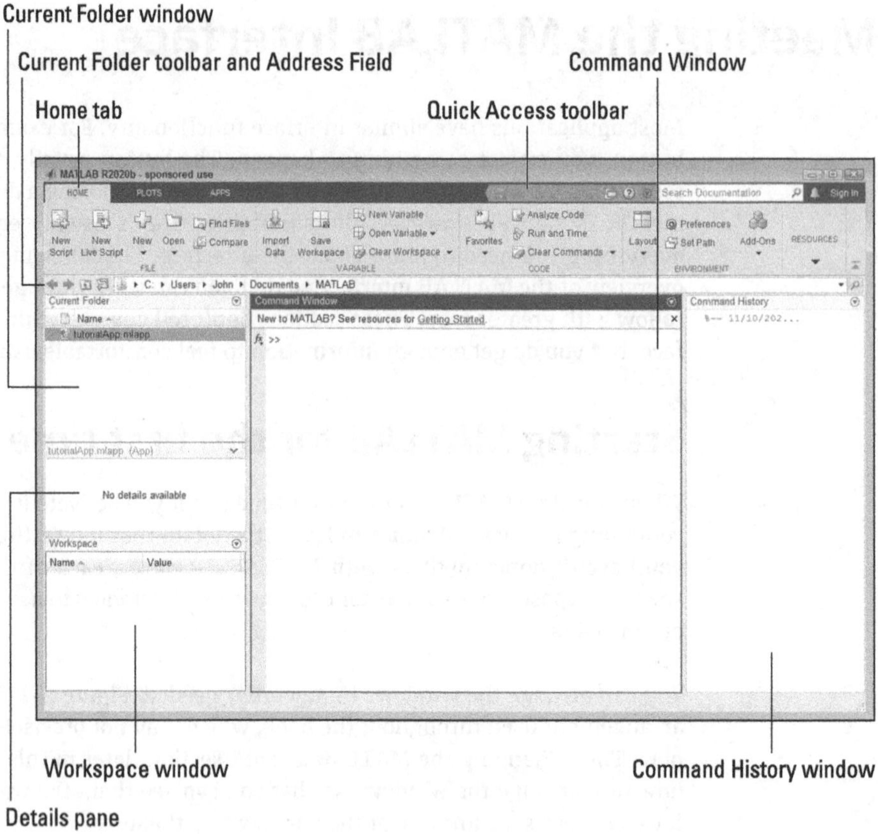

Current Folder window

Current Folder toolbar and Address Field

Command Window

Home tab

Quick Access toolbar

Workspace window

Command History window

Details pane

FIGURE 2-1: The initial view of MATLAB is pretty much empty space.

- **Workspace window:** The Workspace window contains the results of any tasks you ask MATLAB to perform. It provides a scratch pad of sorts that you use for calculations. The Workspace window and Command Window work hand in hand to provide you with a complete view of the work you perform using MATLAB.
- **Command History window:** In some cases, you want to reissue a formula or command. The Command History window acts as your memory and helps you restore formulas and commands that you used in the past. You see the Command History window used later in this chapter.
- **Current Folder window:** The Current Folder window contains a listing of the files you've created in the current folder — files you'd use to store any data you create in MATLAB, along with any scripts or functions you'd use to manipulate data.

- **Details pane:** The Details pane appears in the lower half of the Current Folder window and shows specifics about any file you select. Normally, this pane is hidden from view, and you display it by clicking the up-pointing arrow on the right side of the pane. To hide the pane again, click the down-pointing arrow on the right side.
- **Current Folder toolbar and Address Field:** The Current Folder toolbar helps you manage the current folder location on your storage device. The first four buttons let you move Back, Forward, Up One Level, or Browse for Folder. The Address Field text box appears next. Changing the Address Field text box content also changes the content of the Current Folder window. Clicking the down arrow in the Address Field shows a list of previous destinations from which to choose. The magnifying glass Search icon lets you perform searches for files based on filename, including wildcards (the ?), phrases, and extension.

Considering the default Toolstrip tabs

You always see the three same tabs as a starting point when working with MATLAB: Home, Plots, and Apps. Depending on what you're doing, you may see other task-specific tabs, such as Editor, Publish, and View. The following sections describe general Toolstrip behavior and introduce you to the default tabs. As you perform specific tasks in the book, you also discover the task-specific tabs.

Showing and hiding the Toolstrip

You can minimize the Toolstrip by clicking the up-pointing arrow with a bar on top on the right side. When the Toolstrip is minimized, you can still see the three default tabs: Home, Plots, and Apps.

Click a tab to reveal the Toolstrip long enough to use a MATLAB feature. As soon as you select a Toolstrip feature or click in another MATLAB area, the Toolstrip disappears again. Using this technique allows you full access to the MATLAB features but keeps the Toolstrip hidden to save space.

To maximize the Toolstrip again, click one of the tabs to display the Toolstrip, and then click the thumbtack that appears on the right side after you select one of the tabs. The Toolstrip will appear at full size again.

Home

The Home tab is where you spend most of your time when working with MATLAB to perform common tasks. It contains a number of *sections*, which divide the *controls* used to perform specific tasks. For example, to create a new script, you click the New Script button in the File section. You could also click the New drop-down list and choose Script from the list. Here's how you use the various sections on the Home tab:

- **File:** Contains options for creating new files of various types, opening existing files, locating files on your storage device, and comparing two files.

  TIP

  Notice that there isn't an option for saving a file on the Home tab. This option appears on the Editor tab when you create a new script or other code file that you need to save. MATLAB presents controls only when necessary to keep the interface simple.

- **Variable:** Provides methods for working with variables you create when doing one of the following: working with commands or various kinds of code; importing data from an external source; or clicking New Variable. The Save Workspace button lets you save variables to the storage device between sessions so that it's easier to resume your thoughts.
- **Code:** Allows management of code items you create: functions, scripts, Live Functions, Live Scripts, and so on. You can also perform timings on your code and clear both the Command Window and the Command History window. You use the Editor, Publish, and View tabs that appear automatically when you create a new code item to actually work with the code.
- **Environment:** Helps you control the MATLAB environment by
  - Changing the MATLAB layout
  - Setting preferences
  - Modifying the MATLAB search path used to look for files
  - Managing add-ons such as toolboxes and products
- **Resources:** Links you to standard help options and community support. You can also request support from MathWorks. The Learn MATLAB option opens your browser to a page with links for online courses.

Plots

The Plots tab, shown in Figure 2-2, helps you create visualizations of your data using various plot types. The default plot types appear in the drop-down list in the Plots A section. To use any of the plots, you must first create one or more variables containing data to use in the plot.

Near the bottom of the plot list that appears when you click the down arrow button (not shown in Figure 2-2), you see a Catalog button that shows all the plots applicable to the data you select in a Plot Catalog window when clicked. If you

can't use a particular plot because your data doesn't support it, the plot type appears grayed out in the Plot Catalog window. Don't worry for now about drowning in a sea of plots; Chapters 6 and 7 explain them in detail for you.

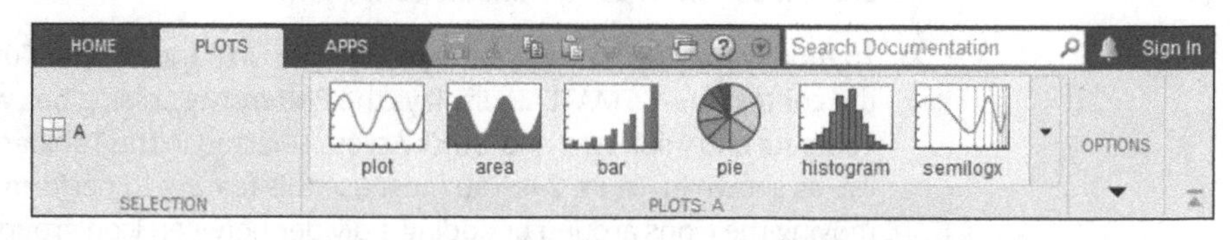

FIGURE 2-2: Use the Plots tab entries to create graphic presentations of your data.

Apps

The Apps tab, shown in Figure 2-3, helps you work with apps, which are simply packaged and portable copies of code that you can type in the Command Window or store in a script, function, Live Script, Live Function, or class. The point is that apps are self-contained and do something interest. Part 3 of the book tells you about the various methods of creating and storing code, including creating apps (Chapter 14) and working with projects (Chapter 15).

FIGURE 2-3: Use the Apps tab entries to develop portable versions of your code.

The Apps section of this tab contains a list of apps you have installed. The Raspberry Pi Resource Monitor app is visible by default. This app lets you monitor resources and processes running on a *Raspberry Pi* (`https://www.raspberrypi.org/`) — an experimental computer that plugs into a wealth of devices and lets you perform a variety of tasks, such as learn about robotics.

Working with the Quick Access toolbar (QAT)

The Quick Access toolbar (QAT) makes commands that you select more accessible because you don't have to select a tab to use them. You can change the QAT to meet your needs:

- To add an icon to the QAT, right-click its entry in the Ribbon/Toolstrip (such as, New Script on the Home tab) and choose Add to Quick Access Toolbar from the context menu.

» To remove an icon from the QAT, right-click its entry in the QAT and, from the context menu, choose Remove from the Quick Access Toolbar.

» To show a label for a particular icon on the QAT, right-click its entry in the QAT and choose Show Label from the context menu.

» To modify the QAT as a whole, right-click the QAT and choose Customize from the context menu. MATLAB displays the Preferences dialog box with the Toolbars entry selected and Quick Access selected in the Toolbar drop-down list, as shown in Figure 2-4. You must use this feature to perform tasks such as moving the icons around or adding a divider between icon groups.

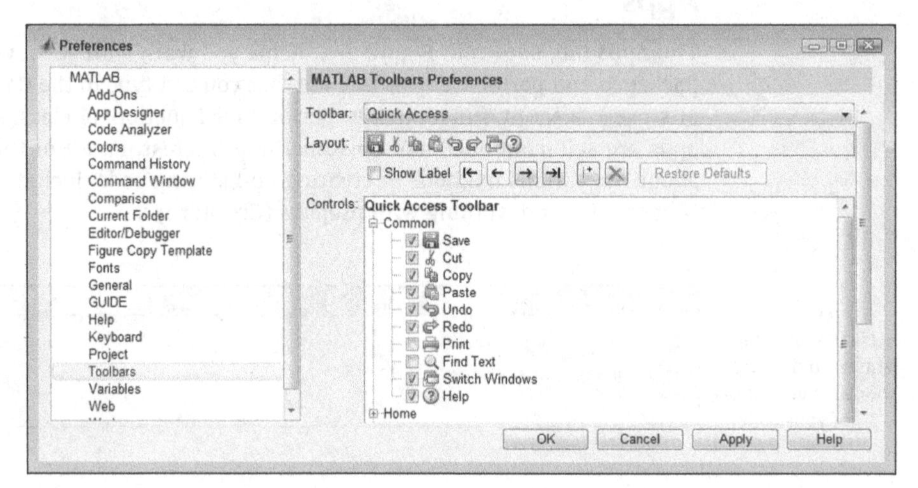

FIGURE 2-4: Use the Preferences dialog to gain full access to the QAT configuration.

Employing the Command Window

The Command Window is where you perform most of your experimentation. This chapter shows how to perform simple tasks using the Command Window, but as the book progresses, you see that the Command Window can do quite a lot for you. You can type any formula or command desired in the Command Window and see a result. Of course, it pays to start with something basic so that you can get a feel for how this window works. Type **2 + 2** and press Enter in the Command Window. You see the results shown in Figure 2-5.

REMEMBER

Typing a formula or command affects more than just the Command Window (which receives the output of the formula `2 + 2` with a result of `ans = 4`, in this example). MATLAB changes other windows as well. The Workspace window now contains a default variable named `ans` that contains the value `4`. *Variables* are boxes (pieces of memory) in which you can place data. The Command History window displays formulas and commands that you type in the Command Window, along with the date and time you typed them. (If you don't see the Command History window, type **command history** and press Enter in the Command Window.) To replay a formula or command in this window, just double-click its entry in the list.

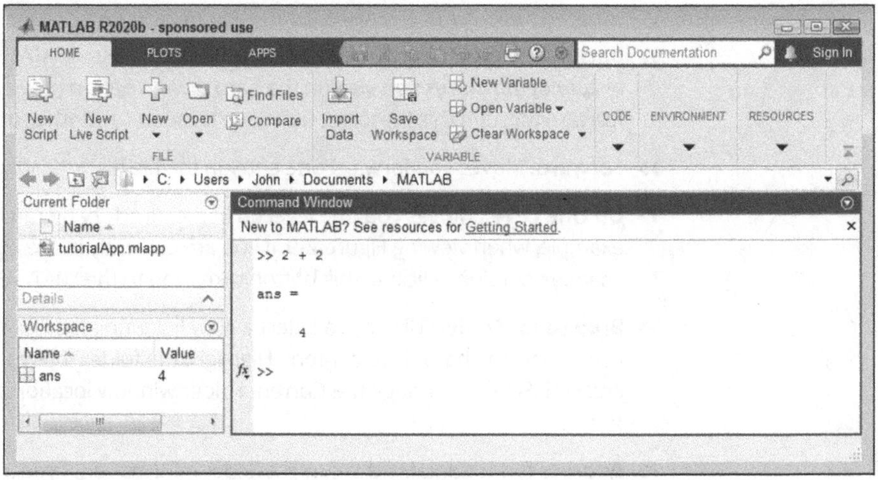

FIGURE 2-5: A very simple command in MATLAB.

Getting additional help with MATLAB

The Help drop-down list in the Resources section of the Home tab contains a number of useful ways to obtain help with your copy of MATLAB besides relying on community support, paid support, and self-paced courses:

- **Documentation:** Provides descriptive, textual help. You can access the general help page through the Help drop-down list. However, you can also highlight a command, function, or other programming construct and press F1 to receive specific help about that MATLAB feature.
- **Examples:** Displays a window containing links to example applications you can try. Each application demonstrates a particular MATLAB feature, such as basic matrix operations. By following the example, you learn how to use MATLAB features in a hands-on environment.
- **Support Web Site:** Navigates to the MATLAB support site online, where you can find a wealth of other help resources, such as videos, apps other people have created, answers to common questions, error reports, and so on. You can access this site directly at `https://www.mathworks.com/support.html`.

Using the Current Folder toolbar

The Current Folder toolbar helps you navigate the Current Folder window with greater precision. Here is a description of each of the toolbar elements when viewed from left to right on the toolbar:

- **Back:** Moves you back one entry in the file history listing. MATLAB retains a history of the places you visit on the hard drive. You can move backward and forward through this list to get from one location to another quickly.
- **Forward:** Moves you forward one entry in the file history listing.
- **Up One Level:** Moves you one level up in the directory hierarchy. For example, when viewing Figure 2-9, if you are currently in the `\MATLAB\Chapter02` folder, clicking this button takes you to the `\MATLAB folder`.
- **Browse for Folder:** Displays a Select a New Folder dialog box that you can use to view the hard drive content. Highlight the folder that you want to use and click Select to change the Current Folder window location to the selected folder.
- **Address field:** Contains the current folder information. Type a new value and press Enter to change the folder.
- **Search:** The Search icon, which looks like a magnifying glass and appears to the right of the Address field, changes the Address field into a search field. In that field, type the search criteria that you want to use and press Return. MATLAB displays the results for you in the Current Folder window.

Viewing the Current Folder window

The Current Folder window (refer to Figure 2-1) really does show the current folder listed in the Address field. The current version of MATLAB provides a tutorial application named `tutorialApp.mlapp` that you can load and view, or run to see what it does. To run the tutorial, type **tutorialApp** and press Enter in the Command Window, or right-click the entry in the Current Folder window and choose Run. The point is that this file is found in your default MATLAB folder.

When you first start MATLAB, the current folder always defaults to the MATLAB folder found in your user folder for the platform of your choice. For Windows users, that means the `C:\Users\<User Name>\Documents\MATLAB` folder (where `<User Name>` is your name). Burying your data way down deep in the operating system may seem like a good idea to the operating system vendor, but you can change the current folder location to something more convenient when desired. The following sections describe techniques for managing data and its storage location using MATLAB.

tutorialApp.MLAPP IS MISSING

You might find that your setup lacks `tutorialApp.mlapp`. Use these steps to create this file:

1. **Select the Apps tab.**

   You see a list of app-related buttons.

2. **Click Design App.**

   The App Designer Start Page appears.

3. **Click the Interactive Tutorial entry in the Examples: General section.**

   A wizard appears to take you through the creation process. Notice that the name of this file is tutorialApp.mlapp.

4. **Choose Save ⇨ Save Copy As in the File section of the Designer tab.**

   You see a Save Copy As dialog box.

5. **Choose a location to save the `tutorialApp.mlapp` file; then click Save.**

   You're ready to work with `tutorialApp.mlapp`.

6. **Close the App Designer window.**

   The MATLAB window appears again.

Temporarily changing the current folder

There are times when you need to change the current folder. Perhaps your data is actually stored on a network drive, or you want to use a shared location so that others can see your data, or you simply want to use a more convenient location on your local drive. The following steps help you change the current folder:

1. **Click Set Path in the Environment section on the Home tab.**

   You see the Set Path dialog box, shown in Figure 2-6.

   This dialog box lists all the places the MATLAB searches for data, with the default location listed first. You can use the Move to Top, Move Up, Move Down, Move to Bottom, and Remove buttons to work with existing folders (go to Step 3 when you're working with existing folders or to Step 2 to add a new folder):

   - To set an existing folder as the default folder, highlight the folder in the list and click Move to Top.
   - To stop using an existing folder, highlight the folder in the list and click Remove.

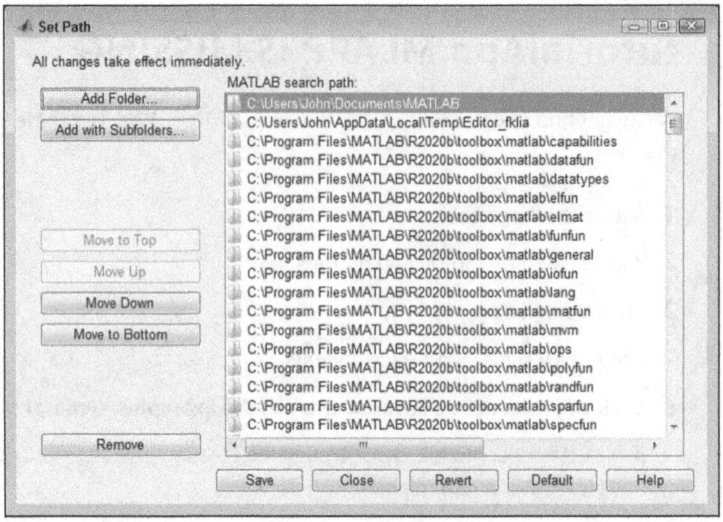

FIGURE 2-6: The Set Path dialog box contains a listing of folders that MATLAB searches for data.

2. **Click Add Folder.**

   You see the Add Folder to Path dialog box, which looks like a standard file open dialog box. This dialog box lets you choose an existing folder that doesn't appear in the current list or add a new folder to use:

   - To use a folder that exists on your hard drive, use the dialog box's tree structure to navigate to the folder, highlight its entry, and then click Select Folder.
   - To create a new folder, highlight the parent folder in the dialog box's tree structure, click New Folder, type the name of the folder, press Enter, and then click Select Folder.

3. **Click Save.**

   MATLAB makes the folder you select the new default folder. (You may see a User Account Control dialog box when working with Windows; click Yes to allow Windows to perform the task.)

4. **Click Close.**

   The Set Path dialog box closes.

5. **Type the new location in the Address field.**

   The Current Folder display changes to show the new location.

Permanently changing the default folder

The default folder is the one that MATLAB uses when it starts. Setting a default folder saves you time because you don't have to remember to change the current folder setting every time you want to work. If you have your default folder set to the location from which you work most of the time, you can usually get right to work and not worry too much about locations on the hard drive.

If you want to permanently change the default folder so that you see the same folder every time you start MATLAB, you must use the `userpath()` command. Even though this might seem like a really advanced technique, it isn't hard. In fact, go ahead and set the `userpath` so that it points to the downloadable source for this book. Simply type **userpath('C:\MATLAB2')** (for *MATLAB For Dummies*, 2nd Edition) in the Command Window and press Enter. You need to change the path to wherever you placed the downloadable source.

To see what the default path is for yourself, type **userpath** and press Enter. MATLAB displays the current default folder.

Creating a new folder

Organizing the files that you create is important so that you can find them quickly when needed. To add a folder to the Current Folder window, right-click any clear area in the window and choose New➪Folder from the context menu. MATLAB creates the new folder for you. Type the name you want to use for the new folder and press Enter.

Each chapter in this book uses a separate folder to store any files you create. When you obtain the downloadable source from the publisher's site (`http://www.dummies.com/go/MATLABFD2e`), you find the files for this chapter in the `\MATLAB2\Chapter02` folder. Every other chapter will follow the same pattern.

Saving a formula or command as a script

After you create a formula or command that you want to use to perform a number of calculations, be sure to save it to disk. Of course, you can save anything that you want to disk, even the simple formula that you typed earlier in this chapter. The following steps help you save any formula or command that you want to disk so that you can review it later:

1. **Choose a location to save the formula or command in the Address field.**
2. **Right-click the formula or command that you want to save in the Command History window and choose Create Script from the context menu.**

   You see the Editor window, as shown in Figure 2-7. The script is currently untitled, so you see the script name as Untitled2*. (Figure 2-7 shows the Editor window undocked so that you can see it with greater ease; the "Changing the MATLAB layout" section, later in this chapter, tells how to undock windows so that you can get precisely the same look.)

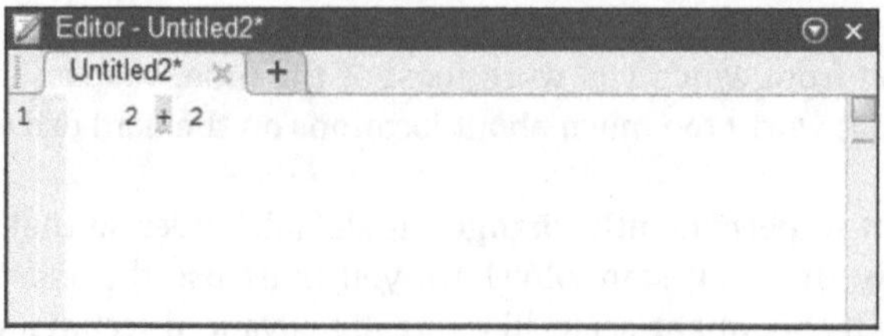

FIGURE 2-7: The Editor turns your formula or command into a script.

TIP

If you want to select multiple commands to place in a script, you can choose them by clicking the first command and then using Ctrl+click to select any additional commands. Each time you Ctrl+click on a command, MATLAB highlights its entry. The commands will appear in the script file in the same order in which they appear in the Command History window, rather than in the order in which you click them. As an alternative, you can click the starting command and then Shift+click the ending command if you want to use all the commands within the group.

Note that you can also create a Live Script from your MATLAB formulas and commands. However, in most cases, you use scripts for simple tasks, so you don't need to worry about Live Script for now. Chapter 11 discusses Live Script in detail and compares a Live Script to a regular MATLAB script.

3. **Click Save on the Editor tab.**

   You see the Select File for Save As dialog box, as shown in Figure 2-8.

4. **In the left pane, highlight the location you want to use to save the file.**

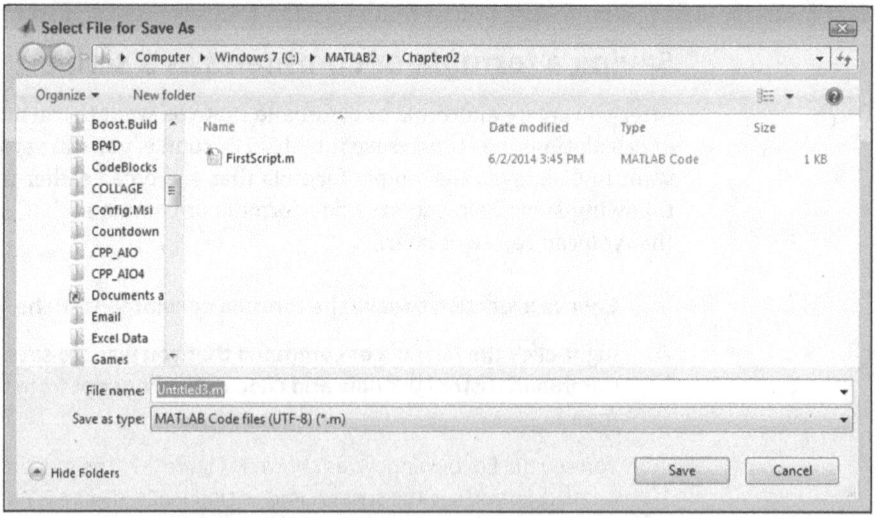

FIGURE 2-8: Choose a location to save your script and provide a filename for it.

5. **Type a name for the script in the File Name field.**

   The example uses `FirstScript.m`. However, when you save your own scripts, you should use a name that will help you remember the content of the file. Descriptive names are easy to remember and make precisely locating the script you want much easier later.

REMEMBER

   MATLAB filenames can start with only letters and numbers. You can't use spaces in a MATLAB filename. However, you can use the underscore in place of a space after the first letter.

6. **Click Save.**

   MATLAB saves the script for you so that you can reuse it later. The title bar changes to show the script name and its location on disk.

7. **Close the Editor window.**

   The Current Folder window displays the folder and script file that you've created using the previous steps in this chapter, as shown in Figure 2-9.

FIGURE 2-9: The Current Folder window always shows the results of any changes you make.

Running a saved script

You can run any script by right-clicking its entry in the Current Folder window and choosing Run from the context menu. When you run a script, you see the script name in the Command Window, the output in the Workspace window, and the actual command in the Command History window.

Saving the current workspace to disk

Sometimes you might want to save your workspace to protect work in progress. The work may not be ready to turn into a script, but you want to save it before quitting for the day or simply to ensure that any useful work isn't corrupted by errors you make later.

To save your workspace, click Save Workspace in the Variable section of the Home tab. You see a Save As dialog box that looks similar to the Select File for Save As dialog box (refer to Figure 2-8). Type a filename for your workspace, such as

FirstWorkspace.mat, and click Save to save it. You see the new file added to the Current Folder window, as shown in Figure 2-9.

REMEMBER

Workspaces use a `.mat` extension, while scripts have a `.m` extension. Make sure that you don't confuse the two extensions. In addition, workspaces and scripts use different icons so that you can easily tell them apart in the Current Folder window.

Changing the MATLAB layout

The MATLAB layout is designed to make experimentation easy and comfortable for you. However, you may find after a while that it really doesn't meet your needs. Fortunately, you can reconfigure the MATLAB layout to any configuration you want. The following sections provide ideas on how you can reconfigure the MATLAB layout.

Minimizing and maximizing windows

Sometimes you need to see more or less of a particular window. It's possible to simply resize the windows, but you may want to see more or less of the window than resizing provides. In this case, you can minimize the window to keep it open but completely hidden from view, or maximize the window to allow it to take up the entire client area of the application.

REMEMBER

On the right side of the title bar for each window, you see a down arrow. When you click this arrow, you see a menu of options for that window, such as the options shown in Figure 2-10 for the Current Folder window. To minimize a window, choose the Minimize option from this menu. Likewise, to maximize a window, choose the Maximize option from the menu.

New ▸
Reports ▸
Compare
Find Files Ctrl+Shift+F
Show ▸
Sort By ▸
Group By ▸
Minimize
Maximize Ctrl+Shift+M
Undock Ctrl+Shift+U
Close Ctrl+W

FIGURE 2-10: The window menus contain options for changing the appearance of the window.

Eventually, you want to change the window size back to its original form. The Minimize or Maximize option on the menu is replaced by a Restore option when you change the window's setup. Select this option to restore the window to its original size.

Opening and closing windows

In some cases, you may no longer need the information found in a particular window. When this happens, you can close the window. MATLAB doesn't actually destroy the window contents, but the window itself is no longer accessible. To close a window that you don't need, click the down arrow on the right side of the window and choose Close from the menu.

After you close a window, the down arrow is no longer accessible, so you can't restore a closed window by using the menu options shown in Figure 2-10. To reopen a window, you click the down arrow on the Layout button in the Environment section of the Home tab. You see a list of layout options like the ones shown in Figure 2-11.

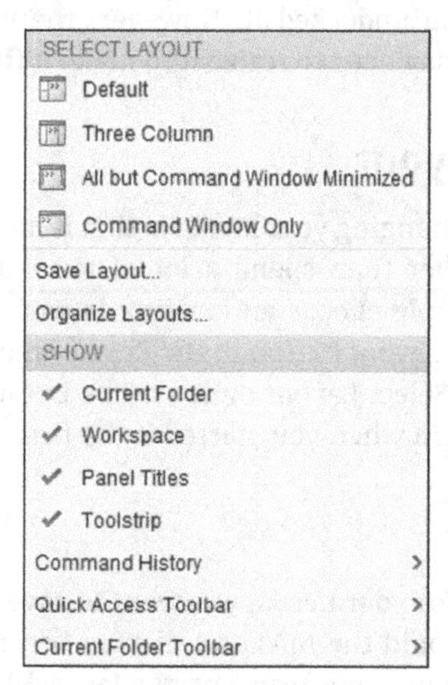

FIGURE 2-11: The Layout menu contains the layout options for MATLAB.

The Show group contains a listing of windows. Each window with a check mark next to it is opened for use (closed windows have no check mark). To open a window, click its entry. Clicking the entry places a check next to that window and opens it for you. The window is automatically sized to the size it was the last time you had it open.

You can also close windows using the options on the Layout menu. Simply click the check next to a window entry to close it.

Docking and undocking windows

Many people have multiple monitors attached to their systems. It's often more efficient to perform the main part of your work on your main monitor and move supplementary windows to a second monitor. However, you really can't move a window until you undock the window from MATLAB so that you can move just that window to another location.

To undock a window, click the down arrow on the right side of its title bar and choose Undock from the menu. The window becomes a separate entity, much like the Current Folder window shown previously in Figure 2-9. You can move the undocked window anywhere you want it, including to a second monitor.

At some point, you may decide that you want MATLAB to have all its windows in one place again. In this case, you click the down arrow on the right side of the window's title bar and choose Dock from the menu. MATLAB places the window precisely where it was before you undocked it. However, the window may not return to its original size — you may need to resize it to make it fit as it did before.

Choosing an existing layout

One of the potential problems of changing your layout is that it may cause MATLAB to become nearly unusable. Rather than spend a lot of time trying to get the original layout back, you can simply choose an existing layout. To perform this task, click the down arrow on the Layout button in the Environment section of the Home tab and choose one of the Select Layout options. The Default entry returns MATLAB to the same state it was in when you started it the first time.

Saving a new layout

After you find the perfect layout for your needs, you want to save it to disk so that you can easily restore it later should the MATLAB display become disorganized (perhaps you've moved things about to perform a particular task). To perform this task, click the down arrow on the Layout button in the Environment section of the Home tab and choose Save Layout. You see the Save Layout dialog box. Type the name of the layout in the space provided and click OK. The layout now becomes available in the Select Layout section of the Layout menu.

To delete a saved layout, click the down arrow on the Layout button in the Environment section of the Home tab and choose Organize Layouts. You see the Organize Layouts dialog box, where you can rename or delete layouts that you've saved, but not the layouts provided with MATLAB.

IN THIS CHAPTER

» Performing basic and more complex calculations

» Interacting with variables

» Using MATLAB functions

» Overcoming errors

Chapter 3
Interacting with MATLAB

You can interact with MATLAB in a lot of ways, and you'll experience quite a few of them as the book progresses. However, it pays to start out slowly to build your skills. This chapter presents an overview of the sorts of things you can do with MATLAB. Use this chapter to get started with a product that can really perform complex tasks with aplomb.

Although you probably won't spend a lot of time using MATLAB as a calculator for even complex calculations, you can do so. Rather than view this kind of use as a waste of time, however, view it as a means of practicing as well as experimentation. Sometimes playing with a product produces unexpected outcomes that can help you in your daily work. To that end, this chapter introduces you to MATLAB through the use of direct calculation entries.

Another type of interaction with MATLAB covered in this chapter occurs through variables. Think of a *variable* as a kind of storage box. You put data into a variable so that you can store that data for a while. Later, when you need the data again, you take it out of the variable, do something with it, and put it back in again. Variables have nothing mystical or difficult about them; in fact, you use variables all the time in real life. For example, you could view your refrigerator as a kind of variable. You put the bag of apples inside to store them for a short time, take the bag out to remove an apple to eat, and put the rest of the bag back into the refrigerator (minus one apple). The point is that developers make a big deal out of fancy terms (that you unfortunately also need to use in order to talk with them), but in reality there isn't anything odd about them. You get a fuller explanation of variables as part of this chapter.

In the process of interacting with MATLAB, you'll make mistakes. Of course, everyone makes mistakes. MATLAB won't blow up if you make a mistake, and your computer won't up and run away. Mistakes are part of the learning process, so you need to embrace them. In fact, most of the greatest people in history made a ton of mistakes (see `https://www.lifehack.org/articles/communication/10-famous-failures-that-will-inspire-you-success.html`). This book assumes that you're going to make mistakes, so part of this chapter discusses how to recover from them. Knowing how to recover means that you don't have to worry about making a mistake, because you can always start fresh.

Using MATLAB as a Calculator

MATLAB performs math tasks incredibly well. Sometimes people get so caught up in "what else" an application can do that they miss the most interesting facts that are staring them right in the face. The following sections help you understand MATLAB as a calculator so that you can use it for experimentation purposes.

Entering information at the prompt

References to using the prompt appear a few times in previous chapters, but those chapters don't fully explain it. The *prompt* is that place where you type formulas, commands, or functions or perform tasks using MATLAB. It appears in the Command Window. Normally, the prompt appears as two greater-than signs (>>). However, when working with some versions of MATLAB, you might see EDU>> (for the student version) or Trial>> (for the trial version) instead. No matter what you see as a prompt, you use it to know where to type the information described in this book.

TIP

Chapter 2 shows you how to use something called the `userpath()` function to alter the permanent path that MATLAB uses when starting up. In this chapter, you discover a useful command known as `clc`. Try it now: Type **clc** and press Enter at the MATLAB prompt. If the Command Window contains any information, MATLAB clears it for you.

The `userpath()` function is called a *function* because it uses parentheses to hold the data — also called *arguments* — that you send to MATLAB. The `clc` command is a *command* because you don't use parentheses with it. Whether something is a function or a command depends on how you use it. The usage is called the function or command *syntax* (the grammar used to tell MATLAB what tasks to perform). It's possible to use `userpath()` in either Function or Command form. To avoid confusion, the book usually relies on function syntax when you need to provide arguments, and command syntax when you don't. So, when you see parentheses, you should also expect to provide input with the *function call* (the act of typing the function and associated arguments, and then pressing Enter).

REMEMBER

MATLAB is also *case sensitive*. That sounds dangerous, but all it really means is that CLC is different from Clc, which is also different from clc. Type **CLC** and press Enter at the MATLAB prompt. You see an error message like the one shown in Figure 3-1. (MATLAB will also suggest the correct command, `clc`, but ignore the advice for right now by highlighting `clc` and pressing Delete, or by simply pressing Esc.) Next, type **Clc** and press Enter at the MATLAB prompt. This time, you see the same error because you made the "same" mistake — at least in the eyes of MATLAB. If you see this error message, don't become confused simply because MATLAB didn't provide a clear response to what you typed — just retype the command, being sure to type the command exactly as written.

```
Command Window
New to MATLAB? See resources for Getting Started.
>> CLC
Unrecognized function or variable 'CLC'.

Did you mean:
fx >> clc
```

FIGURE 3-1: MATLAB is case sensitive, so CLC, Clc, and clc all mean different things.

Notice also the "Did you mean:" text that appears after the error message. Normally, MATLAB tries to help you fix any errors. In some cases, MATLAB can't figure out what's wrong, so it won't provide any alternatives for you. In other cases, MATLAB provides an alternative, so you need to check the prompt to determine whether help is available. Because MATLAB was able to provide the correct command in this case, simply press Enter to clear the Command Window.

TIP

Look in the Command History window. Notice that a red line appears next to each of the errant commands you typed. These red lines tell you when you shouldn't use a command or function again because it produced an error the first time. You should also avoid adding errant commands and functions to any scripts you create.

Entering a formula

To enter a formula, you simply type it. For example, if you type **2 + 2** and press Enter, you get an answer of 4. Likewise, if you type **2 * pi * 6378.1** and press Enter, you get the circumference of the earth in km (see `https://nssdc.gsfc.nasa.gov/planetary/factsheet/earthfact.html` for a list of Earth statistics, including radius). The second formula uses a predefined constant, `pi`, which equals 3.1416. MATLAB actually defines a number of predefined constants that you can use when entering a formula:

- `ans`: Contains the most recent temporary answer. MATLAB creates this special temporary variable for you when you don't provide a variable of your own.
- `eps`: Specifies the accuracy of the floating-point precision (epsilon), which defaults to 2.2204e-16.
- `i`: Contains an imaginary number, which defaults to 0.0000 + 1.0000i.
- `Inf`: Defines a value of infinity, which is any number divided by 0, such as 1 / 0.
- `NaN`: Specifies that the numerical result isn't defined (Not a Number).
- `pi`: Contains the value of pi, which is 3.1416 when you view it onscreen. Internally, MATLAB stores the value to 15 decimal places so that you're assured of accuracy.

Whenever you type a formula and press Enter, you get an output that specifies the value of `ans`, which is a temporary value that holds the answer to your question. For example, try typing **2 * pi * 6378.1** and pressing Enter. You see the circumference of the Earth, 4.0075e+04 km.

Copying and pasting formulas

With MATLAB, you can copy and paste formulas that you create into other documents (such as a script or function file, or to another application). To begin, you highlight the information you want to copy. Use one of these methods to copy the text after you highlight it:

- Click Copy on the Quick Access Toolbar (QAT).
- Right-click the highlighted text and choose Copy from the context menu.
- Rely on a platform-specific method of copying the text, such as pressing Ctrl+C on Windows.

UNDERSTANDING INTEGER AND FLOATING-POINT VALUES

Throughout the book, you see the terms *integer* and *floating point.* These two terms describe kinds of numbers. When most people look at 3 and 3.0, they see the same number: the value three. The computer, however, sees two different numbers. The first is an integer — a number without a decimal portion. The second is a floating-point value — a number that has a decimal portion, even if it's a whole number.

You see these two terms often in this book because the computer works with and stores integer values differently from floating-point values. How the computer interacts differently with them is not important — you just need to know that it does. MATLAB does a great job of hiding the differences from view unless the difference becomes important for some reason, such as when you want to perform integer math — in which you want to work with only whole numbers. For example, 4 divided by 3 is equal to 1 with a remainder of 1 when performing integer math.

Humans also don't pay much attention to the size of a number. Again, the computer must do so because it has to allocate memory to hold the number — and larger numbers require more memory. So, not only do you need to consider the kind of number but also the size of the number when performing some tasks.

Finally, the computer must also consider whether a number has a sign associated with it. The sign takes up part of the memory used to store the number. If you don't need to store a sign, the computer can use that memory to store additional number information. With all these points in mind, here are the kinds of numbers that MATLAB understands:

- **double:** 64-bit floating-point double precision
- **single:** 32-bit floating-point double precision
- **int8:** 8-bit signed integer
- **int16:** 16-bit signed integer
- **int32:** 32-bit signed integer
- **int64:** 64-bit signed integer
- **uint8:** 8-bit unsigned integer
- **uint16:** 16-bit unsigned integer
- **uint32:** 32-bit unsigned integer
- **uint64:** 64-bit unsigned integer

(continued)

(continued)

Sometimes MATLAB won't know what you mean when you type **3**. A value of 3 could be any kind of number. (MATLAB defaults to assuming that all values are doubles unless you specify otherwise.) To specify the kind of number you mean, you enter the type name and place the value in parentheses. For example, double(3) is a 64-bit floating-point number, but int32(3) is a 32-bit signed integer form of the same number.

When you have the text on the Clipboard, you can paste it wherever you want. If you want to paste it somewhere in MATLAB, click wherever you want to put the text, such as after the prompt. Use one of these methods to paste the text:

- Click Paste on the QAT.
- Right-click the insertion point and choose Paste from the context menu.
- Rely on a platform-specific method of pasting text, such as pressing Ctrl+V in Windows.

Changing the Command Window formatting

The Command Window provides the means necessary to change the output formatting. For example, if you don't want the extra space between lines that MATLAB provides by default, you can type **format compact** and press Enter to get rid of it. In fact, try typing that command now. When you type **format compact** and press Enter, you don't see any output. However, the next formula you type shows the difference. Type **2 + 2** and press Enter. You see that the extra spaces between lines are gone, as shown in Figure 3-2.

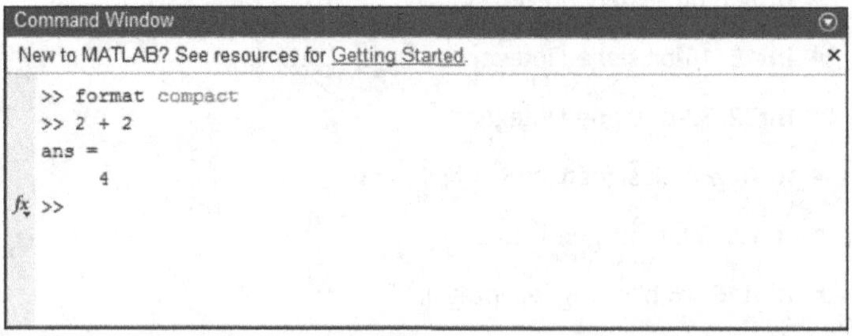

```
Command Window
New to MATLAB? See resources for Getting Started.
>> format compact
>> 2 + 2
ans =
     4
fx >>
```

FIGURE 3-2: Modify the appearance of the Command Window using format commands.

MATLAB provides a number of format commands. Each of them begins with the keyword `format`, followed by an additional instruction. The numeric format instructions, such as `long` and `shortg`, control presentation and the number of decimals when there is a decimal portion to display. (Whole numbers are displayed as whole numbers, rather than as a whole number with zeroes in the decimal portion.) When you type **format** by itself and press Enter, the formatting returns to the default setting. Here is a list of the instructions you can type:

- `short`: All floating-point output has at least one whole number, a decimal point, and four decimal values, such as 4.2000.
- `long`: All floating-point output has at least one whole number, a decimal point, and 15 decimal values, such as 4.200000000000000.
- `shorte` (default): All floating-point output uses exponential format with four decimal places, such as 4.2000e+00.
- `longe`: All floating-point output uses exponential format with 15 decimal places, such as 4.200000000000000e+00.
- `shortg`: All output uses a short general format, such as 4.2, with five digits of space.
- `long`: All output uses a long general format, such as 4.2, with 15 digits of space.
- `shorteng`: All floating-point output uses exponential format with four decimal places and powers in groups of three, such as 4.2000e+000.
- `longeng`: All floating-point output uses exponential format with 14 decimal places and powers in groups of three, such as 4.20000000000000e+000.
- `hex`: All output is in hexadecimal format, such as 4010cccccccccccd.
- `+`: All output is evaluated for positive or negative values, so that the result contains just a + or – sign, such as + when using the formula 2 * 2.1.
- `bank`: All output provides two decimal places, even for integer calculations, such as 4.20.
- `rat`: All output is presented as a ratio of small integers, such as 21/5 for 4.2.
- `compact`: All output appears in single-spaced format.
- `loose`: All output appears in double-spaced format.

Suppressing Command Window output

When performing most experiments, you want to see the result of your actions. However, sometimes you really don't want to keep seeing the results in the Command Window when you can just as easily look in the Workspace window for the result. In these cases, you can follow a command with a semicolon (`;`), and the Command Window output is suppressed. For example, try typing **2 + 2;** and pressing Enter (note the semicolon at the end of the command). You see output in the Workspace window, but not the Command Window.

This technique is often used when you have a complex set of formulas to type and you don't want to see the intermediate results, or when working with large matrices. Of course, you also want to use this approach when you create scripts so that the script user isn't bombarded by the results that will appear as the script runs. Anytime you stop using the semicolon at the end of the command, you start seeing the results again.

Understanding the MATLAB Math Syntax

The MATLAB syntax is a set of rules that you use to tell MATLAB what to do. It's akin to learning another human language, except that the MATLAB syntax is significantly simpler than any human language. In order to communicate with MATLAB, you must understand its language, which is essentially a form of math. Because you already know math rules, you already know many MATLAB rules as well. The following sections get you started with the basics that you use to build an understanding of the MATLAB language. You may be surprised to find that you already know some of these rules, and other rules are simply extensions of those rules.

Adding, subtracting, multiplying, and dividing

MATLAB is a math-based language, so it pays to review the basic rules for telling MATLAB how to perform basic math tasks. Of course, MATLAB performs the basic math functions:

- + or `plus()`: Adds two numbers. For example, you can use `3 + 4` or `plus(3, 4)` to obtain a result of `7`.
- - or `minus()`: Subtracts two numbers. For example, you can use `3 - 4` or `minus(3, 4)` to obtain a result of `-1`.
- * or `times()`: Multiplies two numbers. For example, you can use `3 * 4` or `times(3, 4)` to obtain a result of `12`.
- / or `rdivide()`: Performs right division, which is the form of division you likely learned in school. For example, you can use `3 / 4` or `rdivide(3, 4)` to obtain a result of `0.75`.

» \ or `ldivide()`: Performs left division, which is also called "goes into" or, as you learned in third grade, "guzinta." You know (say this out loud), 5 "guzinta" 5 once, 5 "guzinta" 10 twice, 5 "guzinta" 15 three times, and so on. For example, you can use `3 \ 4` or `ldivide(3, 4)` to obtain a result of `1.3333`.

Most MATLAB operators are binary, which means that they work on two values. For example, `3 + 4` has two values: `3` and `4`. However, some operators are unary, which means that they work on just one value. Here are the basic unary operators:

» + or `uplus()`: Returns the unmodified content of a value or variable. For example, +1 or `uplus(1)` is still equal to `1`.

» - or `uminus()`: Returns the negated content of a value or variable. For example, -1 or `uminus(1)` returns `-1`. However, -1 or `uminus(-1)` returns `1` (the negative of a negative is a positive).

In some cases, you don't want a floating-point result from division. To perform integer division, you have to use special functions — you can't just use operators for the simple reason that no operators are associated with these math tasks. Here are the functions associated with integer math:

» `idivide()`: Performs integer division. You supply two values or variables as input, along with an optional modifier that tells MATLAB how to perform rounding.

REMEMBER

To use the `idivide()` function, you must specify that at least one of the input values is an integer (see the "Understanding integer and floating-point values" sidebar, earlier in this chapter, for details). For example, `idivide(int32(5), int32(3))` provides an output of `1`. (Note that `idivide(int32(5), 3)` also works because the first argument is an `int32`.) Here is a list of the modifiers that you use to provide different rounding effects:

- `ceil`: Rounds toward positive infinity. For example, `idivide(int32(5), int32(3), 'ceil')` produces an output of 2, and `idivide(int32(5), int32(-3), 'ceil')` produces an output of -1.
- `fix`: Rounds toward zero. For example, `idivide(int32(5), int32(3), 'fix')` produces an output of 1, and `idivide(int32(5), int32(-3), 'fix')` produces an output of -1.
- `floor`: Rounds toward negative infinity. For example, `idivide(int32(5), int32(3), 'floor')` produces an output of 1, and `idivide(int32(5), int32(-3), 'floor')` produces a result of -2.
- `round`: Rounds to the nearest integer. For example, `idivide(int32(5), int32(3), 'round')` produces an output of 2, and `idivide(int32(5),`

int32(-3), 'round') produces an output of -2. When the result has decimal portion of .5, round will round the value up, rather than down, so idivide(int32(5), int32(2), 'round') produces an output of 3.

- mod(): Obtains the modulus after division. For example, mod(5, 3) produces an output of 2, and mod(5, -3) produces an output of -1.
- rem(): Obtains the remainder after division. For example, rem(5, 3) produces an output of 2, and rem(5, -3) produces an output of 2. Even though the second example has a negative number as the second argument, the remainder is still positive. However, rem(-5, 3) produces an output of -2 because the first argument is negative.

Rounding can be an important feature of an application because it determines the approximate values the user sees. You can round any formula that you want to produce an integer output. Here are the rounding functions:

- ceil(): Rounds toward positive infinity. For example, ceil(5 / 3) produces an output of 2, and ceil(5 / -3) produces an output of -1.
- fix(): Rounds toward zero. For example, fix(5 / 3) produces an output of 1, and fix(5 / -3) produces an output of -1.
- floor(): Rounds toward negative infinity. For example, floor(5 / 3) produces an output of 1, and floor(5 / -3) produces an output of -2.
- round(): Rounds toward nearest integer. For example, round(5 / 3) produces an output of 2, and round(5 / -3) produces an output of -2. A result with a decimal portion of precisely .5 will round up, rather than down, so round(2.5) outputs a value of 3.

Working with exponents

You use the caret (^) to raise a number to a particular power. MATLAB can handle negative, fractional, and complex number bases as exponents. Here are some examples of exponents:

- 10^3 = 1000
- 2^10 = 1024
- 2.5^2.5 = 9.8821
- 2^-4 = 0.0625
- 2^i = 0.7692 + 0.6390i
- i^i = 0.2079

USING THE LETTER *E* (OR *E*) FOR SCIENTIFIC NOTATION

In the early days of computing, a display would use seven Light Emitting Diode (LED), or Liquid Crystal Display (LCD) segments to display numbers by turning particular segments on or off. Even today, many watches and clocks use this technique. The following figure shows how a seven-segment display works.

When designers made calculators that displayed scientific notation (see `https://www.mathsisfun.com/numbers/scientific-notation.html` for an explanation of scientific notation), they thought of the letter *E*, which reminds users that what follows is an exponent. They could also implement *E* using a seven-segment display, as shown here:

Then designers got lazy and instead of letting uppercase *E* mean scientific notation, they also let a lowercase *e* mean the same thing. In our modern age, designers can use all the pixels that various screens now employ to display the information without using the letter *E*. However, using *E* or *e* caught on, so now we have to use one of them. In addition, seven-segment displays are still commonly used in calculators, watches, clocks, and other devices.

Organizing Your Storage Locker

Computers contain memory, much as your own brain contains memory. The computer's memory stores information that you create using MATLAB. Looking at memory as a kind of storage locker can be helpful. You open the door, put something inside, and then close the door until you need the item again. When that happens, you simply open the door and take the item out. The idea of memory doesn't have to be complex or difficult to understand.

Whenever you tell MATLAB to store something in memory, you're using a *variable.* Developers use the term *variable* to indicate that the content of the memory isn't stable — it can change. The following sections tell you more about the MATLAB storage lockers called variables.

Using ans — the default storage locker

MATLAB always needs a place to store the output of any calculation you perform. For example, when you type **2 + 2** and press Enter, the output tells you that the value is 4. However, it more specifically tells you that `ans = 4`. MATLAB uses `ans` as a storage locker when you don't specify a specific storage locker to use.

MATLAB uses `ans` as a temporary storage locker. The content lasts only as long as you keep MATLAB open and you don't perform another calculation that requires `ans` to hold the output. If you need the result from a calculation for additional tasks, you must store the result in another variable.

If you discover that you need to save the value of `ans` after performing the calculation, you can still save it. Right-click `ans` in the Workspace window and choose Rename from the context menu. The `ans` entry will turn into a text box, where you type the new name you want to use for the storage box. Press Enter. The `ans` entry is now available for another calculation.

Creating your own storage lockers

Whenever you need to use the result of a calculation in future calculations, you must create your own storage locker to hold the information; using the `ans` temporary variable just won't work unless you rename it. Fortunately, creating your own variables is straightforward. The following sections help you create variables that you can use for storing any MATLAB information you want.

Defining a valid variable name

A MATLAB variable name has certain requirements, just as naming other kinds of things must meet specific requirements. For one thing, you can't use a MATLAB keyword like `end` or `break` as a variable name. Here are the rules for creating a MATLAB variable:

- Start with a letter.
- Add:
  - Letters
  - Digits
  - Underscores

With this in mind, naming a variable `7Heaven` doesn't work because this particular variable name begins with a number — and variables must begin with a letter.

Likewise, `Doug'sStuff` doesn't work as a variable name because the apostrophe (`'`) isn't allowed as part of a variable name. However, all the following variable names *do* work:

» `MyVariable`

» `My_Variable`

» `My7Joys`

REMEMBER

In each case, the variable name begins with a letter and is followed by a letter, digit, or underscore. If you violate any of these rules, you see an error message similar to this one:

```
Error: Unexpected MATLAB expression.
```

TIP

Always make variable names meaningful. Even though a variable named `x` is easy to type, remembering what `x` contains isn't so easy. A name such as `CosOutput` is much easier to remember because it has meaning. At least you know that it contains the output from a cosine calculation. The more meaningful you make the name, the easier it will be for you to later determine what a calculation does.

To create your own variable, type the variable name, an equal sign, and the value you want to assign to that variable. For example, to create a variable called `MyName` and assign it a value of `Amy`, you type **MyName = 'Amy'** and press Enter. (The single quotes show that `Amy` is a value [data], rather than another variable with the name of Amy.)

Understanding that variables are case sensitive

The "Entering information at the prompt" section, earlier in this chapter, discusses the need to type command and function names precisely as described in the MATLAB documentation because MATLAB is case sensitive. Variable names are also case sensitive, and this is one of the ways in which many users make mistakes when creating a script. The variable `myVariable` is different from `MyVariable` because the case is different.

Avoiding existing variable names

Avoiding the use of existing MATLAB names such as `pi`, `i`, `j`, `sin`, `cos`, `log`, and `ans` is essential. If you don't know whether a particular name is in use, you can type **exist('*variable_name*')** and press Enter. Try it now with `pi`. Type **exist('pi')** and press Enter. You see an output of `5`, which means that the variable is in use. Now, type **exist('MyVariable')** and press Enter. The output of `0` means that the variable doesn't exist.

MATLAB lets you create case-sensitive variations of existing variables. For example, type **Ans = 'Hello'** and press Enter. You see that the Workspace window now displays two variables, `ans` and `Ans`. Using a variable with the same name but different capitalization as an existing MATLAB variable will cause you problems. You're better off to simply avoid any existing term no matter how you capitalize it.

Operating MATLAB as More Than a Calculator

It's time to take your first steps beyond using MATLAB as a simple calculator. The following sections help you understand some of the MATLAB functions that you eventually use to perform complex tasks.

Learning the truth

Determining whether something is true is an important part of performing most tasks. You determine the truth value of the information you receive almost automatically thousands of times a day. Computers can perform comparisons and report whether something is true (it does compare) or false (it doesn't compare). A man named George Boole (see `https://plato.stanford.edu/entries/boole/`) created a method for quantifying the truth value of information using Boolean logic.

The basic idea is to ask the computer to perform a comparison of two variables. Depending on the values in those variables, the computer will say that it's either true that they compare or false that they compare. Table 3-1 spells out how Boolean logic works within MATLAB (where an output value of 1 means that the statement is true and an output value of 0 means that the statement is false).

It's essential to remember that one equal sign (=) is an assignment operator. It assigns the value you provide to the variable. Two equal signs (==) is an equality operator. This operator determines whether two variables contain the same value.

Using the built-in functions

Previous sections of this chapter introduce you to a number of MATLAB functions, but you've barely scratched the function surface. MATLAB has a lot of other functions, such as `sin()`, `cos()`, `tan()`, `asin()`, `acos()`, `atan()`, `log()`, and `exp()`. Many of these functions appear in other chapters of the book.

TABLE 3-1 Relational Operators

Meaning	Operator	Example
Less than	A < B	A=2; B=3; A<B ans = 1
Less than or equal to	A <= B	A=2; B=3; A<=B ans = 1
Equal	A == B	A=2; B=3; A==B ans = 0
Greater than or equal to	A >= B	A=2; B=3; A>=B ans = 0
Greater than	A > B	A=2; B=3; A>B ans = 0
Not equal	A ~= B	A=2; B=3; A~=B ans = 1

TIP

For an exhaustive list of functions, go to Appendix A or see `https://www.mathworks.com/help/matlab/matlab_prog/function-summary-1.html`. Yes, there really are that many. The site has brief descriptions of each function. Also, you can get additional information by using `help('`*`function_name`*`')`. Try it now. Type **help('sin')** and press Enter. You see output similar to that shown in Figure 3-3.

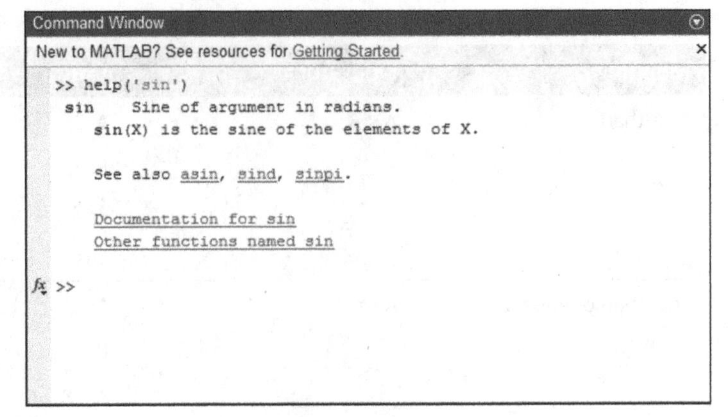

FIGURE 3-3: MATLAB makes it easy for you to learn more about functions you need.

Notice that the help screen contains links. Click any link to receive additional information about that topic.

Accessing the function browser

With all the functions that MATLAB provides, you might think it's impossible to discover what they are without a lot of memorization. Fortunately, help is closer than you might think. Look carefully at the Command Window and you see an fx symbol in the border next to the prompt. Click the down arrow under the symbol and you see the dialog box shown in Figure 3-4. The official name of this dialog box is the Function Browser, and you use it to browse through categories of functions to track down the function you want.

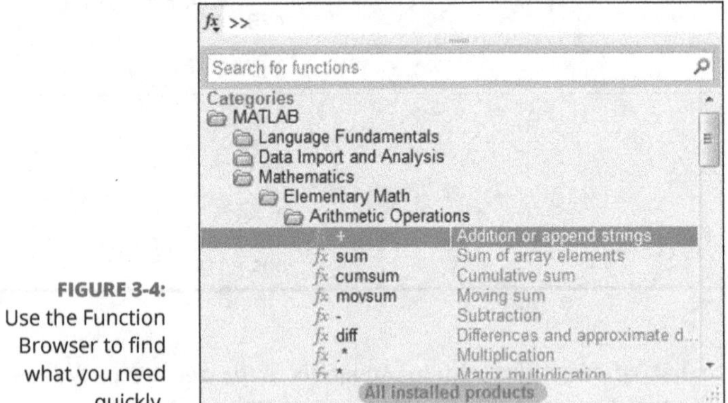

FIGURE 3-4: Use the Function Browser to find what you need quickly.

You can also access the Function Browser using these techniques:

- Right-click the Command Window and choose Function Browser from the context menu.
- Press Shift+F1.

Now that you have a better idea of what the Function Browser is, it's time to look at it in more detail. The following sections provide additional information on using the Function Browser.

Looking through the Function categories

The Function Browser is designed for you to easily drill down into a topic until you find precisely what you need. For example, when you click the Mathematics folder, you see a number of subcategories, such as Elementary Math, Linear Algebra, and Interpolation. When you click Elementary Math, you see yet more subcategories, such as Arithmetic, Trigonometry, and Polynomials. When you finally get to a list of functions, you see the fx symbol next to the entries, as shown in Figure 3-4.

Searching for a particular function

Sometimes you already have a good idea of what you want to find. In such a case, you can type all or part of a function name in the search bar at the top of the Function Browser window. For example, type **sin** (without pressing Enter) in the search field to see all the functions that contain sin in their name or description, such as sine.

Recovering from Mistakes

Everyone makes mistakes. You might think that experts don't make mistakes, but any expert who says so definitely isn't an expert. As Nobel Prize–winning physicist Niels Bohr said, "An expert is a person who has made all the mistakes that can be made in a very narrow field" (see `https://www.goodreads.com/quotes/5-an-expert-is-a-person-who-has-made-all-the`). Making mistakes is part of the learning process. It's also part of the discovery process. If you want to do anything important with MATLAB, you're going to make mistakes. The following sections help you understand what to do when mistakes happen.

Understanding the MATLAB error messages

MATLAB tries to be helpful when you make mistakes. It doesn't always succeed, and you may not always understand the message, but it does try. In most cases, you see an error message that provides enough information for you to at least get started in finding the mistake. For example, if you try to use the `clc` command but type it in uppercase, you get

```
Undefined function or variable 'CLC'.
```

The error message is enough to get you looking for a solution to the problem, even when the problem isn't completely clear. In some cases, MATLAB even provides the correct command for you. All you have to do is press Enter, and it executes.

Some errors are a little harder to figure out than others. For example, Figure 3-5 shows what happens when you try to use `idivide()` without specifying that the inputs are integers.

```
Command Window
New to MATLAB? See resources for Getting Started.
>> idivide(5 / 3)
Error using idivide (line 39)
Not enough input arguments.
>>
```

FIGURE 3-5: Some error messages are a bit complex.

REMEMBER

In this case, you can ignore the links and what looks like gobbledygook. Focus on the second line. It tells you that one of the arguments must belong to the integer class, even though the actual message says `Not enough input arguments`. The Command Window evaluates the expression 5 / 3 and passes the result to idivide. Because that expression creates only a single, idivide is confused because it expects two arguments. When you get past the odd bits of information, you can more easily figure out how to fix the problem.

Stopping MATLAB when it hangs

Most of the time, MATLAB is extremely forgiving. You can make absolutely horrid mistakes, and MATLAB simply provides what it considers a helpful message without destroying anything. However, at times MATLAB has to chew on a bit of code for a while before it discovers the error, such as when you're working with a really large array. In this case, you can talk to your buddy in the next cubicle, get a cup of coffee and read a good book, or press Ctrl+C to stop MATLAB from going any further.

Pressing Ctrl+C always stops MATLAB from performing any additional processing. It's important that you don't use this option unless you really need to do so because MATLAB truly does stop right in the middle of what it's doing, which means that whatever you were doing is in an uncertain state. It's good to know that the option exists, though.

IN THIS CHAPTER

» Understanding the MATLAB file structure

» Working with MATLAB files

» Storing data on disk

Chapter 4

Starting, Storing, and Saving MATLAB Files

Computers have two kinds of storage bins: temporary memory in RAM and permanent memory on a storage device such as a hard drive. (There are also hard drive-like online storage bins, but they'll count as hard drives for most of this chapter.) To make anything you create using MATLAB permanent, you must place it on the hard drive. Unfortunately, hard drives are huge, and if you want to find the data again later, you need to know where you placed the information. That's why knowing something about the MATLAB file structure is important — because you use it to find a place to store your data and to recover that data later.

Data is stored in files, while folders are used to organize the data. To load your data into MATLAB, you must first find the right folder, open it, and then open the file. It works much the same as a filing cabinet. As long as the drawer is closed and the file folder remains inside, the data is inaccessible. Note as well that some of your data may be in the wrong format. When data formatting is a problem, you need to import the data into MATLAB so that MATLAB can make use of it. The same holds true of other applications. When you want to use your MATLAB data with another application, you export it to that application.

The final section of this chapter discusses how to save your work for later use. The act of saving your work moves the data from temporary storage in RAM to permanent storage on the hard drive in a file. Later, when you need to access the data again, you open the file, which moves it from the hard drive into RAM. There is nothing mystical about this process. You perform the same sorts of tasks in the real world every day. Just think about how you use files in a filing cabinet the next time you open one.

REMEMBER

You don't have to type the source code for this chapter manually. In fact, using the downloadable source is a lot easier. You can find the source for this chapter in the `\MATLAB2\Chapter04` folder of the downloadable source. When using the downloadable source, you may not be able to work through some of the hands-on examples because the example files will already exist on your system. In this case, you can create an alternative folder, `Chapter04a`, to hold the results of the hands-on exercises. You can find details of where to find the downloadable source in the Introduction.

Examining MATLAB's File Structure

To keep your data permanently, you must store it on disk. Of course, you could just store it anywhere, but then finding it later would be intensely difficult. In fact, given the size of today's hard drives, you might well retire before you find the data again. (And, if you're storing that data using an online hard drive, just think about how long it would take to search the Internet.) So, relying on some organized method of storing your information is important.

REMEMBER

Applications also rely on specific file types when storing information. The main reason for using a specific file type is to allow the application to recognize its data among all the other data on your drive. Imagine the chaos if every application used the `.txt` file extension for every file, but each application created its own sort of `.txt` file that isn't readable by any other application. Not only you but also the computer would become confused. In addition, using specific file types lets you know what sort of data the file contains.

MATLAB lets you identify the particular kind of information a file holds through the use of unique file extensions. For example, scripts and functions are stored in files with an `.m` extension; Live Scripts and Live Functions appear in `.mxl` files; variables are stored in files with a `.mat` extension; and plots are stored in files with a `.fig` extension. In addition, you can organize your data using a file structure. You can perform all these management tasks from within MATLAB using either the application's GUI or commands. The following sections tell you how all these features work.

Understanding the MATLAB files and what they do

MATLAB provides specific file types for specific needs. The following list tells you about the MATLAB file types and describes their uses:

- `.fig`: Provides access to any plots or other graphics you create. Keep in mind that the file contains all the information required to reconstruct the graphic, but does not contain the graphic itself. This approach means that your image is accessible on any platform that MATLAB supports.

  TECHNICAL STUFF

  A lot of people have asked whether they can access `.fig` files without necessarily having to display the graphic image itself. It turns out that `.fig` files are actually `.mat` files in disguise. The file format is the same (even though the content between the two file types differs). In fact, you can simply rename a `.fig` file with a `.mat` extension and use Import Data on the Home tab to import it. You can also use the `load('untitled.fig', '-mat')` command, where `'untitled.fig'` is the name of the file. Inside the file, you see two variables: `hgM_070000` (contains the graphic objects, which are accessible only when the figure is visible) and `hgS_070000` (contains the user data). Wandering through the two variables will tell you a lot about how MATLAB works with figures. For example, `hgS_070000.children(2).children.properties.XData` displays the x-axis data used to generate the plot. As another example, `hgM_070000.GraphicsObjects.Format3Data.Color = [0.9900, 0, 0]` turns the frame around the figure red.

- `.m`: Holds a MATLAB script. This is a platform-independent file, so you can use the same scripts on whatever platform you're working on at the time. This file also allows you to create a script on one platform and share it with others, even when they use a different platform than you do. MATLAB script files are always written using the MATLAB language.

- `.mat`: Provides access to any data you saved on disk. Opening this file starts the Import Wizard to load the data into the MATLAB workspace.

- `.mdl`: Contains an older version of a Simulink model (see `.slx` below for details on the Simulink model). MATLAB recommends updating these files to the `.slx` format using the procedure at `https://www.mathworks.com/help/simulink/examples/converting-from-mdl-to-slx-model-file-format-in-a-simulink-project.html`.

- `.mex*`: Contains compiled executable code that extends MATLAB functionality in some manner. You execute these files just as you would a script program. The original code is written in either FORTRAN or C++ and then compiled for a specific platform. Each platform has a unique extension associated with it, as shown in the following list:

- .mexa64: Linux
- .mexmaci64: Mac OS X
- .mexw32: 32-bit Windows

MATLAB 32-bit extensions won't run in a 64-bit copy of MATLAB (see https://www.mathworks.com/matlabcentral/answers/225383-32-bit-mex-file-not-working-in-a-64-bit-version-of-matlab). In addition, running a 32-bit version of MATLAB in a 64-bit environment can be problematic (see https://www.mathworks.com/matlabcentral/answers/101743-is-running-32-bit-matlab-on-a-64-bit-platform-supported). The best answer for .mexw32 files is to find a 64-bit replacement. If you own the source code, you can recompile the extension.

- .mexw64: 64-bit Windows

» .mlapp: Defines a MATLAB application created using the MATLAB App Designer, described in Chapter 14. A MATLAB application lets you share your code in a nicely packaged form with other people, and you can create one without having any programming experience (although a little experience does help).

The .mlapp file is actually a kind of .zip file, so you can open it using a file archiver to see what's inside. This knowledge can come in handy when you need to use code found in a .mlapp file with an older version of MATLAB, as described at https://www.mathworks.com/matlabcentral/answers/404815-how-to-get-code-from-a-mlapp-file-using-an-earlier-matlab-version.

» .mlappinstall: Provides the means for installing a MATLAB application on another machine. Because an application can consist of several files, using an application installer makes the task of moving the application to another machine easier and less error prone. You can discover more about the packaging process in the "Packaging Your App" section of Chapter 14.

» .mlpkginstall: Accesses a MATLAB Hardware Support Package Installer, which provides support for various hardware devices within MATLAB. This is actually a type of XML file that describes how to install the hardware support you require. To use .mlpkginstall files, you must have a locally installed copy of MATLAB with a valid license. By default, newer copies of MATLAB ship with the Raspberry Pi Resource Monitor (see https://www.mathworks.com/help/supportpkg/raspberrypi/ug/raspberrypiresourcemonitor-app.html for details). Creating a package is outside the scope of this book, but you can learn how to build one at https://www.mathworks.com/matlabcentral/answers/395889-how-do-i-create-my-own-hardware-support-package.

- `.mltbx`: Allows distribution of MATLAB elements, such as code, data, apps, examples, and documentation, that are designed to work together to perform tasks such as analysis with greater ease. Mathworks provides a host of toolboxes for various tasks and you can also find other toolboxes developed by third parties online. The "Developing a Toolbox" section of Chapter 15 provides more information on creating a toolbox of your own.
- `.mlx`: Contains a Live Script or Live Function. These new ways of writing MATLAB code are a vast improvement over previous techniques because you can see the output in a nicely formatted presentation. The `.mlx` file is actually a kind of `.zip` file, so you can open it using a file archiver to see what's inside. Chapter 11 tells you how to work with Live Scripts, while Chapter 12 tells you how to work with Live Functions.
- `.p`: Performs the same task as an `.m` file, except the content is protected from edits by anyone else. This feature lets you distribute your scripts to other people without fear of giving away programming techniques or trade secrets.
- `.prj`: Defines a MATLAB project. You can create a blank project, start with a folder of MATLAB elements, or obtain code from either Git or Apache Subversion (SVN). Other MATLAB tasks also rely on project files, such as when you create an application or an application installer.
- `.slx`: Contains a Simulink model. Simulink is an add-on product for MATLAB that provides a block diagram environment for performing simulations. You can read more about this product at `https://www.mathworks.com/help/simulink/gs/product-description.html`. This book doesn't discuss the Simulink add-on because it's an advanced product used for higher-end needs.

Exploring folders with the GUI

The GUI method of working with folders in MATLAB requires the Current Folder window shown in Figure 4-1. (To display this window, choose Layout⇨Current Folder in the Environment group of the Home tab.) In this case, the Current Folder toolbar appears at the top of the Current Folder window. You can also place it below the Toolstrip by choosing Layout⇨Current Folder Toolbar⇨Below Toolstrip in the Environment group of the Home tab. (The screenshots in the rest of the book assume that you have selected the Inside Current Folder option.)

FIGURE 4-1: The Current Folder window provides GUI access to the MATLAB folders.

The Current Folder toolbar shows the current folder that the Current Folder window displays. To change locations, simply type the new location in the field provided. You can also select a location by clicking the right-pointing arrow next to each level. The arrow changes to a down-pointing arrow, as shown in Figure 4-2, with a list of destinations below it. Clicking the magnifying glass icon in the field turns it into a Search field, where you can choose the kind of file you want to find.

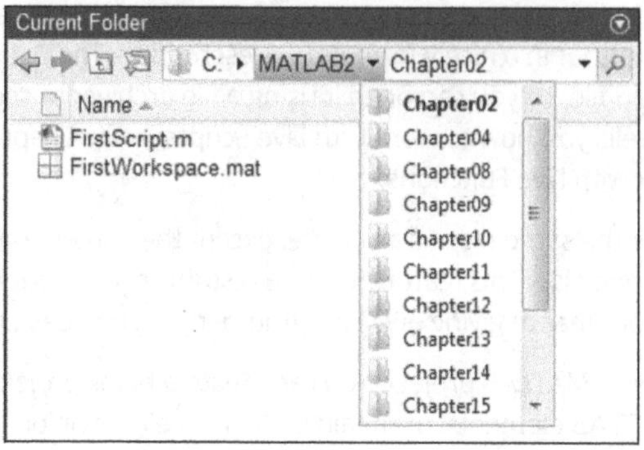

FIGURE 4-2: You can choose new locations by clicking the right-pointing arrow.

The Current Folder toolbar also includes four buttons. Each of these buttons helps you move to another location on the hard drive, as follows:

- **Back:** Moves the location back one position in the history list. The history list is a list that is maintained by MATLAB that tracks the locations you've visited.
- **Forward:** Moves the location forward one position in the history list.
- **Up One Level:** Moves the location up to the parent folder.
- **Browse for Folder:** Displays a Select New Folder dialog box that you can then use to find another location on the hard drive. (See Figure 4-3.) After you find the folder, highlight its entry and click Select Folder to select it.

The Current Folder window provides access to all the folders that you've set up for organizational purposes. In this case, you see the `Chapter02` subfolder (child folder) of the `C:\MATLAB2` folder. The `Chapter02` folder contains two files (refer to Figure 4-2). When you right-click the `Chapter02` folder entry, you see a number of commands on a context menu like the one shown in Figure 4-4.

Note that not all the entries on the context menu have to do with exploring folders or managing them from a file structure perspective. The following list focuses on those commands that *do* help you manage the file structure.

- **Open:** Opens the folder so that it becomes the current folder in the Current Folder toolbar.

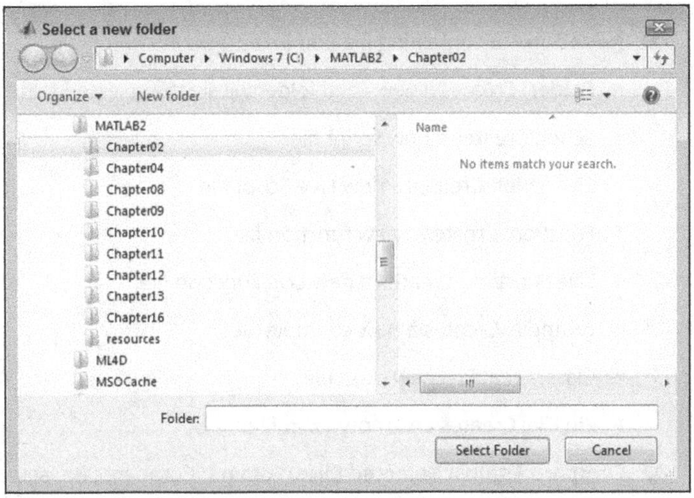

FIGURE 4-3: The Select New Folder dialog box helps you find other locations on the hard drive.

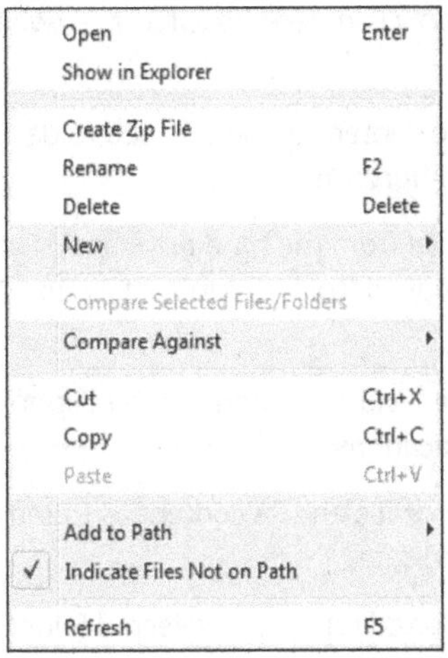

FIGURE 4-4: The context menu associated with a folder contains options for managing the folder content.

- **Show in Explorer (Windows only):** Opens a copy of Windows Explorer so that you can interact with the folder using this Windows tool.
- **Create Zip File:** Creates a new `.zip` file that contains the compressed content of the folder. This feature makes sending the folder to someone else easier.
- **Rename:** Changes the name of the folder.

WARNING

- **Delete:** Removes the folder and its content from the hard drive. Depending on how you have your system configured, this option could permanently destroy any data found in the folder, so use it with care.

- **New:** A menu containing file and folder options.
  - **Folder:** Creates a new child folder within the selected folder.
  - Script: Creates a new script file.
  - Live Script: Creates a new Live Script file.
  - Function: Creates a new function file.
  - Live Function: Creates a new Live Function file.
  - Example: Creates a new example file.
  - Class: Creates a new class file.
  - Zip File: Creates an empty `.zip` file.
- **Compare Against Selected Files/Folders:** Compares the selected files or folders against each other. You select two files or folders to use this option by highlighting the primary file or folder first, then Ctrl+clicking the secondary file or folder next. This command only compares like items — two folders or two files of the same type.
- **Compare Against:** Matches the content of the selected folder against another folder and tells you about the differences.
- **Cut:** Marks the folder for removal from the hard drive and places a copy on the Clipboard. The folder is removed from its current location when you paste the copy in its new location.
- **Copy:** Places a copy of the folder and its content on the Clipboard so that you can paste copies of it in other locations.
- **Paste:** Places a copy of a folder or file and its content as found on the Clipboard in the location you indicate.
- **Add to Path:** Adds a folder, selected folders, or selected folders and subfolders to the path so that MATLAB can locate resources within them when the folder isn't the current folder as described here:
  - **Selected Folders:** Adds only the selected folders even if the folders contain subfolders. Because the subfolders aren't part of the path, MATLAB won't see resources they contain.
  - **Selected Folders and Subfolders:** Adds both the selected folders and any subfolders these folders contain to the path.
- **Indicate Files Not on Path:** Displays folders and files that are on the path in bold when selected.

» **Refresh:** Verifies that the presentation of folders and files in the Current Folder window matches the actual content on the hard drive. Another application may have made modifications to the hard drive content, and with this command you can synchronize MATLAB with the physical device.

Exploring folders with commands

Many people prefer not to use the mouse. In this case, you can duplicate most of the GUI methods of interacting with the current folder using keyboard commands. The results of typing a command and pressing Enter appear in the Command Window. To see how this feature works, try the following steps. (Your folder structure may not look precisely like the one in the book, but you should see appropriate changes as you type the commands.)

1. **Type** cd \MATLAB2 **and press Enter.**

   The Current Folder window changes to show the folder used for the book. The change directory, `cd`, command will always modify the location in the Current Folder window unless you're already there. You may have to change the actual folder information to match your system if you chose not to create the directory structure described in earlier chapters. For example, you might need to type **cd \Users\<*Username*>\MATLAB\Projects** and press Enter instead (and if your path contains spaces, you must enclose the entire path in single quotes).

TIP

   Even though you can't see it in this black-and-white book, MATLAB does provide color coding to make working with commands easier. Notice that the command portion of a command is in black lettering, while the argument part of the command is in purple lettering. The use of color coding helps you better see the commands and how they're structured.

2. **Type** mkdir Chapter04 **and press Enter.**

   MATLAB creates a new folder in the Current Folder window to hold the materials for this chapter. Notice that you don't include a backslash (or slash) when creating a child directory for the current directory.

REMEMBER

   If you're using the downloadable source, you already have a `Chapter04` folder in your `MATLAB` directory. Create a `Chapter04a` folder instead to hold the results of this and any other hands-on exercises in the chapter. Every place you see `Chapter04`, use `Chapter04a` instead.

3. **Type** cd Chapter04 **and press Enter.**

   The directory changes to the one used for this chapter. Notice (again) that you don't include a backslash (or slash) when moving to a subdirectory of the current directory.

4. **Type** copyfile ..\Chapter02\FirstScript.m **and press Enter.**

   You see the copied file appear in the Current Folder window in the `Chapter04` folder.

   a. The `copyfile` command provides the functionality needed to copy a file.

   b. The `..` part of the path statement says to look in the parent folder, which is `\MATLAB`.

   c. The `Chapter02` part of the path says to look in the `Chapter02` subdirectory, which equates to `\MATLAB\Chapter02`.

   d. The `FirstScript.m` part of the path is the name of the file you want to copy to the current folder.

5. **Type** `exist FirstScript.m` **and press Enter.**

   The command used in this case has a number of parts to it:

   MATLAB provides an output value of 2, which means that the file exists. This final step helps you validate that the previous steps all worked as intended. If one of the previous steps had gone wrong, you'd see a failure indicator, such as an error message or a different output value (as shown in the next step), with this step.

6. **Type** `exist MyScript.m` **and press Enter.**

   In this case, the output value of `0` tells you that `MyScript.m` doesn't exist, as shown in Figure 4-5. The procedure didn't tell you to create `MyScript.m`, so this output is completely expected.

Now that you can see how the commands work, it's time to look at a command list. The following list contains an overview of the most commonly used file and folder management commands. (You can get detailed information at `https://www.mathworks.com/help/matlab/file-operations.html`.)

- `cd`: Changes directories to another location.
- `copyfile`: Copies the specified file or folder to another location.
- `delete`: Removes the specified file or object.
- `dir`: Outputs a list of the folder contents.
- `exist`: Determines whether a variable, function, folder, or class exists.
- `fileattrib`: Displays the file or directory attributes (such as whether the user can read or write to the file) when used without attribute arguments. Sets the file or directory attributes when used with arguments.
- `isdir`: Determines whether the input is a folder.

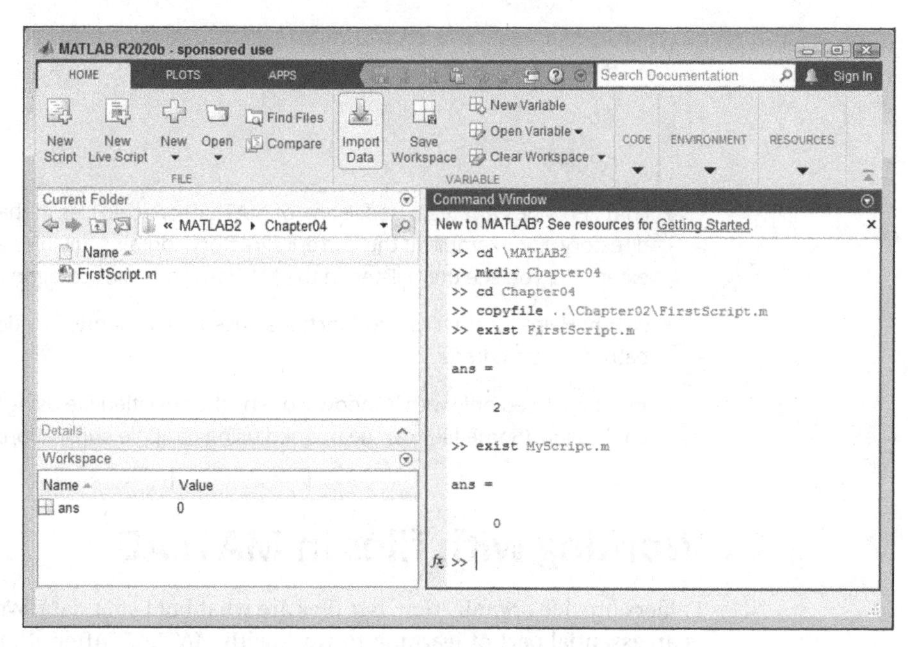

FIGURE 4-5: MATLAB not only allows you to manage the file structure but also to validate it using commands.

- `ls`: Outputs a list of the folder contents.
- `mkdir`: Creates a new directory.
- `movefile`: Moves the specified file or folder to another location.
- `open`: Opens the specified file using the default application. (Some files can be opened using multiple applications.)
- `pwd`: Displays the current path information, including the drive letter.
- `recycle`: Determines whether deleted files or folders are moved to the recycle bin.
- `rmdir`: Deletes the specified directory.
- `type`: Outputs the content of the specified file as text.

TIP

Some commands, such as `type`, can be combined with other commands, such as `disp`, to create well-formatted output. The `disp` command displays text, variables, or arrays. You discover how to use it later in the book (starting with Chapter 8). The point is that you sometimes combine commands to obtain a desired output.

- `visdiff`: Performs a comparison of two files or folders of the following types:
  - Text
  - MAT-Files

* Binary
* Zip
* Folders

» `what`: Provides a categorized listing of MATLAB-specific files in the current directory. For example, if the current directory contains any files with an `.m` extension, you see them listed in the MATLAB code files category.

» `which`: Helps locate files and functions based on filename, function name path, or other criteria.

» `winopen`: Used only with Windows; opens the specified file using the default application. (Some files can be opened using multiple applications.)

Working with files in MATLAB

Folders provide organization, but files are what hold your data. Working with files is an essential part of learning to work with MATLAB. After all, if you can't find your data, you can't do anything with it. Managing the data to ensure that it remains safe, secure, and reliably accessible is important. The following sections describe how to perform common tasks with files.

Using the right-click to your advantage

Every file and folder shown in the Current Folder window has a context menu associated with it. A context menu always displays just the actions that you can perform with that file or folder. By right-clicking various files and folders, you see the context menu and might discover new tasks that you can perform with the file or folder you highlighted.

REMEMBER

Right-clicking a file or folder can never damage it. The only time you might damage the file or folder is if you select an item from the context menu. To close the context menu after you view it, click in an empty area outside the context menu or press Escape.

TIP

Depending on your platform, you may also see shortcut keys when viewing the context menu. For example, when working with Windows, you can highlight a file and press Ctrl+C to copy it to the Clipboard — all without using the context menu at all. Pasting is just as easy: Select the folder you want to use to store the file and press Ctrl+V. As mentioned, these shortcut keys are platform specific, which is why they aren't used in the book.

Copying and pasting

Copying and pasting creates a copy of an existing data file and places that copy in another location. You use this process in a number of ways. For example, you might want to make a backup of your data before you modify it or share the data with a friend. Even though the following steps use a specific file and locations, you can easily use them for any file you want to copy and paste and with any location. In this case, you copy `FirstWorkspace.mat` found in the `Chapter02` folder to the `Chapter04` folder (or `Chapter04a` if you're using the alternative setup with the downloadable source).

1. **Open the `\MATLAB\Chapter02` folder in the Current Folder window.**

   You see two files: `FirstScript.m` and `FirstWorkspace.mat`.

2. **Right-click `FirstWorkspace.mat` and choose Copy from the context menu.**

   This action copies the file onto the Clipboard. You won't actually see anything happen in the window.

3. **Click the Up One Level button in the Current Folder toolbar.**

   You return to the `\MATLAB` folder. This folder should show two subdirectories: `Chapter02` and `Chapter04`, unless you're working with the downloadable source, in which case you see all the folders for this book. If you don't see both subdirectories, make sure to create the `Chapter04` subdirectory using the steps found in the "Exploring folders with commands" section, earlier in this chapter.

TIP

   The `Chapter02` subdirectory is darker than the `Chapter04` subdirectory, and the file folder appears open rather than closed. The reason for this difference is that the `Chapter02` subdirectory is on your MATLAB path, while the `Chapter04` subdirectory isn't. To add `Chapter04` to the path, right-click its entry and choose Add To Path ➪ Selected Folders or Add To Path ➪ Selected Folders and Subfolders from the context menu.

4. **Double-click the `Chapter04` folder to open it.**

   This folder should contain a single existing file, `FirstScript.m`.

5. **Right-click anywhere within the folder area and choose Paste.**

   MATLAB copies the file to the new location for you. At this point, the `Chapter04` folder should look precisely like the `Chapter02` folder.

Cutting and pasting

The process for cutting and pasting a file is almost the same as copying and pasting it. (See the previous section for details.) The only difference is that you select Cut rather than Copy from the context menu. However, the results are slightly different. When you cut and paste a file, the file is actually moved from one location to another. You use Cut and Paste when you don't want to create multiple copies of a file and simply want to place the file in another location.

Dragging

Dragging a file or folder moves it from one location to another. All you need to do is click the file. While you hold the mouse button down, you drag the file to a new location. MATLAB moves the file to the location you specify.

If the location already has a file with that name, MATLAB displays a message asking whether you're sure you want to move the file. You must confirm the move before MATLAB performs the task. The new file replaces the existing file, so you could experience data loss.

Accessing and Sharing MATLAB Files

To make data useful, you need to be able to open the files containing it. Otherwise, there isn't any point in saving the data. Likewise, not all your colleagues will have a copy of MATLAB, or they may want to use a different application to interact with the MATLAB data. For you to use their data, you must be able to *import* data files created by other applications. When you want to share your data with others, you must *export* your data to files that are understood by other applications. MATLAB provides great support for both imported and exported data.

Opening

The fastest way to open any MATLAB file is to double-click its entry in the folder found in the Current Folder window. You can also right-click the entry and choose Open from the context menu. MATLAB automatically opens the file using a default application or method.

It's important to realize that MATLAB always uses a default application or method. Data files are sometimes associated with other applications. In addition, some data files can be opened in more than one way.

When you want to use an alternative method of opening a file, you must rely on the underlying platform. For example, when working with Windows, right-click the file and choose Show in Explorer from the context menu. A copy of Windows Explorer opens, and you can work with alternative applications in that copy. Right-click the file in Windows Explorer and choose one of the alternative applications shown in the Open With menu of the context menu. Figure 4-6 shows an example for `FirstScript.m` that presents several choices of application in which to open the file using (with the default shown at the top of the list).

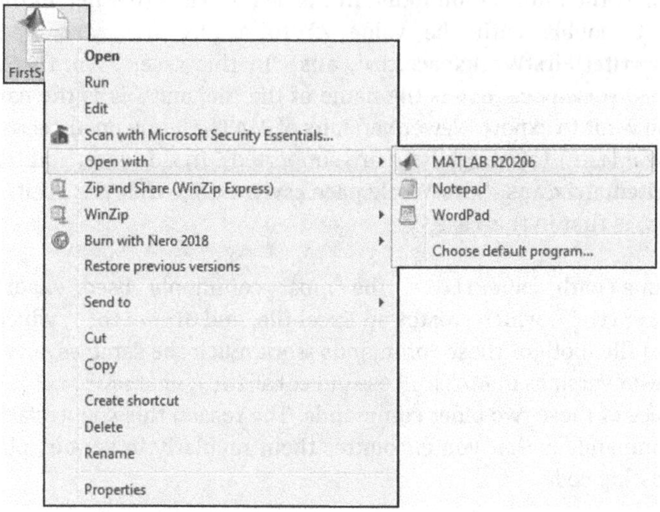

FIGURE 4-6: Use a platform-specific means of opening files using alternative applications.

MATLAB also uses different techniques for interacting with files when you work with commands. The default action for a `.mat` file is to load it into MATLAB and execute it — showing the result in the Workspace window. However, you can either load it or open it, as needed. Here are the two commands you use (assuming that you want to work with `FirstWorkspace.mat`):

```
» open('FirstWorkspace.mat')
» load('FirstWorkspace.mat')
```

The first command actually opens the workspace so that you can see a result in the Command Window in addition to the Workspace window. However, the results aren't loaded into the Workspace window as they normally would be if you double-clicked the file. To achieve this same effect, you must use the second command, which loads the workspace into MATLAB. Try loading `FirstWorkspace.mat` now so that you have a value in `ans` to use for the sections that follow.

Exporting

You rely on commands to export data from MATLAB. The list of data formats at `https://www.mathworks.com/help/matlab/import_export/supported-file-formats.html` includes commands in the Export Function column for each format that MATLAB supports. To export the data in this section, you need to first open the `FirstWorkspace.mat` file using the technique found in the previous section.

Most of the commands work with a single variable. For example, if you want to export the information found in the `ans` variable (defined in the previous section as a double with the value 4) to a CSV file, you type something like **csvwrite('FirstWorkspace.csv', ans)**. In this case, `csvwrite()` is the function, `FirstWorkspace.csv` is the name of the file, and `ans` is the name of the variable you want to export. Newer versions of MATLAB rely on `writematrix()` instead of `csvwrite()` (although both are supported). In this case, you type something like **writematrix(ans, 'FirstWorkspace.csv')**. Notice that the position of the variable, `ans`, is first in this case.

Along with `csvwrite()`, the most commonly used export commands are `xlswrite()`, which creates an Excel file, and `dlmwrite()`, which creates a delimited file. Both of these commands work much the same as `csvwrite()`. Note that newer versions of MATLAB use `writetable()`, `writematrix()`, or `writecell()` in place of these two older commands. The reason this chapter also covers the older commands is that you encounter them regularly in various places online and in existing code.

Some file formats require quite a bit of extra work. For example, to create an eXtensible Markup Language (XML) file, you must first build a document model for MATLAB to use. You can see the procedure for performing this task at `https://www.mathworks.com/help/matlab/ref/xmlwrite.html`.

Importing

MATLAB makes importing whatever data you need from an external source easy. To use the procedure in this section, you must create `FirstWorkspace.csv` using the technique found in the previous section. The following steps show you how to import any file using `FirstWorkspace.csv` as an example:

1. **Click Import Data in the Variable group of the Home tab.**

   You see the Import Data dialog box. Note that MATLAB defaults to showing every file it can import.

   If you find that the list of files is too long, you can click the Recognized Data Files drop-down list and choose just one of the common file types. The list displays just those files, making a selection easier.

2. **Highlight the file you want to import and click Open.**

   The example assumes that you're using `FirstWorkspace.csv`, but any file that MATLAB supports for import will work. MATLAB displays an Import dialog box that contains import information about the file, as shown in Figure 4-7.

This dialog box contains settings that you use to import the data and ensure that it's useful in MATLAB. Figure 4-7 shows the settings for a comma-separated value (CSV) file, and the rest of the procedure assumes that you're working with such a file. However, the process is similar for other file types.

FIGURE 4-7: The Import dialog box lets you tweak the import settings.

3. **(Optional) Modify the settings as needed so that the data appears as it should appear in the Workspace window.**

   You can choose to limit the amount of data imported by changing the range. You can also select a different delimiter (which changes how the data appears onscreen).

4. **Verify that the Unimportable Cells group has no entries.**

   Cells that MATLAB can't import might reflect an error or simply mean that you have some settings wrong.

5. **Click Import Selection.**

   MATLAB imports the data using the Import Data option (the default) on the Import Selection drop-down list. As alternatives, you can also choose to generate a Live Script, script, or function based on the data, rather than actually import the data into the workspace.

6. **Close the Import window.**

   Notice that the Workspace window includes entries for

   - ans: Obtained by loading the `FirstWorkspace.mat` file
   - `FirstWorspace`: Obtained by importing the data from `FirstWorkspace.csv`

TIP

You can read about the data formats that MATLAB can import at `https://www.mathworks.com/help/matlab/import_export/supported-file-formats.html`. This site also contains commands that you can use to import the files rather than relying on the GUI to do the work. However, the GUI is always faster and easier to use, so it's the recommended course.

Saving Your Work

An essential part of ending any session with MATLAB is saving your work. Otherwise, you could lose everything you've worked so hard to achieve. In fact, smart users save relatively often to avoid the power-failure penalty. How often you save depends on your personal work habits, the value of the work, and the potential need to use time and system resources efficiently. No matter how you save, or when, the following sections help you get the job done.

Saving variables with the GUI

Although Chapter 2 does show you how to save the entire workspace, sometimes you need to save just one variable. You can perform this task using the GUI and the following steps:

1. **Right-click the variable that you want to save in the Workspace window and choose Save As from the context menu.**

   You see the Save to MAT-File dialog box.

2. **Type a name for the file in the File Name field.**

   Choose something that will help you remember the purpose of the variable.

TIP

   You can use the tree structure in the left pane to choose a different folder if you don't want to use the current folder to store the file containing the variable information.

3. **Click Save.**

   MATLAB saves the variable to the file you choose.

Saving variables using commands

You can use commands to save your variables to disk. In fact, the command form is a little more flexible than the GUI. The basic command you use is `save('filename')`, where *filename* is the name of the file you want to use. Note that `save()` overwrites any existing file of the same name without telling you first.

When you want to save specific variables, you must add a list of them after the filename. For example, `save('MyData.mat', 'ans')` would save a variable named `ans` to a file named `MyData.mat` in the current folder. You can include path information as part of the filename if you want to save the data in a different folder. For example, `save('C:\Temp\MyData.mat', 'ans')` would save the data in the `C:\Temp` folder. If you want to save multiple variables, simply create a comma-delimited list of them. To save `Var1` and `Var2` to `MyData.mat`, you type **save('MyData.mat', 'Var1', 'Var2')**.

These initial commands save the output in MATLAB format. However, you can also specify a format. The formats are listed at `https://www.mathworks.com/help/matlab/ref/save.html#inputarg_fmt`. For example, to save the previous variables in ASCII format, you type **save('MyData.txt', 'Var1', 'Var2', '-ASCII')**.

Saving commands with the GUI

You can't save commands that you type directly into the Command Window using the GUI. What you do instead is save them using the Command History window. The "Saving a formula or command as a script" section of Chapter 2 describes how to save both formulas and commands.

Saving commands using commands

MATLAB does let you save commands to disk using a command: `diary`. A *diary* is simply an on-disk record of the commands that you type in the Command Window. Recording starts when you start the diary — the resulting file doesn't contain all the entries in the Command History window. Later, you can review the file and edit it just as you would a script. The `diary` command actually has a number of forms, as follows:

- `diary`: Creates a diary file with the filename `diary`. Because this file has no extension, it isn't associated with anything. The output is ASCII, and you can open it with any text editor.
- `diary('`*`filename`*`')`: Creates a diary file that has the name *filename*. You can give the output file an `.m` extension, which means that you can open it as a script using the MATLAB editor. This approach is actually better than using `diary` by itself because the resulting file is easier to work with.
- `diary off`: Turns off recording of your commands so that they aren't recorded to the file. Setting the diary to `off` lets you experiment before committing commands that you don't want to the file on disk.
- `diary on`: Resumes recording of your commands.

Using online storage

Many individuals and organizations now use online storage facilities such as GitHub (https://github.com/) to store data. This makes it possible to use your data anywhere and from multiple devices. In addition, you can share data with any number of individuals who may need to collaborate with you. Using online storage means downloading and installing additional software, configuring it, and then using it to make a connection. The process for performing these steps is outside the scope of this book, but you can find a procedure for it at https://www.mathworks.com/help/matlab/matlab_prog/set-up-git-source-control.html.

2 Manipulating and Plotting Data in MATLAB

IN THIS PART . . .

Defining the MATLAB memory structures

Manipulating data in memory

Performing basic plotting tasks

Using advanced plotting features

IN THIS CHAPTER

» Interacting with vectors and matrices

» Performing addition, subtraction, multiplication, and division

» Working with more than two dimensions

» Getting help with matrixes

Chapter 5

Embracing Vectors, Matrices, and Higher Dimensions

The previous chapters of this book introduce you to MATLAB and its interface. Starting in this chapter, you become immersed in math a little more serious than 2 + 2. Of course, in this "more serious" math, many problems revolve around vectors and matrices, so these are good topics to start with. This chapter helps you understand how MATLAB views both vectors and matrices, and how to perform basic tasks with these structures. The chapter then takes you from two-dimensional matrices to matrices with three or more dimensions. All this material gives you a good idea of just how MATLAB can help you solve your vector and matrix problems.

Of course, you might still have questions. In fact, a single chapter of a book can't answer every question on this topic. That's why you also need to know how to obtain additional help. The last section of the chapter provides insights into how you can get additional matrix-oriented help from MATLAB and force it to do more of your matrix work for you. (After all, MATLAB is there to serve your needs, not the other way around.)

Working with Vectors and Matrices

A *vector* is simply a row of data elements. The length of this row has no limit and the values have no specific interval. A *matrix* is a two-dimensional table of data elements. Again, the size of this table has no limit (either in rows or columns), and the numbers have no specific interval. Both structures are well understood by mathematicians and engineers. They are used extensively by MATLAB to perform tasks that might otherwise require the use of complex structures not understood by these groups, which would unnecessarily complicate MATLAB usage.

The following sections describe how MATLAB uses vectors and matrices to make creating programs easier and demonstrates some of the ways in which MATLAB uses them. (Note that this chapter's discussion assumes that you're coming to the table with a basic understanding of linear algebra. If you find that you need to brush up on this particular area, check out the "Locating linear algebra resources online" sidebar.)

Understanding MATLAB's perspective of linear algebra

Linear algebra deals with vector spaces and linear mappings between those spaces. You use linear algebra when working with lines, planes, and subspaces and their intersections. When working with linear algebra, vectors are viewed as coordinates of points in space, and the algebra defines operations to perform on those points. MATLAB divides linear algebra into these major areas:

- Matrix analysis
  - Matrix operations
  - Matrix decomposition
  - Matrix properties
- Linear equations
- Eigenvalues
- Singular values
- Matrix functions
  - Logarithms
  - Exponentials
  - Factorization

LOCATING LINEAR ALGEBRA RESOURCES ONLINE

This chapter doesn't furnish a tutorial on linear algebra. Of course, not everyone remembers that college course in linear algebra, and some things that you don't use every day are likely to be a little hard to remember. With this in mind, you might want to locate a linear algebra tutorial to jog your memory. Many good sources of information about linear algebra are available online.

One of the more interesting places to get some information about linear algebra is the Khan Academy at `https://www.khanacademy.org/math/linear-algebra`. Most of the information is relayed through videos, so you get the benefit of a classroom-like presentation. The presentations are short, for the most part — usually shorter than ten minutes — so you can watch segments as time allows. In addition, you can pick and choose among the videos to watch.

If all you really want is a quick brush-up on linear algebra, you might not need something as time-consuming as what the Khan Academy provides. In that case, you might want to check out the linear algebra tutorial in four pages at `https://minireference.com/blog/linear-algebra-tutorial/`. A number of people using this resource complained that it went really fast. After reviewing it, I can report that the four pages are well done, but they really do assume that you need a light refresher and already know how to use linear algebra quite well.

A middle-ground tutorial is found on Kardi Teknomo's Page at `https://people.revoledu.com/kardi/tutorial/LinearAlgebra/index.html`. The interesting thing about this tutorial is that it's interactive. You get somewhat detailed text instruction and then get to try your new skills right there on the site. The act of reading the information and then practicing what you learn makes the information stick better.

The point is that you're likely to find a tutorial that meets your specific needs. You just need to invest a few minutes in trying out the various tutorials until you find one that fits your particular learning style. Offering such diversity in a single chapter of a book simply isn't possible, so that's why the online resources are so important.

This chapter doesn't cover absolutely every area, but you are exposed to enough linear algebra to perform most tasks effectively using MATLAB. As the book progresses, you see additional examples of how to make MATLAB perform tasks using linear algebra. Think of this chapter as a really good start toward the goal of making MATLAB perform linear algebra tasks for you at a level of speed and accuracy you couldn't achieve otherwise.

Entering data

Chapter 3 shows you how to import data from a spreadsheet or another data source. Of course, that's fine if you have a predefined data source. However, you'll often need to create your own data, so knowing how to type it yourself is important.

Think about how you use data when working with math: The data appears as a list of numbers or text. MATLAB uses a similar viewpoint. It also works with lists of numbers and text that you create through various methods. The following sections describe how to enter data as lists by using assorted techniques.

Entering values inside square brackets

The left square bracket, [, starts a list of numbers or text. The right square bracket,], ends a list. Each entry in a list is separated by a comma (,). To try this technique yourself, open MATLAB, type **b=[5, 6]** in the Command Window, and press Enter. You see

```
b =
     5     6
```

The information is stored as a list of two numbers. Each number is treated as a separate value. Double-click `b` in the Workspace window and you see two separate entries, as shown in Figure 5-1. Notice that the Variables window shows `b` as a 1-x-2 list of doubles in which the entries flow horizontally.

FIGURE 5-1: Typing comma-separated numbers in square brackets produces a list of numbers.

REMEMBER

You can type **format compact** and press Enter to save display space. If you want to clear space in the Command Window for typing additional commands, type **clc** and press Enter. When you get too many variables in the Workspace window, type **clear all** and press Enter. Chapter 3 provides additional details on configuring MATLAB output.

Starting a new line or row with the semicolon

The comma creates separate entries in the same row. You use the semicolon (;) to produce new rows. To try this technique yourself, type **e=[5; 6]** in the Command Window and press Enter. You see

```
e =
     5
     6
```

As in the previous section, the information is stored as a list of two numbers. However, the arrangement of the numbers differs. Double-click e in the Workspace window and you see two separate entries, as shown in Figure 5-2. Notice that the Variables window shows e as a 2-x-1 list in which the entries flow vertically.

FIGURE 5-2: Typing semicolon-separated numbers produces rows of values.

Separating values with a comma or a semicolon

It's possible to create a matrix by combining commas and semicolons. The commas separate entries in the same row and the semicolons create new rows. To see this for yourself, type **a=[1, 2; 3, 4]** in the Command Window and press Enter. You see

```
a =
     1     2
     3     4
```

REMEMBER

Notice how the output looks like the linear algebra you're used to. MATLAB makes every effort to use a familiar interface when presenting information so that you don't have to think about how to interpret the data. If the output doesn't appear as you expect, it could be a sign that you didn't create the information you expected, either.

Finding dimensions of matrices with the Size column

Figures 5-1 and 5-2 show one way to obtain the size of a numeric list (it appears in the upper-left corner of the window). However, you can use an easier method. Click the down arrow on the right side of the Workspace window and select Choose Columns ⇨ Size from the context menu.

TIP

You may also find it helpful to display the minimum and maximum values for each entry. This information comes in handy when working with large vectors or matrices where the minimum and maximum values aren't obvious. To obtain this information, Choose Columns ⇨ Min, and then choose Choose Columns ⇨ Max.

Depending on your computer screen, you may need to click and drag the Size, Min, and Max columns more to the left so that you can see them. You can also resize the window. Figure 5-3 shows the results of the entries you created in the previous sections.

Workspace

Name	Value	Size	Min	Max
a	[1,2;3,4]	2x2	1	4
b	[5,6]	1x2	5	6
e	[5;6]	2x1	5	6

FIGURE 5-3: The Size column tells you the dimensions of your matrix or vector.

Creating a range of values using a colon

Typing each value in a list manually would be time-consuming and error-prone because you'd eventually get bored doing it. Fortunately, you can use the colon (:) to enter ranges of numbers in MATLAB. The number on the left side of the colon specifies the start of the range, and the number on the right side of the colon specifies the end of the range. To see this for yourself, type **g=[5:10]** and press Enter. You see

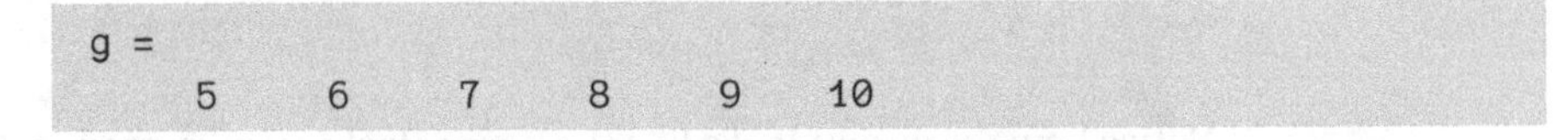

```
g =
     5     6     7     8     9    10
```

Creating a range of values using linspace()

Using the colon to create ranges has a problem. MATLAB assumes that the *step* (the interval between numbers) is 1. However, you may want the numbers separated by some other value. For example, you might want to see 11 values between the range of 5 and 10, instead of just 6.

The `linspace()` function solves this problem. You supply the starting value, the ending value, and the number of values you want to see between the starting and ending value. To see how `linspace()` works, type **g=linspace(5,10,11)** and press Enter. You see

```
g =
  Columns 1 through 5
    5.0000    5.5000    6.0000    6.5000    7.0000
  Columns 6 through 10
    7.5000    8.0000    8.5000    9.0000    9.5000
  Column 11
   10.0000
```

In this case, the step value is 0.5. Each number is 0.5 higher than the last, and there are 11 values in the output. The range is from 5 to 10, just as in the colon example in the previous section. In short, using `linspace()` is a little more flexible than using the colon, but using the colon requires less typing and is easier to remember.

Adding a step to the colon method

It turns out that you can also specify the step when using the colon method. However, in this case, you add the step between the beginning and ending of the range when defining the range. So, you type the beginning number, the step, and the ending number, all separated by colons. To try this method for yourself, type **g=[5:0.5:10]** and press Enter. You see

```
g =
  Columns 1 through 5
    5.0000    5.5000    6.0000    6.5000    7.0000
  Columns 6 through 10
    7.5000    8.0000    8.5000    9.0000    9.5000
  Column 11
   10.0000
```

This is precisely the same output as that of the `linspace()` example. However, when using this method, you specify the step directly, so you don't control the number of values you receive as output. When using the `linspace()` approach, you specify the number of values you receive as output, but MATLAB computes the step value for you. Each technique has advantages, so you need to use the one that makes sense for your particular need.

Transposing matrices with an apostrophe

Using the colon creates row vectors. However, sometimes you need a column vector instead. To create a column vector, you end the input with an apostrophe, which is also called a prime in MATLAB parlance. To see how this works for yourself, type **h=[5:0.5:10]'** and press Enter. You see

```
h =
     5.0000
     5.5000
     6.0000
     6.5000
     7.0000
     7.5000
     8.0000
     8.5000
     9.0000
     9.5000
    10.0000
```

When you look at the Workspace window, you see that `g` is a `1-x-11` vector, while `h` is an `11-x-1` vector. The first entry is a row vector and the second is a column vector.

You can transpose matrices as well. The rows and columns change position. For example, earlier you typed **a=[1,2;3,4]**, which produced

```
a =
     1     2
     3     4
```

To see how this matrix looks transposed, type **i=[1,2;3,4]'** and press Enter. You see

```
i =
     1     3
     2     4
```

NON-NUMERIC VECTORS AND MATRICES

Even though this chapter focuses almost exclusively on numeric vectors and matrices, MATLAB enables you to use other kinds of data to create vectors and matrices as well. In many cases, you see these other forms referred to as an *array*, rather than a *vector*, because the term *vector* has special meaning in math. For example, you can create a character array like this:

```
A = ['Hello, There!']
```

Notice that all the letters appear together in a single-quoted string, which is viewed by MATLAB as a 1-x-13 character array. However, when working with a character array, you can also create precisely the same array using `A = ['Hello, ', 'There!']`. However, if you place each of the character arrays in an individual array, separated by a semicolon, like this:

```
A = [['Hello,']; ['There!']]
```

you obtain a 2-x-6 character array, a matrix-like presentation, instead. Notice the use of a semicolon to separate dimensions. Again, MATLAB uses the term *array*, rather than *matrix*, because the term *matrix* has special meaning to mathematicians. Note that if you were to use `A = ['Hello, '; 'There!']` instead, you'd receive an error message telling you that the array dimensions aren't consistent, so the extra square brackets really are needed. A string array is somewhat different because you create it using a double-quoted series of character entries, like this:

```
B = ["One", "Two", "Three", "Four"]
```

The result is a 1-x-4 string array. Each string is separate. If you want a matrix-like presentation, you might use the following code:

```
B = [["One", "Two"]; ["Three", "Four"]]
```

which results in a 2-x-2 string array. Note that when working with string arrays, you can get by using `B = ["One", "Two"; "Three", "Four"]` instead (without the additional square brackets). Boolean arrays follow the same pattern as character and string arrays. You might create one like this:

```
C = [false, true, false, true]
```

The result is a 1-x-4 logical array. The point is that you need to maintain an open mind when working with vectors and matrices — it's not all about the numbers. You see other examples of non-numeric forms as the book progresses.

Adding and Subtracting

Now that you know how to enter vectors and matrices in MATLAB, it's time to see how to perform math using them. Adding and subtracting is a good place to start.

REMEMBER

The essential rule when adding and subtracting vectors and matrices is that they must be the same size. You can't add or subtract vectors or matrices of different sizes because MATLAB will display an error message. Use the following steps to see how to perform this task:

1. **Type** a=[1,2;3,4] **and press Enter.**

   You see

```
a =
     1     2
     3     4
```

2. **Type** b=[5,6;7,8] **and press Enter.**

   You see

```
b =
     5     6
     7     8
```

3. **Type** c = a + b **and press Enter.**

   This step adds matrix a to matrix b. You see

```
c =
     6     8
    10    12
```

4. **Type** d = b - a **and press Enter.**

   This step subtracts matrix b from matrix a. You see

```
d =
     4     4
     4     4
```

5. **Type** e=[1,2,3;4,5,6] **and press Enter.**

   You see

```
e =
     1     2     3
     4     5     6
```

If you attempt to add or subtract matrix e from either matrix a or matrix b, you see an error message. However, the following step tries to perform the task anyway.

6. **Type** f = e + a **and press Enter.**

   As expected, you see the following error message:

   ```
   Matrix dimensions must agree.
   ```

   The error messages differ a little between addition and subtraction, but the idea is the same. The matrices must be the same size in order to add or subtract them.

NON-NUMERIC ADDITION AND SUBTRACTION

It's possible to perform addition and subtraction with non-numeric data. In fact, when performing certain character and string processing tasks, you do just that. For example, if you have the matrix a=[1,2;3,4] and the matrix b=["a", "b"; "c", "d"], then c = a+b would create the following output:

```
c =
  2×2 string array
    "1a"    "2b"
    "3c"    "4d"
```

In this case, MATLAB converts the values in matrix a to strings, and then concatenates the values in matrix b to produce matrix c. However, the rules are different for character arrays, such as when you create d=['a', 'b'; 'c', 'd'] and then enter **e=a+d** to produce this output:

```
e =
    98   100
   102   104
```

This time, MATLAB converts the characters in matrix d to their numeric values (a=97, b=98, c=99, and d=100) and adds them to matrix a. You can also perform addition and subtraction with other non-numeric data types, such as logical arrays. In fact, you can also perform multiplication and division, as described in the next section, "Understanding the Many Ways to Multiply and Divide."

Understanding the Many Ways to Multiply and Divide

After addition and subtraction come multiplication and division. MATLAB is just as adept in meeting these needs as it is in every other area. The following sections describe the many ways in which you can use multiplication and division in MATLAB.

Performing scalar multiplication and division

A *scalar* is just technobabble for ordinary numbers. When you multiply ordinary numbers by vectors and matrices, you get a result where every element is multiplied by the number. To try this for yourself, type **a = [1,2;3,4] * 3** and press Enter. You see the following output:

```
a =
     3     6
     9    12
```

The example begins with the matrix, `[1,2;3,4]`. It then multiplies each element by `3` and places the result in `a`.

Division works in the same manner. To see how division works, type **b = [6, 9; 12, 15] / 3** and press Enter. You see the following output:

```
b =
     2     3
     4     5
```

Again, the example begins with a matrix, `[6, 9; 12, 15]`, and right divides it by `3`. The result is stored in `b`.

REMEMBER

MATLAB supports both *right division*, in which the left side is divided by the right side (what most people would consider the standard way of doing things), and *left division*, in which the left side is divided by the right side. When working with scalars, whether you use right division or left division doesn't matter as long as you divide the matrix by the scalar. To see this fact for yourself, type **c = 3 \ [6, 9; 12, 15]** and press Enter. (Notice the use of the backslash, \, for left division.) You get the same result as before:

```
c =
     2     3
     4     5
```

Employing matrix multiplication

Multiplication occurs at several different levels in MATLAB. The following sections break down the act of matrix multiplication so that you can see each level in progression.

Multiplying two vectors

Vectors are just matrices of only one row or column. Remember that you create a row vector by separating values using a comma, such as `[1, 2]`. To create column vectors, you use a semicolon, such as `[3; 4]`. You can also use prime to create a row or column vector. For example, `[3, 4]'` is equivalent to `[3; 4]`. (Pay particular attention to the use of commas and semicolons.)

REMEMBER

When you want to multiply one vector by another, you must have one row and one column vector. Try it for yourself by typing **d = [1, 2] * [3; 4]** and pressing Enter. You get the value `11` for output. Of course, the method used to perform the multiplication is to multiply the first element in the row vector by the first element of the column vector, and add the result to the multiplication of the second element of the row vector and the second element of the column vector. What you end up with is `d = 1 * 3 + 2 * 4`. This form of multiplication is also called an *inner product* or a *dot product.*

You can also use the `dot()` function to obtain a dot product by typing **d = dot([1, 2], [3; 4])** and pressing Enter. The `dot()` function is smarter than using the `*` operator because it will automatically transpose a vector when necessary. For example, typing **d = [1, 2, 3] * [4, 5, 6]** and pressing Enter produces an error, but typing **d = dot([1, 2, 3], [4, 5, 6])** and pressing Enter produces an output of `32`. To obtain the same value using the `*` operator, you need to type **d = [1, 2, 3] * [4; 5; 6]** and press Enter (note the semicolons), which transposes the second vector.

You can also create an *outer product* using MATLAB. In this case, each element in the first vector is multiplied by every element of the second vector (technically matrix multiplication), and the results of each multiplication are placed in a separate element. To put this in perspective, you'd end up with a 2-x-2 matrix consisting of `[3 * 1, 3 * 2; 4 * 1, 4 * 2]`. The easiest way to see how this works is by trying it yourself. Type **e = [3; 4] * [1, 2]** and press Enter. You see

```
e =
     3     6
     4     8
```

TIP

Another way to perform this task is to use the `bsxfun()` function: `e = bsxfun (@times, [1, 2], [3; 4]`. You supply a function name (see `https://www.mathworks.com/help/matlab/ref/bsxfun.html` or type **help('bsxfun')** and press Enter) to perform an element-by-element math operation on two objects (vectors, in this case). This example uses the `@times` function, which performs multiplication. The two inputs are a row vector and a column vector. The output is a 2-x-2 matrix using the same technique as the multiplication approach.

Multiplying a matrix by a vector

When performing multiplication of a matrix by a vector, the order in which the vector appears is important. Row vectors appear before the matrix, but column vectors appear after the matrix. To see how the row vector approach works, type **f = [1, 2] * [3, 4; 5, 6]** and press Enter. You see an output of

```
f =
    13    16
```

The first element is produced by 1 * 3 + 2 * 5. The second element is produced by 1 * 4 + 2 * 6. However, the number of rows in the matrix must agree with the number of elements in the vector. For example, if the vector has three elements in a row, the matrix must have three elements in a column to match. To see how this works, type **g = [1, 2, 3] * [4, 5; 6, 7; 8, 9]** and press Enter. The result is

```
g =
    40    46
```

The number of matrix columns controls the number of output elements. For example, if the matrix were to have three columns, the output would have three elements. To see this principle in action, type **h = [1, 2, 3] * [4, 5, 6; 7, 8, 9; 10, 11, 12]** and press Enter. The result is

```
h =
    48    54    60
```

Working with a column vector is similar to working with a row vector, except that the position of the vector and matrix are exchanged. For example, if you type **i = [4, 5, 6; 7, 8, 9; 10, 11, 12] * [1; 2; 3]** and press Enter, you see this result:

```
i =
    32
    50
    68
```

Notice that the output is a column vector instead of a row vector. The result is produced by these three equations:

```
1 * 4 + 2 * 5 + 3 * 6
1 * 7 + 2 * 8 + 3 * 9
1 * 10 + 2 * 11 + 3 * 12
```

The order of the multiplication differs because you're using a column vector instead of a row vector. MATLAB produces the same result as you would get when performing the task using other means, but you need to understand how the data-entry process affects the output.

Multiplying two matrices

When working with matrices, the number of columns in the first matrix must agree with the number of rows in the second matrix. For example, if the first matrix contains two rows containing three entries each, the second matrix must contain three rows and two entries each. To see this for yourself, type **j = [1, 2, 3; 4, 5, 6] * [7, 8; 9, 10; 11, 12]** and press Enter. You see the output as

```
j =
    58    64
   139   154
```

Notice that the number of rows in the first matrix determines the number of rows in the result. Likewise, the number of columns in the second matrix determines the number of columns in the result. The output of the first column, first row is defined by 1 * 7 + 2 * 9, + 3 * 11. Likewise, the output of the second column, first row is defined by 1 * 8 + 2 * 10 + 3 * 12. The matrix math works just as you would expect.

REMEMBER

Order is important when multiplying two matrices (just as it is when working with vectors). You can create the same two matrices, but obtain different results depending on order. If you reverse the order of the two matrices in the previous example by typing **k = [7, 8; 9, 10; 11, 12] * [1, 2, 3; 4, 5, 6]** and pressing Enter, you obtain an entirely different result:

```
k =
    39    54    69
    49    68    87
    59    82   105
```

Again, it pays to know how the output is produced. In this case, the output of the first column, first row is defined by 7 * 1 + 8 * 4. Likewise, the output of the second column of the first row is defined by 7 * 2 + 8 * 5.

Dividing two vectors

MATLAB produces an output if you try to divide two vectors, even though the result isn't useful. You can read the details at `https://van.physics.illinois.edu/qa/listing.php?id=24304` and many other places online. For example, if you type **l = [2, 3, 4] / [5, 6, 7]** and press Enter, you receive a result of

```
l =
    0.5091
```

Likewise, you could try typing **l = [2, 3, 4] \ [5, 6, 7]** and press Enter. The results would be different:

```
l =
         0         0         0
         0         0         0
    1.2500    1.5000    1.7500
```

You get the same reproducible results every time, but you can see that they're interesting, at best.

Effecting matrix division

As with matrix multiplication, matrix division takes place at several different levels. The following sections explore division at each level.

Dividing a vector by a scalar

Dividing a vector by a scalar and producing a usable result is possible. For example, type **m = [2, 4, 6] / 2** and press Enter. You see the following result:

```
m =
     1     2     3
```

Each of the entries is divided by the scalar value. Notice that this is right division. Using left division (`m = [2, 4, 6] \ 2`) would produce an unusable result; however, using `m = 2 \ [2, 4, 6]` would produce the same result as before. MATLAB would do its best to accommodate you with a result, just not one you could really use. (See the "Dividing two vectors" section for an explanation.)

Dividing a matrix by a vector

When dividing a matrix by a vector, defining the sort of result you want to see is important. For example, if you use `n = [2, 4] / [2, 4; 6, 8]`, you receive the following output, which is actually solving a system of linear equations (`xb = a`):

```
n =
     1     0
```

You can obtain the same result using the `mrdivide()` function (see `https://www.mathworks.com/help/matlab/ref/mrdivide.html`), and it's the equivalent of `n = [2, 4] * inv([2, 4; 6, 8])`.

However, most people want to perform an element-by-element division. In this case, you use *array* division: `n = [2, 4; 6, 8] ./ [2, 4]` (notice the dot before the division symbol). You see the following output:

```
n =
     1     1
     3     2
```

In this case, the element in column 1, row 1 is defined by 2 / 2. Likewise, the element in column 1, row 2 is defined by 6 / 2. You can also use the `bsxfun()` function with the `@rdivide` function to obtain the same result.

Dividing two matrices

When dividing two matrices, the dimensions of the two matrices must agree. For example, you can't divide a 3-x-2 matrix by a 2-x-3 matrix — both matrices must be the same dimensions, such as 3-x-2. To see how this works, type **o = [2, 4; 6, 8] / [1, 2; 3, 4]** and press Enter. You see the following result:

```
o =
     2     0
     0     2
```

Performing left division of two matrices is also possible. To see the result of performing left division using the same matrices, type **p = [2, 4; 6, 8] \ [1, 2; 3, 4]** and press Enter. Here's the result you see:

```
p =
    0.5000         0
         0    0.5000
```

REMEMBER

As with vector division in the previous section, matrix division isn't array division. What you really do is multiply one matrix by the inverse of the other. For example, using the two matrices in this section, you can accomplish the same result of left division by typing **q = [2, 4; 6, 8] * inv([1, 2; 3, 4])** and pressing Enter. To perform right division, you simply change the inverted matrix by typing **r = inv([2, 4; 6, 8]) * [1, 2; 3, 4]** and pressing Enter. The `inv()` function always returns the inverse of the matrix that you provide as input (assuming an inverse exists), so you can use it to help you understand precisely how MATLAB is performing the task. However, using the `inv()` function is computationally inefficient. To make your scripts run faster, dividing is always better.

You can use the `inv()` function in many ways. For example, multiplying any matrix by its inverse, such as by typing **s = [1, 2; 3, 4] * inv([1, 2; 3, 4])**, yields the identity matrix:

```
s =
    1.0000         0
    0.0000    1.0000
```

Element-by-element (array) division follows the same pattern with matrixes as it does with vectors. So, when you type **t = [2, 4; 6, 8] .\ [1, 2; 3, 4]** and press Enter, you obtain the following result:

```
t =
    0.5000    0.5000
    0.5000    0.5000
```

Likewise, you can perform right division using **u = [2, 4; 6, 8] ./ [1, 2; 3, 4]** to obtain the following result:

```
u =
     2     2
     2     2
```

Creating powers of matrices

Sometimes you need to obtain the power or root of a matrix. MATLAB provides several different methods for accomplishing this task. The most common method is to use the circumflex (^) to separate the matrix from the power to which you want to raise it. To see how this works, type **v = [1, 2; 3, 4]^2** and press Enter. The output is the original matrix squared, as shown here:

```
v =
     7    10
    15    22
```

To obtain the root of a matrix, you use a fractional value as input. For example, to obtain the square root of the previous example, you use a value of 0.5. To see this feature in action, type **x = [1, 2; 3, 4]^0.5** and press Enter. You see the following output:

```
x =
   0.5537 + 0.4644i   0.8070 - 0.2124i
   1.2104 - 0.3186i   1.7641 + 0.1458i
```

To verify your values, type **int32(x^2)** and press Enter. You should see the original array values. The reason you use `int32()` is to convert the complex numbers to integer values, just like the original matrix.

It's even possible to obtain the inverse of a matrix by using a negative power. For example, try typing **z = [1, 2; 3, 4]^(−1)** and pressing Enter (notice that the −1 is enclosed in parentheses to avoid confusion). You see the following output:

```
z =
   -2.0000    1.0000
    1.5000   -0.5000
```

CHECKING MATRIX RELATIONS

This chapter discusses a number of techniques to perform any given task. For example, you can create the inverse of a matrix using the `inv()` function, or you can simply raise it to a power of `-1`. The problem is that you don't really know that they are equal outputs. Fortunately, you can use relational operators with matrixes, as in: `inv([1, 2; 3, 4]) == [1, 2; 3, 4]^(-1)`. The output you see is a logical array:

```
ans =
  2×2 logical array
   1   1
   1   1
```

A value of `1` indicates true. Consequently, if you were to ask the opposite question, `inv([1, 2; 3, 4]) ~= [1, 2; 3, 4]^(-1)`, the output would be a logical array of `0`s, indicating false for each comparison. It shouldn't surprise you, then, that when you compare `[1, 3; 6, 8] >= [2, 4; 6, 8]`, to determine which elements in the first matrix are greater than or equal to the elements in the second matrix, you see a logical array output of

```
ans =
  2×2 logical array
   0   0
   1   1
```

TIP

MATLAB also provides the means for performing array powers and roots. For example, `aa = [1, 2; 3, 4].^2` produces the following output, in which each element is multiplied by itself (note the addition of the period before the circumflex):

```
aa =
     1     4
     9    16
```

Using complex numbers

Complex numbers consist of a real part and an imaginary part (see `https://www.mathsisfun.com/numbers/imaginary-numbers.html` for a quick overview of imaginary numbers). MATLAB uses the `i` and `j` constants to specify the imaginary part of the number. For example, when you compute the square root of the matrix `[1, 2; 3, 4]`, you obtain an output that contains imaginary numbers. To see this for yourself, type **ad = [1, 2; 3, 4]^0.5** and press Enter. You see the following result:

```
ad =
    0.5537 + 0.4644i   0.8070 - 0.2124i
    1.2104 - 0.3186i   1.7641 + 0.1458i
```

REMEMBER

The first column of the first row contains a real value of `0.5537` and an imaginary value of `0.4644i`. The `i` that appears after the value `0.4644` tells you that this is an imaginary number. The `j` constant means the same thing as the `i` constant, except that the `j` constant is used in electronics work (`i` is already used to represent current).

You can perform tasks with imaginary numbers just as you would any other number. For example, you can square the `ad` matrix by typing **ae = ad^2** and pressing Enter. The result might not be what you actually wanted, though:

```
ae =
    1.0000 + 0.0000i   2.0000 + 0.0000i
    3.0000 - 0.0000i   4.0000 + 0.0000i
```

After a matrix includes imaginary numbers, you need to convert them to obtain a desired format. For example, if you type **af = int32(ad^2)** and press Enter, you obtain the desired result, shown here:

```
af =
  2x2 int32 matrix
   1   2
   3   4
```

The `int32()` function performs the required conversion process for you. Of course, using `int32()`, or any other function of the same type, at the wrong time can result in data loss. For example, if you type **ag = int32([1, 2; 3, 4]^0.5)** and press Enter, you lose not only the imaginary part of the number but the fractional part as well. The output looks like this:

```
ag =
  2x2 int32 matrix
   1   1
   1   2
```

MATLAB assumes that you know what you're doing, so it doesn't stop you from making critical errors. The output conversion functions are

» double()

» single()

» int8()

» int16()

» int32()

» int64()

» uint8()

» uint16()

» uint32()

» uint64()

» complex()

Working with exponents

You use the matrix exponential to perform tasks such as solving differential equations (read about them at `http://www.sosmath.com/matrix/expo/expo.html`). MATLAB provides two functions for working with exponents directly (the article at `https://www.mathworks.com/help/matlab/math/matrix-exponentials.html` offers interesting reading if you want to see other ways to perform the task). The first is the `expm()` function, which calculates a standard matrix exponential. For example, when you type **ah = expm([1, 2; 3, 4])** and press Enter, you see this result:

```
ah =
   51.9690   74.7366
  112.1048  164.0738
```

MATLAB also makes it easy to calculate an element-by-element exponential using the `exp()` function. To see how this works, type **ai = exp([1, 2; 3, 4])** and press Enter. You see the following output:

```
ai =
    2.7183    7.3891
   20.0855   54.5982
```

Working with Higher Dimensions

A vector is one dimensional — just one row or one column. A matrix is a two-dimensional table, much like the kind you're used to with Excel spreadsheets, with rows being one dimension and columns being the other. You can go as high as you want. If a matrix is like a page in a book (two dimensions), three dimensions is like the book itself, and four like a shelf of books. In fact, there is no limit to the number of dimensions you can use to express an idea or data element. Images are an example of computational objects that rely on more than one dimension:

- The first dimension is the x coordinate of a pixel.
- The second dimension is the y coordinate of a pixel.
- The third dimension is the pixel color.

Now that you have a better idea of how you might use more than just two dimensions, it's time to see how you can implement them. The following sections describe how to work with multiple dimensions when using MATLAB.

Creating a multidimensional matrix

MATLAB provides a number of ways to create multidimensional arrays. The first method is to simply tell MATLAB to create it for you and fill each of the elements with zeros. The `zeros()` function helps you perform this task. To create a 2-x-3-x-3 matrix, you type **aj = zeros(2, 3, 3)** and press Enter. You see the following output:

```
aj(:,:,1) =
     0     0     0
     0     0     0
aj(:,:,2) =
     0     0     0
     0     0     0
aj(:,:,3) =
     0     0     0
     0     0     0
```

This output tells you that there are three stacked 2-x-3 matrices, and each one is filled with zeros. Of course, you might not want to start out with a matrix that's filled with zeros, so you can use another approach. The following steps help you create a 2-x-3-x-3 matrix that is already filled with data:

1. **Type** ak(:,:,1) = [1, 2, 3; 4, 5, 6] **and press Enter.**

   You see the following result:

   ```
   ak =
        1     2     3
        4     5     6
   ```

   This step creates the first page of the three-dimensional matrix. You want three pages, so you actually need to perform this step three times.

2. **Type** ak(:,:,2) = [7, 8, 9; 10, 11, 12] **and press Enter.**

   MATLAB adds another page, as shown:

   ```
   ak(:,:,1) =
        1     2     3
        4     5     6
   ak(:,:,2) =
        7     8     9
       10    11    12
   ```

   If you look at the Workspace window at this point, you see that the size column for ak is now 2 x 3 x 2. It's at this point that you see the third dimension added. Before you added this second page, MATLAB simply treated ak as a 2-x-3 matrix, but now it has the third dimension set.

3. **Type** ak(:,:,3) = [13, 14, 15; 16, 17, 18] **and press Enter.**

   The output now looks much like the `aj` output, except that the elements have values, as shown here:

   ```
   ak(:,:,1) =
        1     2     3
        4     5     6
   ak(:,:,2) =
        7     8     9
       10    11    12
   ak(:,:,3) =
       13    14    15
       16    17    18
   ```

You don't have to define assigned values using multiple steps. The `cat()` function lets you create the entire three-dimensional matrix in one step. The first entry that you make for the `cat()` function is the number of dimensions. You then add the data for each dimension, separated by commas. To see how this works, type **al = cat(3, [1, 2, 3; 4, 5, 6], [7, 8, 9; 10, 11, 12], [13, 14, 15; 16, 17, 18])** and press Enter. You see this output (which looks amazingly like the ak matrix):

```
al(:,:,1) =
     1     2     3
     4     5     6
al(:,:,2) =
     7     8     9
    10    11    12
al(:,:,3) =
    13    14    15
    16    17    18
```

REMEMBER

You may also decide that you don't want to type that much but still don't want zeros in the matrix. In this case, use the randn() function for random normally distributed data, or use the rand() function for uniformly distributed data. This function works just like the zeros() function, but it fills the elements with random data.

CREATING NON-NUMERIC MULTIDIMENSIONAL ARRAYS

You can create multidimensional arrays of data types other than numbers. For example, if you want a 2-x-3-x-3 array of strings, you can create it using strings(2, 3, 3). You can then assign values to the array elements much as you do with numbers.

In some cases, when working with non-numeric arrays, it's actually easier to start with a 2D array and then extend it. For example, you might start with a logical array like this: c = [true, false; true, false]. You could then extend it to a third dimension like this: c(:, :, 2) = [false, true; false, true], with an output like this:

```
  2x2x2 logical array
c(:,:,1) =
   1  0
   1  0
c(:,:,2) =
   0  1
   0  1
```

To see how this function works, type **am = randn(2, 3, 3)** and press Enter. You see a three-dimensional array filled with random data. Your output isn't likely to look precisely like the following output, but this output does give you an idea of what to expect:

```
am(:,:,1) =
    1.4090    0.6715    0.7172
    1.4172   -1.2075    1.6302
am(:,:,2) =
    0.4889    0.7269    0.2939
    1.0347   -0.3034   -0.7873
am(:,:,3) =
    0.8884   -1.0689   -2.9443
   -1.1471   -0.8095    1.4384
```

Accessing a multidimensional matrix

No matter how you create the matrix, eventually you need to access it. To access the entire matrix, you simply use the matrix name, as usual. However, you might not need to access the entire matrix. For example, you might need to access just one page. The examples in this section assume that you created matrix `ak` in the previous section. To see just the second page of matrix `ak`, you type **ak(:, :, 2)** and press Enter. Not surprisingly, you see the second page, as shown here:

```
ans =
     7     8     9
    10    11    12
```

The colon (`:`) provides a means for you to tell MATLAB that you want the entire range of a matrix dimension. The values are rows, columns, and pages in this case. So the request you made was for the entire range of page 2. You could ask for just a row or column. To get the second row of page 2, you type **ak(2, :, 2)** and press Enter. The output looks like this:

```
ans =
    10    11    12
```

The second column of page 2 is just as easy. In this case, you type **ak(:, 2, 2)** and press Enter. The output appears in column format, like this:

```
ans =
     8
    11
```

Accessing an individual value means providing all three values. When you type **ak(2, 2, 2)** and press Enter, you get 11 as the output because that's the value in row 2, column 2, of page 2 for matrix ak.

You also have access to range selections for multidimensional matrices. In this case, you must provide a range for one of the entries. For example, if you want to obtain access to row 2, columns 1 and 2, of page 2 for matrix ak, you type **ak(2, [1:2], 2)** and press Enter. Notice that the range appears within square brackets, and the start and end of the range are separated by a colon. Here is the output you see in this case:

```
ans =
    10    11
```

REMEMBER

The use of ranges works wherever you need them. For example, say that you want rows 1 and 2, columns 1 and 2, of page 2. You type **ak([1:2], [1:2], 2)** and press Enter. The result looks like this:

```
ans =
     7     8
    10    11
```

Replacing individual elements

As you work through problems and solve difficulties, you might find changing some of the data in a matrix necessary. The problem is that you don't want to have to re-create the matrix from scratch just to replace one value. Fortunately, you can replace individual values in MATLAB. The examples in this section assume that you created matrix ak in the "Creating a multidimensional matrix" section, earlier in this chapter.

The previous section tells you how to access matrix elements. You use this ability to change values. For example, the value in row 2, column 2, of page 2 in matrix ak is currently set to 11. You may decide that you really don't like the number 11 there and want to change it to 44 instead. To perform this task, type **ak(2, 2, 2) = 44** and press Enter. You see the following result:

```
ak(:,:,1) =
     1     2     3
     4     5     6
ak(:,:,2) =
     7     8     9
    10    44    12
```

```
ak(:,:,3) =
    13    14    15
    16    17    18
```

TIP

Notice that MATLAB displays the entire matrix. Of course, you may not want to see the entire matrix every time you replace a single value. In this case, end the command with a semicolon. When you type **ak(2, 2, 2) = 44;** and press Enter, the change still takes place, but you don't see the result onscreen. For now, continuing to display the information is a good idea so that you can tell whether you have entered the commands correctly and have obtained the desired result.

Replacing a range of elements

If you have a number of values to replace in a matrix, replacing them one at a time would become boring. More important, you start to make mistakes after a while, and your results don't come out as you thought they would. Replacing a range of values with a single command is the best idea in this case. The examples in this section assume that you created matrix ak in the "Creating a multidimensional matrix" section, earlier in this chapter.

You have many different ways to make replacements to a range of elements in your existing matrix. Of course, before you can replace a range of elements, you need to know how to access them. The "Accessing a multidimensional matrix" section, earlier in this chapter, tells you how to access matrix elements.

You can make a single value replacement for a range. Say that you want to replace row 2, columns 1 and 2, of page 2 with the number 5. To perform this task, type **ak(2, [1:2], 2) = 5** and press Enter. The single value appears in both places, as shown in this output:

```
ak(:,:,1) =
     1     2     3
     4     5     6
ak(:,:,2) =
     7     8     9
     5     5    12
ak(:,:,3) =
    13    14    15
    16    17    18
```

Of course, a single value replacement might not work. You can also create range replacements in which you replace each element with a different value. For example, you might want to replace row 2, column 1, of page 2 with the number 22, and row 2, column 2, of page 2 with the number 33. To perform this task, you type **ak(2, [1:2], 2) = [22, 33]** and press Enter. Here is the output you see:

```
ak(:,:,1) =
    1     2     3
    4     5     6
ak(:,:,2) =
    7     8     9
   22    33    12
ak(:,:,3) =
   13    14    15
   16    17    18
```

Column changes work the same way. In this case, you might want to replace row 1, column 3, of page 2 with the number 44, and row 2, column 3, of page 2 with the number 55. To perform this task, you type **ak([1:2], 3, 2) = [44, 55]** and press Enter. Notice that you didn't have to define the input vector using a column format. Here's the result you see:

```
ak(:,:,1) =
    1     2     3
    4     5     6
ak(:,:,2) =
    7     8    44
   22    33    55
ak(:,:,3) =
   13    14    15
   16    17    18
```

When replacing a rectangular range, you need to use a proper matrix for input. For example, you might want to replace a rectangular range between columns 1 and 3, rows 1 and 2, of page 2 with the values 7 through 12. To perform this task, you type **ak(1:2, 1:3, 2) = [7:9;10:12]** and press Enter. Notice that the input uses ranges in this case, too. Here's the result you see:

```
ak(:,:,1) =
    1     2     3
    4     5     6
ak(:,:,2) =
    7     8     9
   10    11    12
```

```
ak(:,:,3) =
    13    14    15
    16    17    18
```

Modifying the matrix size

You might not think that resizing a matrix is possible, but MATLAB can do that, too. It can make the matrix larger or smaller. The technique for making the matrix smaller is a bit of a trick, but it works well, and you likely will have a need for it at some point. The examples in this section assume that you created matrix ak in the "Creating a multidimensional matrix" section, earlier in this chapter.

As with range replacement, you need to know how to access ranges before you start this section. The "Accessing a multidimensional matrix" section, also earlier in this chapter, tells you how to access matrix elements.

The current ak matrix is 2 x 3 x 3. You might want to add another row, even if that row consists only of zeros, to make the matrix square for some advanced task you need to perform. Some tasks work properly only with square matrices, so this is a real concern. To add another row to the existing matrix, type **ak(3, :, :) = 0** and press Enter. You see the following result:

```
ak(:,:,1) =
     1     2     3
     4     5     6
     0     0     0
ak(:,:,2) =
     7     8     9
    10    11    12
     0     0     0
ak(:,:,3) =
    13    14    15
    16    17    18
     0     0     0
```

TIP

All three pages now have another row. However, you might decide that you really don't want that extra row after all. To delete the row, you need to perform a bit of a trick — you set the row to a null (empty) value using an empty matrix ([]). To see how this works, type **ak(3, :, :) = []** and press Enter. You see the following result:

```
ak(:,:,1) =
     1     2     3
     4     5     6
ak(:,:,2) =
     7     8     9
    10    11    12
ak(:,:,3) =
    13    14    15
    16    17    18
```

At this point, you probably wonder what would happen if you added a column or row to just a single page. Try typing **ak(:, 4, 1) = [88, 99]** and pressing Enter. This command adds a fourth column to just page 1 and fills it with the values 88 and 99. MATLAB provides the following output:

```
ak(:,:,1) =
     1     2     3    88
     4     5     6    99
ak(:,:,2) =
     7     8     9     0
    10    11    12     0
ak(:,:,3) =
    13    14    15     0
    16    17    18     0
```

Notice that the other pages also have a fourth column now. The column is filled with zeros, but MATLAB automatically adds it for you to keep things tidy. Note that if you try to get rid of a column in the same way by typing **ak(:, 4, 1) = []** and pressing Enter, you see a `A null assignment can have only one non-colon index.` error message. You must type **ak(:, 4, :) = []** and press Enter instead.

Using cell arrays and structures

The matrices you have created so far all contain the same data type, such as `double` or `uint8`. Every matrix you create will contain data of the same type — you can't mix types in a matrix. When performing certain tasks, such as machine learning, you need other structures:

- A *cell array* works much like a spreadsheet.
- A *structure* works much like a database record.

These two containers let you store other kinds of data, and mix and match types as needed. You can use them to create a small database or some sort of alternative storage on your machine without resorting to another application. The following sections provide an introduction and point you to more help in case you need to know more.

Understanding cell arrays

Cell arrays are naturals for spreadsheets because an individual cell in a cell array is like a cell in a spreadsheet. In fact, when you import a spreadsheet into MATLAB, each cell in the spreadsheet becomes a cell in a MATLAB cell array. Because spreadsheets are so popular, you're more likely to encounter a cell array than a structure.

You use the `cell()` function to create a new cell array. For example, to create a 2-x-2-x-2 cell array, you type **an = cell(2, 2, 2)** and press Enter. You see this result:

```
  2×2×2 cell array
an(:,:,1) =
    {0×0 double}    {0×0 double}
    {0×0 double}    {0×0 double}
an(:,:,2) =
    {0×0 double}    {0×0 double}
    {0×0 double}    {0×0 double}
```

The cells are empty at this point. Cell arrays rely on a different kind of brackets, the curly braces ({}), to provide access to individual elements. In order to make the `an` cell array useful, begin by typing the following lines of code, pressing Enter after each line:

```
rng(5)
an{1,1,1}='George';
an{1,2,1}='Smith';
an{2,1,1}=rand();
an{2,2,1}=uint16(1953);
an{1,1,2}=true;
an{1,2,2}=false;
an{2,1,2}=14.551+2.113i;
an{2,2,2}='The End!'
```

Because all the lines except for the last one ended with a semicolon, you didn't see any output. However, after you type the last line, you see the following output from MATLAB:

```
  2x2x2 cell array
an(:,:,1) =
    {'George'}     {'Smith'}
    {[0.2220]}     {[ 1953]}
an(:,:,2) =
    {[                1]}    {[       0]}
    {[14.5510 + 2.1130i]}    {'The End!'}
```

TIP

Note that this example sets the seed value for the random number generator using `rng(5)` so that the first value produced by `rand()` is the same unless you specify a different generator (see the `rng()` documentation at `https://www.mathworks.com/help/matlab/ref/rng.html` for details). The advantage of this approach is that you obtain the same numbers in the same sequence when you perform testing. However, you want to use a unique seed when working with production applications so that the `rand()` output truly does appear random.

The output looks a little different from a multidimensional matrix. You can access it the same way, except that you use curly braces. For example, type **an{1, :, 2}** and press Enter to see the first row of page 2. The result looks like this:

```
ans =
  logical
   1
ans =
  logical
   0
```

REMEMBER

Each of the entries is treated as a separate item, but you can select ranges and work with individual values, just as you do when working with a multidimensional matrix. However, you must use the curly braces when working with cell arrays.

TIP

You can distinguish between cell arrays and matrices in the Workspace window by the icons they use. The cell array icon contains a pair of curly braces, so it contrasts well with the matrix icon, which looks like a mini table. The Value column also specifically tells you that the entry is a cell rather than a specific data type, such as a double. You can find additional information about cell arrays at `https://www.mathworks.com/help/matlab/cell-arrays.html`.

Understanding structures

Structures are more closely related to SQL database tables than spreadsheets. Each entry consists of a field name and value pair. The field names are generally descriptive strings, but the values can be anything that relates to that field name. To get a better idea of how a structure works, type **MyStruct = struct('FirstName', 'Amy', 'LastName', 'Jones', 'Age', 32, 'Married', false)** and press Enter. You see the following output:

```
MyStruct =
  struct with fields:

    FirstName: 'Amy'
     LastName: 'Jones'
          Age: 32
      Married: 0
```

Notice how the field names are paired with their respective values. A structure is designed to reside in memory like a database. Currently, `MyStruct` has just one record in it. You can access this record by typing **MyStruct(1)** and pressing Enter. The results are as follows:

```
ans =
  struct with fields:

    FirstName: 'Amy'
     LastName: 'Jones'
          Age: 32
      Married: 0
```

Dealing with an entire record probably isn't what you had in mind, though. To access a particular field, you type a period, followed by the field name. For example, type **MyStruct.LastName** and press Enter to access the `LastName` field. You get the following answer:

```
ans =
    'Jones'
```

Because `MyStruct` contains just one record, you don't have to include the record number to access the `LastName` field. A single record structure isn't very useful. You might have quite a few records in a real structure. To add another record to `MyStruct`, type **MyStruct(2) = struct('FirstName', 'Harry', 'LastName', 'Smith',**

'Age', 35, 'Married', true) and press Enter. The output might surprise you this time. You see

```
MyStruct =
1x2 struct array with fields:
    FirstName
    LastName
    Age
    Married
```

The output tells you how many records are in place. You can test for the second record by typing **MyStruct(2)** and pressing Enter. The output is precisely as you expect:

```
ans =
  struct with fields:

    FirstName: 'Harry'
     LastName: 'Smith'
          Age: 35
      Married: 1
```

TIP

Don't limit your input of structures to the common data types. Structure data may contain a matrix, even multidimensional matrices, and you can mix sizes. In addition, structures and cell arrays can contain each other. An element in a structure can be a cell array, and a cell in a cell array can be a structure. The point is that these are extremely flexible ways to store information when you need them; however, you shouldn't make things overly complex by using them when you don't need them. If you can create storage that uses one common data type, matrices are the way to go. You can find additional information about structures at `https://www.mathworks.com/help/matlab/structures.html`.

Using the Matrix Helps

As you work with matrices, you may need to test your code, and MATLAB has provided some help in the form of ways to create a matrix (Table 5-1), test matrices (Table 5-2), and diagnose matrix problems (Table 5-3). The tables in this section help you work more productively with matrices and get them working considerably faster. You can find additional matrix help at `https://www.mathworks.com/help/matlab/matrices-and-arrays.html`.

TIP

When working with functions such as `rand()` and `randn()`, MATLAB provides a pseudo-random value based on a constantly changing seed value, which may not prove effective for testing. To obtain the same series of values every time you start your application, you can use the `rng()` function to set the seed value at the outset. In this way, the output of `rand()` or `randn()` will be the same every time you start the testing over. You can discover more about `rng()` at `https://www.mathworks.com/help/matlab/ref/rng.html`.

TABLE 5-1 Matrix Creation

Function	What It Does	Generic Call	Example

`zeros()`	Creates a matrix of all zeros	`zeros(`*`<mat_size>`*`)`, where *`<mat_size>`* is a positive integer number, two number arguments, or a vector of numbers.	`>> zeros(3)`
			`ans =`
			`0 0 0`
			`0 0 0`
			`0 0 0`

`ones()`	Creates a matrix of ones	`ones(`*`<mat_size>`*`)`, where *`<mat_size>`* is a positive integer number, two number arguments, or a vector of numbers.	`>> ones(3)`
			`ans =`
			`1 1 1`
			`1 1 1`
			`1 1 1`

`eye()`	Creates an identity matrix with one on the main diagonal and zero elsewhere	`eye(`*`<mat_size>`*`)`, where *`<mat_size>`* is a positive integer number, two number arguments, or a vector of numbers. This call doesn't allow you to create N-dimensional arrays, which have more than two dimensions. The `rand()` and `randn()` functions do allow for N-dimensional arrays.	`>>eye(3)`
			`ans=`
			`1 0 0`
			`0 1 0`
			`0 0 1`

`rand()`	Creates a matrix of uniformly distributed random numbers	`rand(`*`<mat_size>`*`)`, where *`<mat_size>`* works similarly to the argument(s) of `eye()`. You could also use `rand(3, 3, 3)` to produce a 3-x-3-x-3 array.	`>>rand(3)`
			`ans=`
			`0.8147 0.9134 0.2785`
			`0.9058 0.6324 0.5469`
			`0.1270 0.0975 0.9575`

Function	What It Does	Generic Call	Example

`randn()`	Creates a matrix of normally distributed random numbers (mean=0, SD=1)	`randn(`*<mat_size>*`)`, where *<mat_size>* works similarly to the argument(s) of `eye()`. You could also use `randn(3, 3, 3)` to produce a 3-x-3-x-3 array.	`>> randn(3)`
			`ans =`
			`0.5377 0.8622 -0.4336`
			`1.8339 0.3188 0.3426`
			`-2.2588 -1.3077 3.5784`

`blkdiag()`	Makes a block diagonal matrix	`blkdiag(`*a*`,`*b*`,`*c*`,...)`, where *a*, *b*, *c*, ... are matrices.	`>> blkdiag(ones(2), ones(2))`
			`ans =`
			`1 1 0 0`
			`1 1 0 0`
			`0 0 1 1`
			`0 0 1 1`

TABLE 5-2 **Test Matrices**

Function	What It Does	Generic Call	Example

`magic()`	Creates a magic square matrix — the sum of rows, columns, and diagonals are the same.	`magic(n)`, where *n* is the number of rows and columns.	`>> magic(3)`
			`ans =`
			`8 1 6`
			`3 5 7`
			`4 9 2`

`gallery()`	Produces a wide variety of test matrices for diagnosis of your code	`gallery(...'`*<option>*`',...`*<mat_size>*`,`*j*`)`,where *'option'* is a string that defines what task to perform, such as binomial, which creates a binomial matrix. *<mat_size>* is a positive integer number, two number arguments, or a vector of numbers. Each different positive integer *j* produces a different matrix.	`>> gallery ('normaldata',3,3)`
			`ans =`
			`0.9280 -0.7230 0.2673`
			`0.1733 -0.5744 1.3345`
			`-0.6916 -0.3077 -1.3311`

TABLE 5-3 Helpful Commands

Function	What It does	Generic Call	Example
rng()	Controls the random number generator	rng(<my_seed>, '*<my_option>*'),where *<my_seed>* is a numeric value used to define the starting point for random values and *'<my_option>'*is the option used to set the random number generator.	rng('default') resets the random number generator to a known value. This command is useful to reproduce random matrices.
size()	Returns the size of a matrix	size(*<your_matrix>*)	>> size(zeros([2,3,4])) ans = 2 3 4
length()	Returns the length of a vector	length(*<your_matrix>*)	>> length(0:50) ans = 51
spy()	Produces a figure identifying where zeros are in a matrix	spy(*<your_matrix>*)	>> spy(blkdiag(ones(100),... ones(200),ones(100)))

IN THIS CHAPTER

» Defining and understanding plots

» Working with the plot function

» Changing plot specifics

» Creating 2D plots

Chapter 6
Understanding Plotting Basics

MATLAB includes fabulous routines for plotting (or graphing) the data and expressions that you supply to the software. Using MATLAB's familiar interface, you can produce visual representations of various functions and data sets, including 2D x-y graphs, log scales, bar, and polar plots, as well as many other options. The visuals that MATLAB produces resemble anything from the graph of an algebraic equation to pie charts often used in business and to specialized graphs.

In this chapter, you find out how to use 2D plotting functions to create expression and data plots and how the same process works with other plotting routines in MATLAB. You also discover the commonly used visual styles for representing various types of data, how to combine plots, and how to modify the plots to match specific data sets.

Considering Plots

A *plot* is simply a visualization of data. Most people see a series of numbers in a table and can't really understand its meaning. Interpreting what the data means is hard to do without thinking about the relationship between data points. A plot makes the relationships between data points more obvious to the viewer and helps the viewer see patterns in the data. The following sections help you discover how MATLAB plots are special and can make the visualization of your data interesting and useful.

Understanding what you can do with plots

People are visually oriented. You could create a standard table showing the data points for a sine wave and have no one really understand that it was a sine *wave* at all or that the data points *move* in a certain way. However, if you plot that information, it becomes apparent to everyone that a sine wave has a particular presentation and appearance. The pattern of the sine wave becomes visible and understandable.

A sine wave consists of a particularly well-known set of data points, so some people might recognize the data for what it is. As your data becomes more complex, however, recognizing the patterns becomes more difficult — to the point at which most people won't understand what they're seeing. So the first goal of a plot is to make the pattern of data readily apparent.

Presentation is another aspect of plotting. You can take the same data and provide multiple views of it to make specific points — the company hasn't lost much money on bad widgets, for example, or the company has gained quite a few new customers due to some interesting research. Creating the right plot for your data defines a specific view of the data: It helps you make your point about whatever the data is supposed to represent.

Creative interaction with the data is another reason to use plots. People see not only the patterns that are present in plots but also see the ones that could be present given the right change in conditions. It's the creative interaction that makes plotting data essential for scientists and engineers. The ability to see beyond the data is an important part of the plotting process.

Comparing MATLAB plots to spreadsheet graphs

Although it might seem obvious at first, spreadsheet graphs are generally designed for use in business. As a result, the tools you find are better suited to making a point about some business need, such as this quarter's sales or the project production rate in the factory. A spreadsheet graph includes the tools of business, such as the need to add trend lines of various sorts to show how the numbers are changing over time.

MATLAB plots are more suited to scientific and engineering needs. A MATLAB plot does include some of the same features as a spreadsheet graph. For example, you can create a pie chart in either environment and assign data points to the chart in about the same manner. However, MATLAB includes plots that you can't find in the business environment, such as a semilogx (used to plot logarithmic data; see `https://www.mathworks.com/help/matlab/ref/semilogx.html`). A business user probably wouldn't have much need for a stem plot — the plot that shows the frequency at which certain values appear.

The way in which the two environments present information differs as well. A spreadsheet graph is designed to present an overview in an aesthetically pleasing manner. The idea is to convince a viewer of the validity of the data by showing general trends. Business users tend not to have time to dig into the details; they need to make decisions quickly based on trends. MATLAB graphs are all about the details. With this in mind, you can zoom in on a graph, examine individual data points, and work the plot in ways that a business user doesn't require.

REMEMBER

No best approach to presenting information in graphic form exists. The only thing that matters is displaying the information in a manner that most helps the viewer. The essential difference in the two environments is that one allows the viewer to make decisions quickly and the other allows the viewer to make decisions accurately. Each environment serves its particular user's needs.

Creating a plot using commands

MATLAB makes creating a plot easy. Of course, before you can create any plot, you need a source of data to plot. The following steps help you create a data source and then use that data source to generate a plot. Even though MATLAB's plotting procedure looks like a really simplistic approach, it's actually quite useful for any data you want to plot quickly. In addition, it demonstrates that you don't even have to open any of the plotting tools to generate a plot in MATLAB.

1. **Type** x = -pi:0.01:pi; **and press Enter in the Command Window.**

   MATLAB generates a vector, x, and fills it with a range of data points for you. The data points begin at -pi and end at pi, using 0.01 steps. The use of the semicolon prevents the output of the data points to the Command Window, but if you look in the Workspace window, you see that the vector has 629 data points.

2. **Type** plot(x, sin(x)), grid on **and press Enter.**

   The plot shown in Figure 6-1 appears. It's a sine wave created by MATLAB using the input you provided.

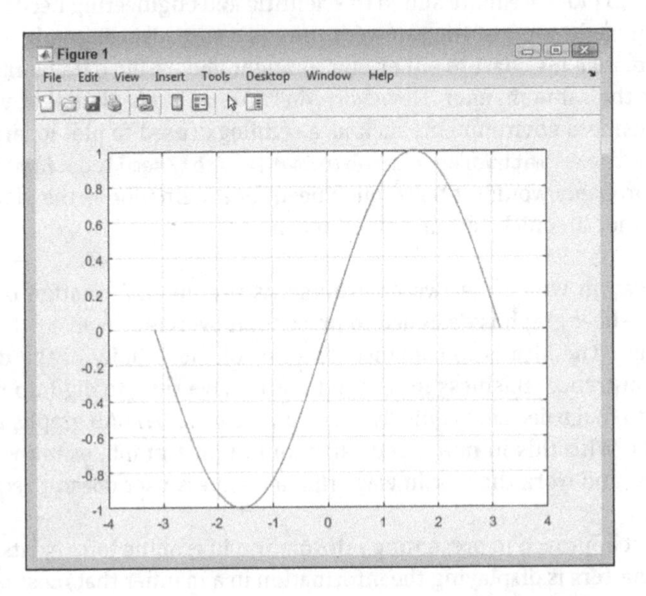

FIGURE 6-1: The plot uses all the defaults that MATLAB provides, except for turning the grid on.

The `plot()` function accepts the data point entries that you provide. The vector `x` contains a series of values between –pi and pi. Taking the sine of each of these values using the `sin()` function creates the values needed to generate the plot shown. This version of the `plot()` function shows the minimum information that you can provide. The `x` value that appears first contains the information for the x-axis of the plot. The `sin(x)` entry that appears second contains the information for the y-axis of the plot.

TIP

The addition of the `grid` on command displays a grid on the plot to make it easier to see line values. Normally, MATLAB displays a plot without a grid to keep from hiding values from view or potentially causing misinterpretations. You can add a graphing command, like `grid` on, by adding to the same line as `plot()` with a comma separation.

REMEMBER

You can create any sort of plot using commands, just you can use the graphic aids (such as the GUI shown in Figure 6-1) that MATLAB provides. For example, type **area(x,sin(x)), grid** and press Enter. You see the plot shown in Figure 6-2. However, this time the sine wave is shown as an area plot. MATLAB also has methods for modifying the appearance of the plot using commands.

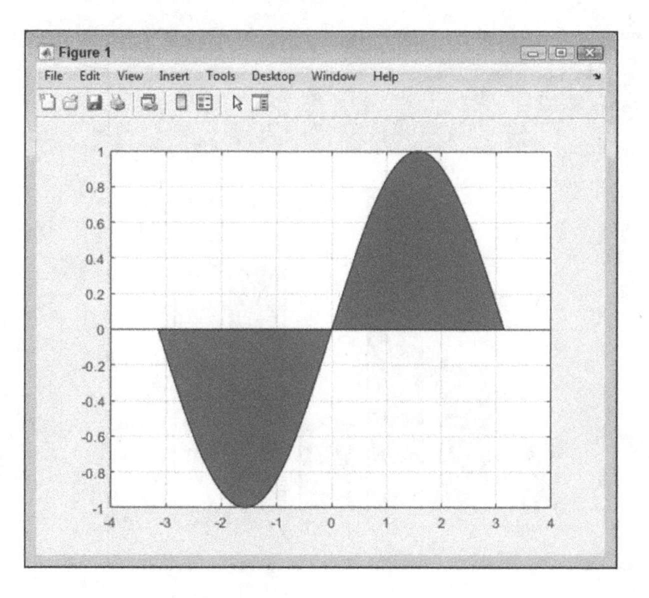

FIGURE 6-2: You can do anything with commands that you can do with the GUI.

Creating a plot using the Workspace window

The Workspace window displays all the variables that you create, no matter what type they might be. What you may not realize is that you can right-click any of these variables and create a plot from them. (If you don't see your plot listed, select the Plot Catalog option on the context menu to see a full listing of the available plots.) The following steps help you create a variable and then plot it using the Workspace window functionality.

1. **Type** `y = [5, 10, 22, 6, 17];` **and press Enter in the Command Window.**

   You see the variable `y` appear in the Workspace window.

2. **Right-click** `y` **in the Workspace window and choose** `bar(y)` **from the context menu that appears.**

   MATLAB creates a bar graph using the default settings, as shown in Figure 6-3. Notice that MATLAB labels this bar graph Figure 1.

TIP

Even though this method might seem really limited, it's a great way to create a quick visualization of data so that you can see patterns or understand how the various data points interact. The advantage of this method is that it's quite fast. In addition, the output varies based on the data you provide. For example, try the example using `y1 = [5, 10, 22, 6, 17; 2, 8, 7, 19, 21];` (row-oriented) and `y2 = [5, 10, 22, 6, 17; 2, 8, 7, 19, 21]';` (column-oriented) to see a difference.

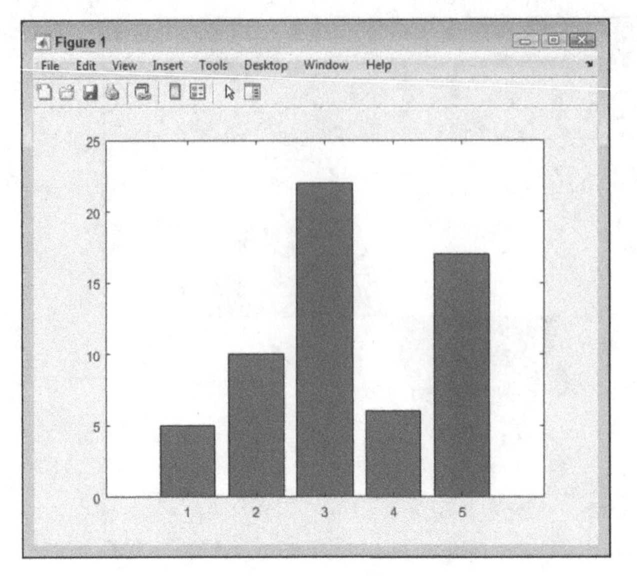

FIGURE 6-3: Bar graphs are best used for a few discrete values that you want to compare.

WARNING

MATLAB overwrites the previous plot you create when you create a new plot, unless you use the `hold` command that is described later in the chapter. If you created the examples in the previous section, you should note that all the plots have appeared in the Figure 1 window and that no new plot windows have been created. Your old plot is immediately overwritten when you create a new one unless you save the old plot to disk or use the `hold` command.

Creating a plot using the Plots tab options

When you view the Plots tab in MATLAB, you see a gallery of the kinds of plots you can create. You initially see just a few of the available plots. However, if you click the downward-pointing arrow button at the right side of the gallery, you see a selection of plot types like the one shown in Figure 6-4.

TIP

Note the option buttons at the bottom with Plots for `y` currently selected. The default is to show only the plots you can use with the currently selected data. If you choose All Plots instead, the list will also show plots that need additional input, such as a contour plot in this case. Unusable plots appear grayed out. Hover your mouse over the plot entry for a few seconds to determine what you need to use the plot, such as a matrix for the contour plot. The All Plots option is helpful when you try to remember a plot and can't seem to find it in the standard list.

To use this feature, select a variable in the Workspace window and then choose one of the plots from the gallery list. This is the technique to use if you can't quite remember what sort of plot you want to create (making the command option less convenient), or if the option doesn't appear in the Workspace window context menu. For example, you might want to create a horizontal bar plot using variable `y`.

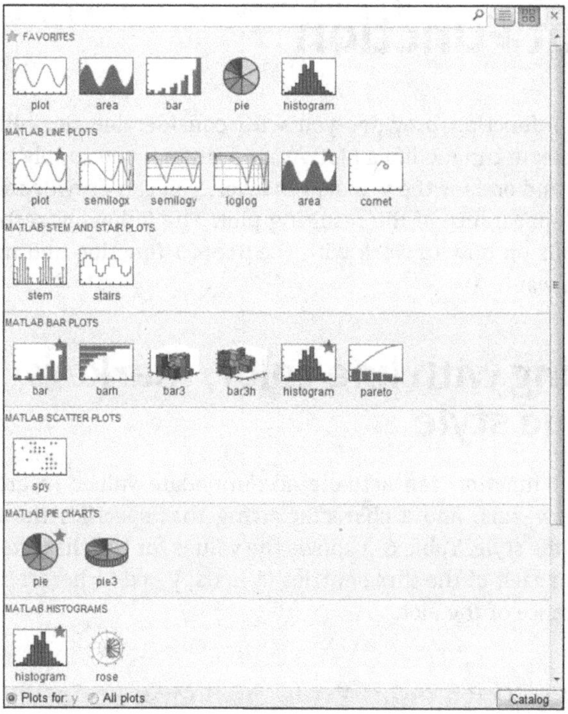

FIGURE 6-4: MATLAB comes with a large number of plot types that you can use.

To perform this task, simply click variable `y` in the Workspace window and then choose `barh` in the MATLAB Bar Plots section of the gallery. The output that MATLAB comes up with looks like Figure 6-5.

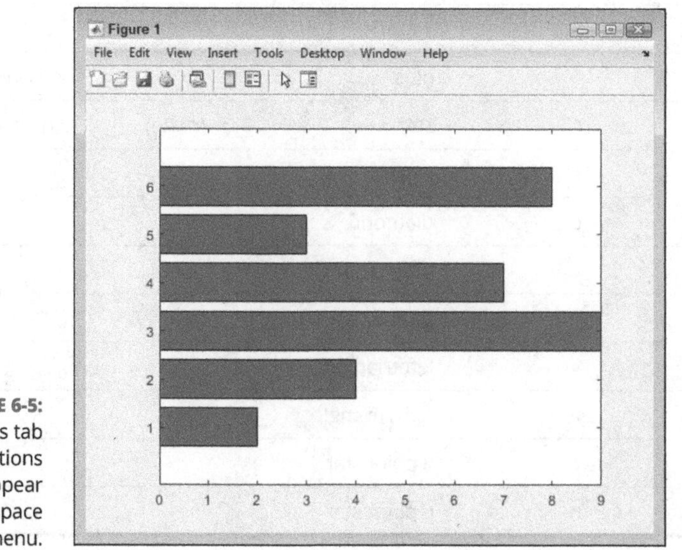

FIGURE 6-5: The Plots tab contains options that don't appear on the Workspace context menu.

Using the Plot Function

The plot() function provides you with considerable flexibility in using commands to create and modify a plot. As a minimum, you supply two vectors: one for the x-axis and one for the y-axis. However, you can provide more information to adjust the appearance of the resulting plot. The following sections provide additional details on how to work with the plot() function and make it provide the output you want.

Working with line color, markers, and line style

The plot() function can actually accommodate values in groups of three: the x-axis, the y-axis, and a character string that specifies the line color, marker style, and line style. Table 6-1 shows the values for the character string; you'd use values from each of the three entries (x-axis, y-axis, character string) to change the appearance of the plot.

TABLE 6-1 Line Color, Data Point Style, and Line Style

Color		Marker		Style	
Code	Line Color	Code	Marker Style	Code	Line Style
b	blue	.	point	-	Solid
g	green	o	circle	:	Dotted
r	red	x	x-mark	-.	dash dot
c	cyan	+	plus	--	Dashed
m	magenta	*	star	(none)	no line
y	yellow	s	square		
k	black	d	diamond		
w	white	v	down triangle		
		^	up triangle		
		<	left triangle		
		>	right triangle		
		p	5 point star		
		h	6 point star		

You can combine the entries in various ways. For example, type **plot(1:length(y), y, 'r+--')** and press Enter to obtain the plot shown in Figure 6-6. Note that the `1:length(y)` argument creates a vector with the values 1 through 5. Even though you can't see it in the book, the line is red. The markers show up as plus signs, and the line is dashed, as you might expect. Note that the vectors must be the same length, or you see an error message, so creating a range of `1` to `length(y)` prevents the error message.

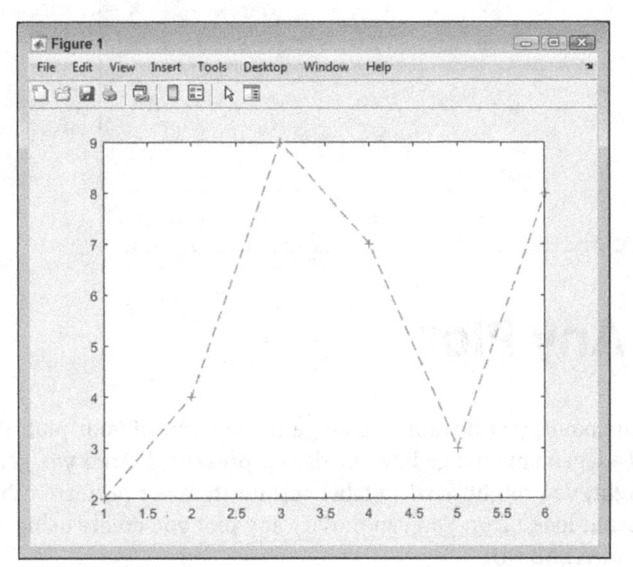

FIGURE 6-6: Spruce up your plot using styles.

Creating multiple plots in a single command

In many cases, you need to plot more than one set of data points when working with a plot. The `plot()` function can accommodate as many series as needed to display all your data. For example, you might want to plot both sine and cosine of x to compare them. To perform this task, you type **plot(x, sin(x), 'g-', x, cos(x), 'b-')** and press Enter (remember that `x` was defined earlier as `x = -pi:0.01:pi;`). Figure 6-7 shows the result.

In this case, sine appears as a green solid line. The value of cosine is in blue with a solid line. Notice that each series appears in the `plot()` command as three values: x-axis, y-axis, and format string. You can add as many series as needed to complete your plot.

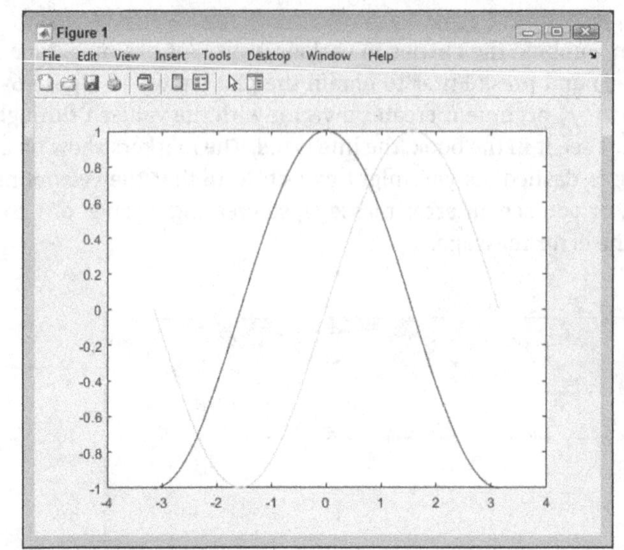

FIGURE 6-7: Plot multiple series when necessary.

Modifying Any Plot

At some point, you'll want to change the content of your plot. Perhaps you want to add a legend or change how the data is presented. After you get the data looking just right, you might need to label certain items or perform other tasks to make the output look nicer. You can modify any plot you create using either commands or the MATLAB GUI.

REMEMBER

The modification method that you use is entirely up to you. Some people work better at the keyboard, others using the mouse, and still others using a combination of the two. Working at the keyboard is a lot faster but requires that you memorize the commands to type. The GUI provides you with great memory aids, but working with the mouse is slower, and you might not be able to find a particular property you want to change when it becomes buried in a menu somewhere. The following sections describe techniques to use for modifying any plot.

Making simple changes

You can make a number of simple changes to your plot that don't require any special handling other than to type the command. For example, to add a grid to an existing plot, you simply type **grid on** and press Enter. (MATLAB has a number of `grid` commands. For example, `grid MINOR` toggles the minor grid lines. Type **help grid** and press Enter to obtain additional information.)

Adding a legend means typing a name for each of the plots. For example, if you want to add a legend to the plot in Figure 6-7, you type **legend('Sine', 'Cosine')** and press Enter. You can also change items such as the legend orientation. The default orientation is vertical, but you can change it to horizontal by typing **legend('orientation', 'horizontal')** and pressing Enter. Notice that the property name comes first, followed by the property value. There are other properties that control location; you access these by typing commands like **legend('location', 'northwest')** and pressing Enter. The idea is to keep the legend out of the way.

MATLAB also lets you add titles to various parts of the plot. For example, to give the plot a title, type **title('Sine and Cosine')** and press Enter. You can also provide labels for the x-axis using `xlabel()` and for the y-axis using `ylable()`. The point is that you have full control over the appearance of the plot. Figure 6-8 shows the effects of the commands that you have tried so far. (Compare it to Figure 6-7.)

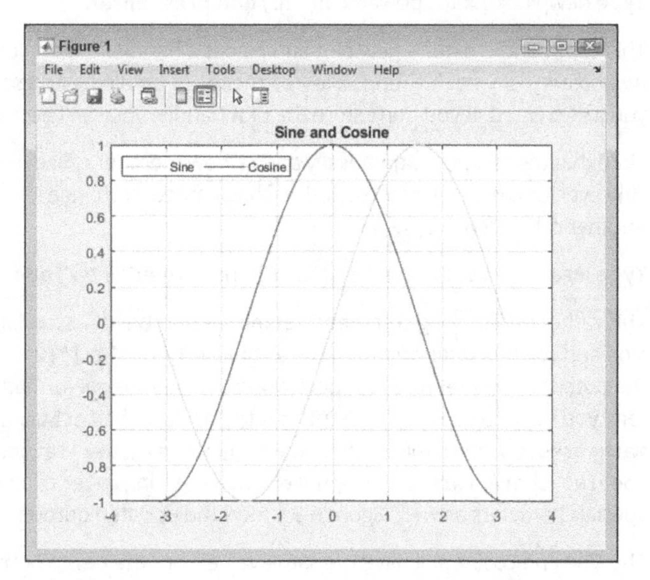

FIGURE 6-8: Change your plot setup using commands.

If you make a mistake, you can always clear the current plot by using the `clf` command (don't use it now, you'll mess up your plot for future sections). The `clf` command does for the plot what the `clc` command does for the Command Window. Make sure that you actually want to clear the plot before using the `clf` command because there isn't any sort of undo feature to restore the plot.

Adding to a plot

You may decide that you want to add another plot to an existing plot. For example, you might want to plot the square of x for each of the values used in the previous examples. To make this technique work, you need to perform the three-step process described here:

1. **Type** `hold on` **and press Enter.**

REMEMBER

If you try to add another plot without placing a hold on the current plot, MATLAB simply creates a new plot and gets rid of the old one. The `hold on` command lets you retain the current plot while you add something to it. Using `hold` by itself toggles the hold state, and you can use `hold off` to return MATLAB to its default state of overwriting the current plot each time you generate a new one.

2. **Type** newplot = plot(x, power(x, 2), 'm:') **and press Enter.**

This command creates a new plot and places a handle to that plot in `newplot`. A *handle* is just what it sounds like — a means of obtaining access to the plot you just created. If you don't store the plot handle, you can't access it later.

Note that the legend automatically adds a new member called `data1`. However, `data1` isn't very descriptive, so you need to change it by issuing another call to `legend()`.

3. **Type** legend('Sine', 'Cosine', ['${x}^{2}$'], 'interpreter', 'latex') **and press Enter.**

This rather odd-looking command contains a number of essential parts. The sine and cosine values are the same as before, but `['${x}^{2}$']` produces the output x^2, where the 2 is superscripted. To see special formatted characters, you need to use the LaTeX interpreter, as described at `https://www.mathworks.com/help/matlab/creating_plots/greek-letters-and-special-characters-in-graph-text.html`. Using LaTeX allows you to add special characters as well. Figure 6-9 shows the updated output.

The sine and cosine still have the same values, but the new plot has much larger values, so the previous plot lines appear to have shrunk. However, compare the values in Figures 6-8 and 6-9 and you see that the values of sine and cosine are the same.

4. **Type** hold off **and press Enter.**

The `hold off` command releases the plot. To create new plots, you must release your hold on the existing plot.

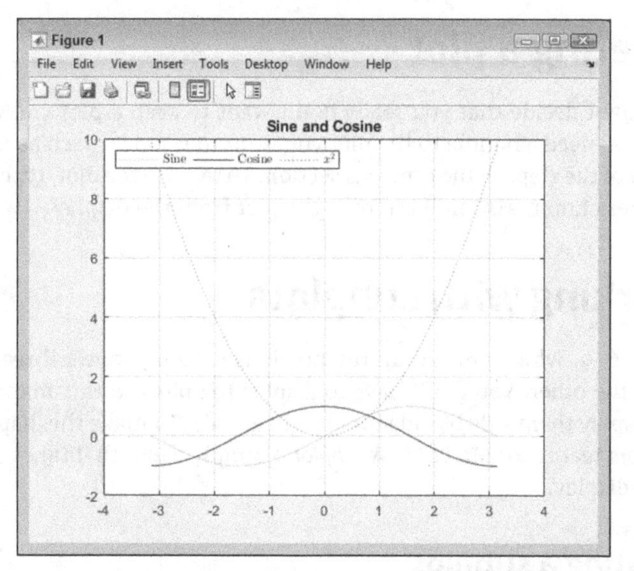

FIGURE 6-9: Add a new plot to the existing setup.

USING THE FIGURE() FUNCTION

This chapter concentrates on various sorts of plots because plots provide you with output. However, the `figure()` function can be an important part of your toolbox when you start creating scripts. You use the `figure()` function alone to create a new *figure* (an object containing graphics, such as plots, that might be displayed in a separate window) doesn't have any sort of information in it. The advantage is that you can then fill the new figure with anything you want. In addition, the `figure()` function creates a new figure without overwriting the old one. The `figure()` function returns a handle to the figure rather than to the plot inside the figure. If you have multiple plots inside a figure, you can use the figure handle to select all the plots rather than just one of them.

You use the `figure()` function with a handle to make the figure associated with a particular handle the current figure. For example, the `figure(MyFigure)` command would make the figure pointed to by `MyFigure` the current figure. When working with multiple figures, you need some method of selecting between them, and the `figure()` function provides the best method of doing that.

Of course, you might have created the figure as a plot rather than as a figure. The plot handle doesn't work with the `figure()` function. Use the `gcf()` (*Get Current Figure*) function to obtain the figure handle for any figure you create using a plot. You can then save the figure handle in a variable for later use.

Deleting a plot

You might decide that you really don't want to keep a plot you've added. In this case, you need a handle to the plot you want to remove, such as the handle stored as part of the steps in the previous section. To remove the plot, type **delete(newplot)** and press Enter. MATLAB removes the plot from the display.

Working with subplots

Figure 6-9, which appears in the previous section, shows three plots — one on top of the other. You don't have to display the plots in this manner. Instead, you can display them side by side (or even in a grid). To make this happen, you use the subplots feature of MATLAB. A *subplot* is simply a plot that takes up only a portion of the display.

Creating a subplot

The best way to understand subplots is to see them in action. The following steps help you create the three previous plots as subplots:

1. **Type** clf **and press Enter.**

   MATLAB clears any previous plot you created.

2. **Type** subplot(1, 3, 1) **and press Enter.**

   This function creates a grid consisting of one row and three columns. It tells MATLAB to place the first plot in the first space in the grid. You see the blank space for the plot.

3. **Type** p1 = plot(x, sin(x), 'g-') **and press Enter.**

   You see the first plot added to the display.

   REMEMBER

   Notice that the example is creating the plots one at a time. You can't combine plots in a single call when using subplots. In addition, you need to maintain a handle to each of the plots in order to configure them.

4. **Type** subplot(1, 3, 2) **and press Enter.**

   MATLAB selects the second area for the next plot.

5. **Type** p2 = plot(x, cos(x),'b-') **and press Enter.**

   You see the second plot added to the display.

6. **Type** subplot(1, 3, 3) **and press Enter.**

   MATLAB selects the third area for the next plot.

7. **Type** p3 = plot(x, power(x, 2), 'm:') **and press Enter.**

   You see the third plot added to the display, as shown in Figure 6-10.

REMEMBER

Each plot takes up the entire area allocated to it. You can't compare plots easily because each plot is in its own space and uses its own units of measure. However, this approach does have the advantage of letting you see each plot clearly.

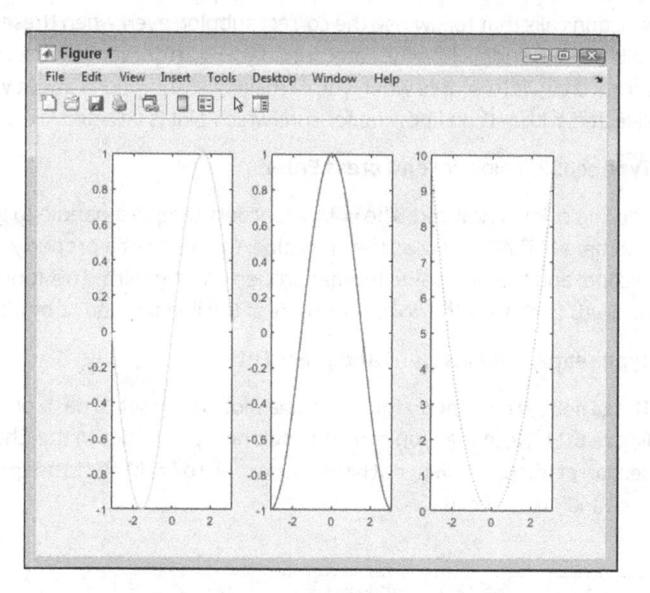

FIGURE 6-10: Each plot appears in its own area.

Changing subplot information

The `subplot()` function doesn't change anything — it merely selects something. For example, the plots in Figure 6-10 lack titles. To add a title to the first plot, follow these steps:

1. **Type** subplot(1, 3, 1) **and press Enter.**

   MATLAB selects the first subplot.

2. **Type** title('Sine') **and press Enter.**

   You see a title added to the first subplot.

Configuring individual plots

To work with a subplot in any meaningful way, you need to have a handle to the subplot. The following steps describe how to change the color and line type of the second plot:

1. **Type** subplot(1, 3, 2) **and press Enter.**

   MATLAB selects the second subplot. Even though the handle used with the `set()` command in the following step will select the subplot for you, this step is added so that you can actually see MATLAB select the subplot. In some cases, performing this task as a separate step is helpful to ensure that any function calls that follow use the correct subplot, even when these function calls don't include a handle. Later, when you start creating scripts, you find that errors creep into scripts when you're making assumptions about which plot is selected, rather than knowing for sure which plot is selected.

2. **Type** set(p2, 'color', 'r') **and press Enter.**

   The line color is now red. The `set()` function accepts a handle to a plot or another MATLAB object as the first value, the name of a property as the second, and the new value for that property as the third. This function call tells MATLAB to change the `color` property of the line pointed at by `p2` to red.

3. **Type** set(p2, 'LineStyle', ' -.') **and press Enter.**

   The `LineStyle` property for the cosine plot is now set to dash dot, as shown in Figure 6-11. Note that you can find additional properties in the Chart Objects section at `https://www.mathworks.com/help/matlab/line-plots.html?s_tid=CRUX_lftnav`.

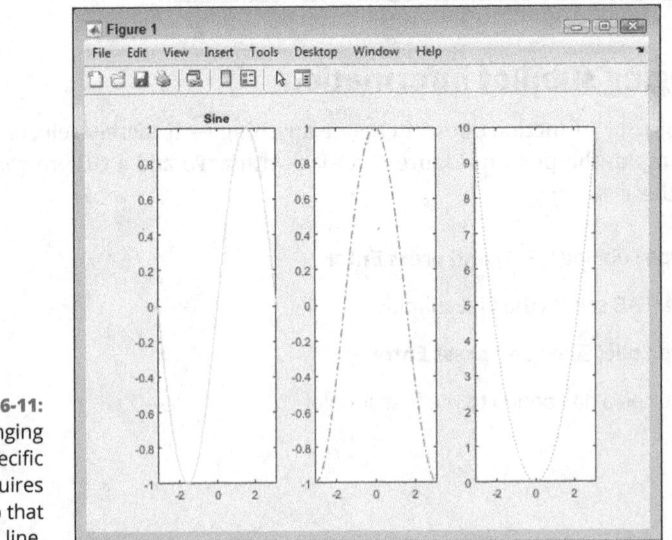

FIGURE 6-11: Changing line-specific features requires a handle to that line.

Plotting with 2D Information

MATLAB has built-in plotting routines that are suitable for many types of data and applications. Table 6-2 gives you an overview of various 2D plotting functions, including what they plot and how they're commonly used. You use these functions in place of the `plot()` function used throughout the chapter to create plots. The output will contain the kind of plot you have requested, such as a pie chart when using the `pie()` function. (MATLAB also supports 3D plotting; for more on that aspect of plotting, check out Chapter 7.)

TABLE 6-2 MATLAB Plotting Routines

Routine	What It Plots	Used By
`plotyy()`	Data with two y axes	Business users rely on this plot to show two sets of units, for example, quantity sold and money.
`loglog()`	Data with both x and y axes as log scales	Science, Technology, Engineering, and Mathematics (STEM) users rely on this plot to show power or root dependence of y versus x.
`semilogx()`	Data with x-axis log scale	STEM users rely on this plot to show logarithmic dependence of y versus x.
`semilogy()`	Data with y-axis log scale	STEM and social science users rely on this plot to show exponential dependence of y versus x and population growth (as an example).
`scatter()`	Data in x-y pairs	Experimentalists and statisticians rely on this plot to show patterns created by the individual data points.
`hist()`	Frequency of occurrence of particular values of data	Experimentalists and statisticians rely on this plot to understand imprecision and inaccuracy.
`area()`	x-y data with areas filled in	Business and STEM users rely on this plot to see (and understand) the contributions of parts to a whole.
`pie()`	Set of labeled numbers	Business users rely on this plot to see (and understand) the fractional contributions of each part to a whole.
`ezpolar()`	Data in terms of radius and angle	STEM users rely on this plot to show the angular dependence of information.

IN THIS CHAPTER

» Working with 3D plots

» Creating enhanced plots

» Employing plot extras

Chapter 7
Using Advanced Plotting Features

Chapter 6 helps you create plots that convey 2D data in visual form. Using plots in this manner helps you present the data in a way that most humans understand better than abstract numbers. Visual presentations are concrete and help the viewer see patterns that might be invisible otherwise. The 3D plots described in this chapter do the same thing as those 2D plots, only with a 3D data set. The viewer sees depth as well as height and width when looking at that data. Using a 3D data set can greatly improve the amount of information the user obtains from a plot. For example, a 3D plot could present the variation of a data set over time so that the user gains insights into how the data set changes.

If you worked through Chapter 6, you focused mostly on small changes to improve the aesthetics of your plots. This chapter looks at some of the fancier things you can do to make plots even more appealing. In many cases, nontechnical viewers require these sorts of additions in order to appreciate the data you present. Making data as interesting as possible can only help to improve your presentation and convince others to accept your interpretation of the data. Of course, making plots that look nice is also just plain fun, and everyone could use a little more fun in their creation and presentation of data.

The last part of this chapter helps you work with plots at a lower level to create better visual effects or to perform manipulations that would be hard otherwise.

For example, you discover more about working with the plot axes to enhance plot visualizations. These more advanced plot techniques are good to know, especially if you work with complex data, but you can also skip this last section until you actually need the advanced functionality it discusses.

Plotting with 3D Information

A 3D plot has an x-, y-, and z-axis (height, width, and depth, if you prefer). The addition of depth lets you present more information to the viewer. For example, you could present historical information about a plot so that each element along the z-axis is a different date. Of course, the z-axis, like the x and y axes, can represent anything you want. The thing to remember is that you now have another method of presenting information to the viewer.

REMEMBER

It's also important to consider that you're presenting 3D information on a 2D surface — the computer screen or a piece of paper. Some users forget this fact and find that some of their data hides behind another plot object that is greater in magnitude. When working with 3D plots, you need to arrange the information in such a manner that you can see it all onscreen.

The following sections describe various kinds of plots and how to create them. Each plot type has specific uses and lends itself to particular kinds of data display. Of course, the kind of plot you choose depends on how you want to present the data as well.

Using the bar() function to obtain a flat 3D plot

The bar chart is a standard form of presentation that is mostly used in a business environment. You can use a bar chart to display either 2D or 3D data. When you feed a bar chart a vector, it produces a 2D bar chart. Providing a bar chart with a matrix produces a 3D chart. The following steps help you create a 3D bar chart:

1. **Type** SurveyData = [8, 7, 6; 13, 21, 15; 32, 27, 32] **and press Enter.**

   MATLAB creates a new matrix named `SurveyData` that is used for many of the examples in this chapter. You see the following output:

```
SurveyData =
     8     7     6
    13    21    15
    32    27    32
```

2. **Type** bar(SurveyData) **and press Enter.**

   You see a flat presentation of `SurveyData`, as shown in Figure 7-1. The x-axis shows each of the columns in turn. (If you could see colors in the book, you would see that the first column is blue, the second is red, and the third is yellow.) The y-axis presents the value of each cell (such as 8, 7, and 6 for the first `SurveyData` row). The z-axis presents each row in a group, and each group corresponds to a number between 1 and 3.

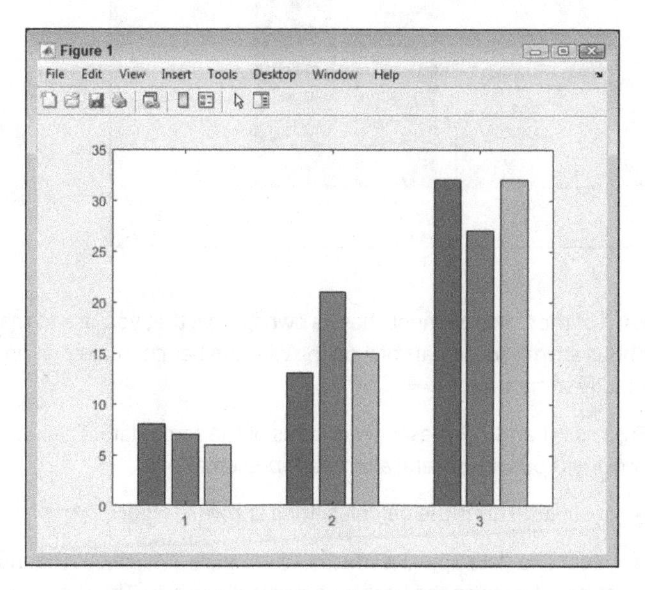

FIGURE 7-1: A flat presentation of the x, y, and z axes of SurveyData.

3. **Type** Bar1 = bar(SurveyData, 'stacked') **and press Enter.**

   You see the same `SurveyData` matrix presented as a stacked bar chart, as shown in Figure 7-2. In this case, the x-axis elements are shown stacked one on top of the other.

   The example also outputs information about the bar chart handles (a means of obtaining access to the plot). The values may differ, but you should see three handles output like the following (each handle is named Bar; previous versions of MATLAB used a number to represent the handle in the output):

```
Bar1 =
  1x3 Bar array:
    Bar     Bar     Bar
```

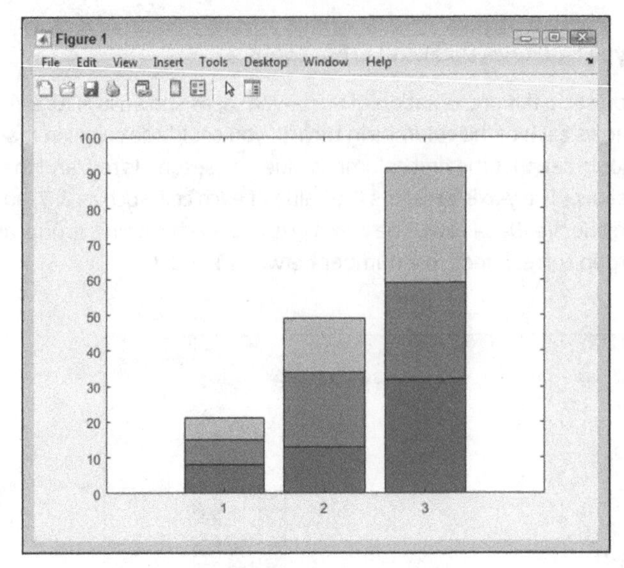

FIGURE 7-2: A stacked presentation of the SurveyData matrix.

REMEMBER

Each of the z-axis elements has its own handle that you use to manipulate it. This is an important part of working with the bar chart later when you want to modify something.

Figures 7-1 and 7-2 present two forms of the same data. The `bar()` function provides you with several alternative presentations:

- `grouped`: This is the default setting shown in Figure 7-1.
- `hist`: The data appears much like in Figure 7-1, except that no spaces appear between the bars for a particular group. The groups do still have spaces between them.
- `hisc`: The groups are positioned so that each group starts at a number on the x-axis, rather than being centered on it.
- `stacked`: This is the stacked appearance shown in Figure 7-2.

4. **Type** get(Bar1(1)) **and press Enter.**

The `get()` function obtains the properties you can work with for a particular object. In this case, you request `Bar1(1)`, which is the first group in Figure 7-2. In other words, this would be the first member of the z-axis. You see the following output:

```
   Annotation: [1×1 matlab.graphics.
                eventdata.Annotation]
    BarLayout: 'stacked'
     BarWidth: 0.8000
```

```
        BaseLine: [1×1 Baseline]
       BaseValue: 0
    BeingDeleted: off
      BusyAction: 'queue'
   ButtonDownFcn: ''
           CData: [3×3 double]
        Children: [0×0 GraphicsPlaceholder]
        Clipping: on
     ContextMenu: [0×0 GraphicsPlaceholder]
       CreateFcn: ''
 DataTipTemplate: [1×1 matlab.graphics.
                   datatip.DataTipTemplate]
       DeleteFcn: ''
     DisplayName: ''
       EdgeAlpha: 1
       EdgeColor: [0 0 0]
       FaceAlpha: 1
       FaceColor: [0 0.4470 0.7410]
   FaceColorMode: 'auto'
HandleVisibility: 'on'
         HitTest: on
      Horizontal: off
   Interruptible: on
       LineStyle: '-'
       LineWidth: 0.5000
          Parent: [1×1 Axes]
   PickableParts: 'visible'
        Selected: off
SelectionHighlight: on
     SeriesIndex: 1
    ShowBaseLine: on
             Tag: ''
            Type: 'bar'
        UserData: []
         Visible: on
           XData: [1 2 3]
       XDataMode: 'auto'
     XDataSource: ''
      XEndPoints: [1 2 3]
           YData: [8 13 32]
     YDataSource: ''
      YEndPoints: [8 13 32]
```

REMEMBER

After you know the properties that you can modify for any MATLAB object, you can use those properties to start building scripts. (You created your first script in Chapter 2.) Just creating and then playing with objects is a good way to discover just what MATLAB has to offer. Many of these properties will appear foreign to you and you don't have to worry about them, but notice that the `YData` property contains a vector with the three data points for this particular bar.

TIP

It's also possible to obtain individual property values. For example, if you use the `get(Bar1(1), 'YData')` command, you see the current `YData` values for just the first bar.

5. **Type** set(Bar1(1), 'YData', [40, 40, 40]) **and press Enter.**

   The `set()` function lets you modify the property values that you see when using the `get()` function. In this case, you modify the `YData` property for the first bar — the blue objects when you see the plot onscreen. Figure 7-3 shows the result of the modification.

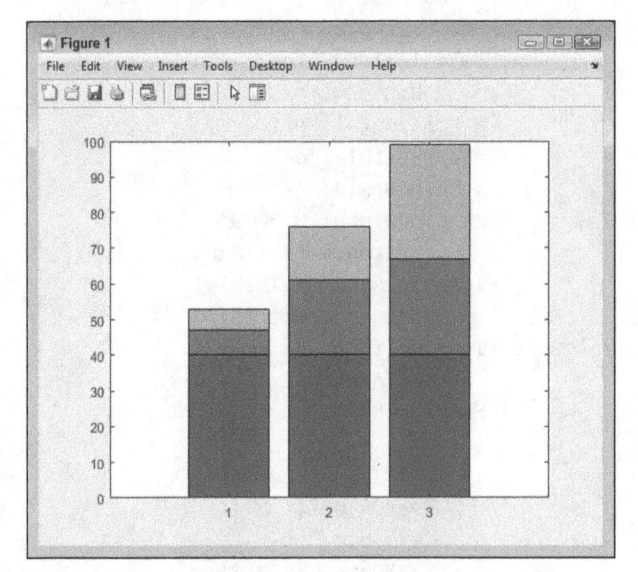

FIGURE 7-3: Rather than re-create a plot, you can simply modify values to obtain the result you want.

Using bar3() to obtain a dimensional 3D plot

The flat form of the 3D plot is nice, but it lacks pizzazz. When you present your information to other engineers and scientists, the accuracy of the flat version is welcome. Everyone can see the 3D data clearly and work with it productively. A business viewer might want something a bit different. In this case, presenting a pseudo-3D look is better because the business user gets a better overall view of the data. Precise measurements aren't quite as useful in this case — but seeing how the data relate to each other is. To create a dimensional plot of the data that appears in the previous section, type **Bar2 = bar3(SurveyData)** and press Enter. You see a result similar to the one shown in Figure 7-4.

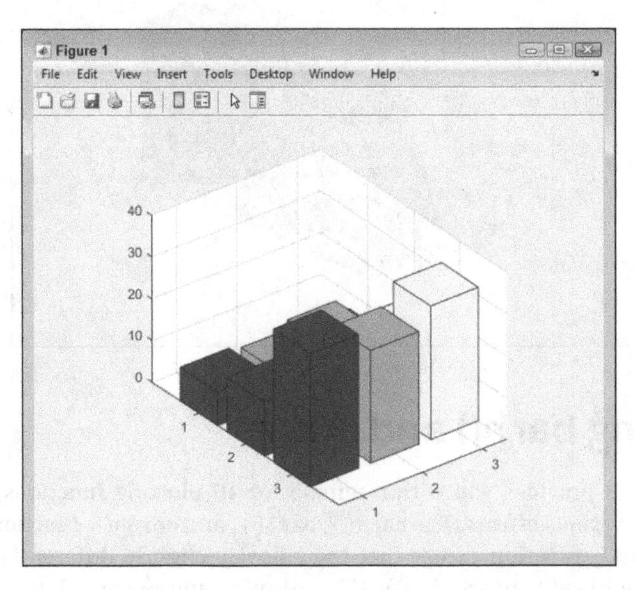

FIGURE 7-4: Dimensional plots display the relationships between data well.

The two problems with the presentation in Figure 7-4 are that you can't see some of the data, and none of it is presented to best effect. To rotate the image, you use the `view()` function. The `view()` function can accept either x, y, and z rotation in degrees, or a combination of azimuth and elevation. Using x, y, and z rotation is easier for most people than trying to figure out azimuth and elevation. To change Figure 7-4 so that you can more easily see the bars, type **view([-45, 45, 30])** and press Enter. Figure 7-5 shows the result.

REMEMBER

The `view()` function uses *absolute* rotation rather than *relative* rotation, in which one change would affect the next. As a result, if you type **view([-45, 45, 30])** and press Enter a second time, you obtain the same result as before. To obtain a new view, you must provide different values.

TIP

As an alternative to using the `view()` function, you can also click the Rotate 3D button, shown in Figure 7-5. It's the button with the circular arrow that appears to the left of the hand (pan) icon. Although the `view()` function is more precise and lets you make changes to the view without moving your hands from the keyboard, the Rotate 3D button can be faster and easier.

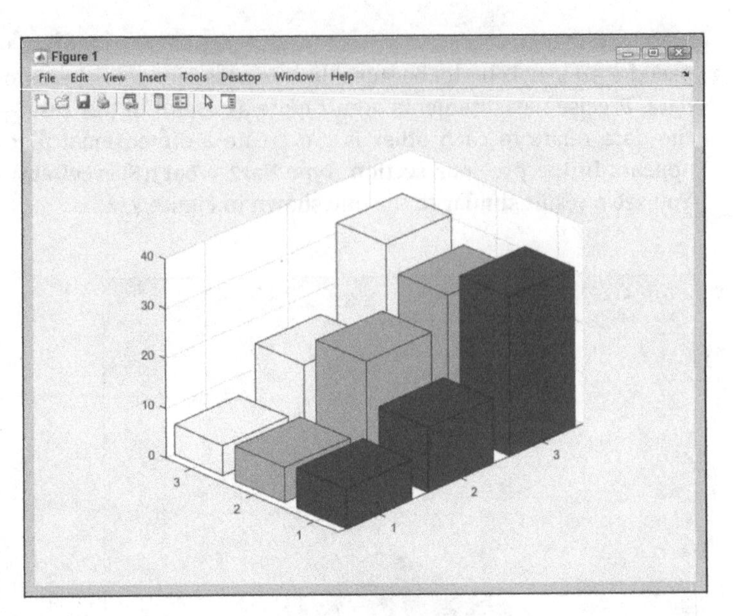

FIGURE 7-5: Changing the view makes seeing the data easier.

Using barh() and more

MATLAB provides you with a number of 3D plotting functions that you use to obtain various effects. The `barh()`, `bar3()`, and `bar3h()` functions work just like the `bar()` function except that they display slightly differently. Closely related are the `hist()`, `histc()`, `rose()`, `polar()`, and `pareto()` functions. Table 7-1 lists the various plotting functions that you have at your disposal as well as a brief description of how they work.

TABLE 7-1 Bar Procedures and Other Related Plotting Procedures

Function	What It Does	Examples
`bar()`	Plots a flat bar chart that relies on color and grouping to show the z-axis.	`bar(SurveyData)` `bar(SurveyData', 'stacked')`
`bar3()`	Plots a dimensional bar chart that uses color and perspective to show the z-axis.	`bar3(SurveyData)` `bar3(SurveyData','stacked')`
`bar3h()`	Plots a horizontal dimensional bar chart that uses color and perspective to show the z-axis.	`bar3h(SurveyData)` `bar3h(SurveyData','stacked')`
`barh()`	Plots a horizontal flat bar chart that relies on color and grouping to show the z-axis.	`barh(SurveyData)` `barh(SurveyData','stacked')`

Function	What It Does	Examples
`hist()`	Plots frequency of occurrence for bins given raw data and, optionally, bin centers.	`hist(randn(1,100), 5)` creates 100 normally distributed random numbers and places them in five equally spaced bins. `hist(randn(1,100),[-3.5,-2.5,-1.5,-.5,.5,1.5,2.5,3.5])` creates 100 normally distributed numbers and places them in specific bin centers.
`histc()`	Obtains frequency data for each bin and displays it as text (rather than as a plot). The advantage is that you can specify bin edges.	`histc(randn(1,100), [-4:1:4])` specifies bins that are 1 unit wide, with edges on integers starting at -4. You could use this information in a plot as `bar([-4:1:4],ans,'histc')`.

`pareto()`	Plots a bar chart ordered by highest bars first — used in business to identify factors causing the greatest effect.	`histc(randn(1,100),[-4:1:4])`
		
		`pareto(ans)`

`polar()`	Plots a polar display of data in which the rings of the circle represent individual data values.	`histc(randn(1,100),[-4:1:4])`
		
		`polar(ans)`

`rose()`	Plots data bars versus angles in a polar-like display. As with the `hist()` function, you may also specify bin centers.	`rose(randn(1,100), 5)` creates 100 normally distributed numbers and places them in five equally spaced bins.

Enhancing Your Plots

For visual information to be meaningful and more informative, you need to add titles, labels, legends, and other enhancements to plots of any type (both 3D and 2D). (The greater visual appeal of 3D plots only makes the plot prettier, not more informative.) The following sections of the chapter won't make you into a graphic designer, but they will let you create more interesting plots that you can use to help others understand your data. The goal of these sections is to help you promote better communication. The examples in the following sections rely on the 3D plot you created in the "Using bar3() to obtain a dimensional 3D plot" section, earlier in this chapter.

Getting an axes handle

Before you can do anything, you need a handle to the current axes. The best way to obtain such a handle (assuming that you have a figure displayed, such as the one in Figure 7-5) is to type **Bar2Axes = gca()** (Get Current Axes) and press Enter. The `gca()` function returns the handle for the current plot. When you type **get(Bar2Axes)** and press Enter, you see the properties associated with the current plot.

TRICKS OF THE TRADE FOR WORKING WITH FIGURES

Knowing a few tricks is helpful when working with plots. The tricks help you perform work faster and more efficiently. In addition, they make working with plots more fun. The following list contains tips that you can try on your own figures, but don't use them with the current examples.

- Start over by using the `clf` command, which stands for Clear Figure (which is precisely what it does).
- Stay organized by obtaining a handle (a method to gain access to the figure) using the `gcf()` (*G*et *C*urrent *F*igure) function. Don't confuse the figure handle with the plot handle mentioned earlier — a figure contains one or more plots, so the figure handle is different.
- Make a particular figure the current figure, use the `figure()` function with the variable containing the figure handle.
- Reset figures to default values using the `reset()` function with the variable containing the figure handle. This feature comes in handy when the changes you make produce undesirable results.
- See the properties associated with the current figure by using the `get()` function with the variable containing the figure handle. MATLAB displays a list of the properties and their current values. If a property doesn't have a value, the value is left blank.

Modifying axes labels

MATLAB automatically creates labels for some of the axes for you. However, the labels are generic and don't really say anything. To modify anything on the axes, you need an axes handle (as described in the previous section).

After you have the handle, you use the appropriate properties to modify the appearance of the axes. For example, to modify the x-axis label, you type **xlabel(Bar2Axes, 'X Axis')** and press Enter. Similarly, for the y-axis, you type **ylabel(Bar2Axes, 'Y Axis')** and press Enter. You can also use the `zlabel()` function for the z-axis.

Each of the ticks on an axis can have a different label as well. The default is to simply assign them numbers. However, if you want to assign meaningful names to the x-axis ticks, you can type **set(Bar2Axes, 'XTickLabel', {'Yesterday',**

'Today', 'Tomorrow'}) and press Enter. Notice that the labels appear within a cell array using curly brackets ({}). Likewise, to set the y-axis ticks, you can type **set(Bar2Axes, 'YTickLabel', {'Area 1', 'Area 2', 'Area3'})** and press Enter. You can also use a `ZTickLabel` property, which you can modify. The tick mark labels will repeat if you don't provide enough labels for each tick mark.

To control the tick values, you type **set(Bar2Axes, 'ZTick', [0, 5, 10, 15, 20, 25, 30, 35, 40])** and press Enter. Those two axes also have `XTick` and `YTick` properties. Of course, in order to see the z-axis ticks, you also need to change the limit (the size of the plot in that direction). To perform this task, you type **set(Bar2Axes, 'ZLim', [0 45])** and press Enter.

TIP

Many of the `set()` function commands have alternatives. For example, you can change the `ZLim` property by using the `zlim()` function. The alternative command in this case is `zlim(Bar2Axes, [0 45])`. Using a `set()` function does have the advantage of making it easier to enter the changes because you have to remember only one function name. However, the result is the same no matter which approach you use, so it's entirely a matter of personal preference.

Use the `get()` function whenever necessary to discover additional interesting properties to work with. Properties are available to control every aspect of the axes' display. For example, if you want to change the color of the axes' labels, you use the `XColor`, `YColor`, and `ZColor` properties. Figure 7-6 shows the results of the changes in this section.

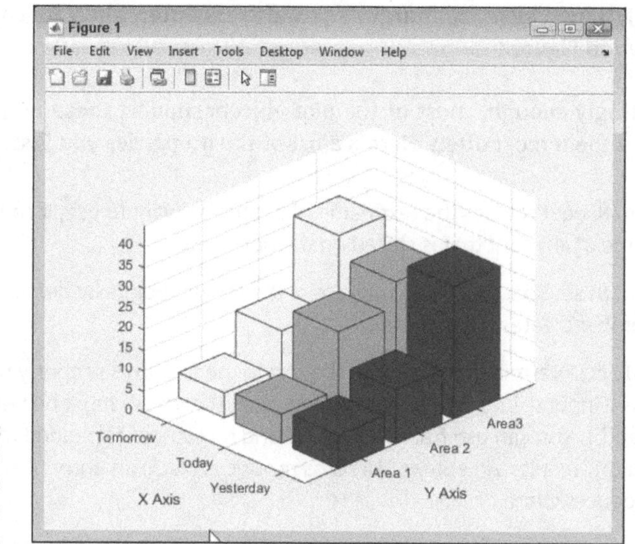

FIGURE 7-6: Properties control the appearance of the axes in your plot.

REMEMBER

Many properties have an automatic setting. For example, to modify the ZLim property so that it uses the automatic setting, you type **zlim(Bar2Axes, 'auto')** and press Enter. The alternative when using a set() function is to type **set(Bar2Axes, 'ZLimMode', 'auto')** and press Enter. Notice that when you use the zlim() function, you can set either the values or the mode using the same command. When using the set() function, you use different properties (ZLim and ZLimMode) to perform the task. However, the important thing to remember is that the auto mode tells MATLAB to configure these items automatically for you.

TIP

Using commands to change plot properties is fast and precise because your hands never leave the keyboard and you don't spend a lot of time searching for a property to change in the GUI. However, you can always change properties using the GUI as well. Click the Edit Plot button (the one that looks like a hollow arrow on the Figure Toolbar in Figure 7-5) to put the figure into edit mode. Click the element you want to modify to select it. Right-click the selected element and choose Open Property Editor to modify the properties associated with that particular element.

Adding a title

Every plot should have a title to describe what the plot is about. You use the title() function to add a title. However, the title() function accepts all sorts of properties so that you can make the title look precisely the way you want. To see how this function works, type **title(Bar2Axes, 'Sample Plot', 'FontName', 'Times', 'FontSize', 22, 'Color', [.5, 0, .5], 'BackgroundColor', [1, 1, 1], 'EdgeColor', [0, 0, 0], 'LineWidth', 2, 'Margin', 4)** and press Enter. MATLAB changes the title, as shown in Figure 7-7.

REMEMBER

Interestingly enough, most of the plot objects support these properties, but the title uses them most often. Here's a list of the properties you just changed:

- FontName: Provides the text name of a font you want to use. It can be the name of any font that is stored on the host system.
- FontSize: Specifies the actual size of the font (in points by default). A larger number creates a larger font.
- Color: Determines the color of the text in the title. This property requires three input values for red, green, and blue. The values must be between 0 and 1. You can use fractional values and mix colors as needed to produce specific results. An entry of all zeros produces black; an entry of all ones produces white.
- BackgroundColor: Determines the color of the background behind the text in the title. It uses the same color scheme as the Color property.

- **EdgeColor**: Determines the color of any line surrounding the title. It uses the same color scheme as the `Color` property.
- **LineWidth**: Creates a line around the title of a particular width (in points by default).
- **Margin**: Adds space between the line surrounding the title (the edge) and the text (in points by default).

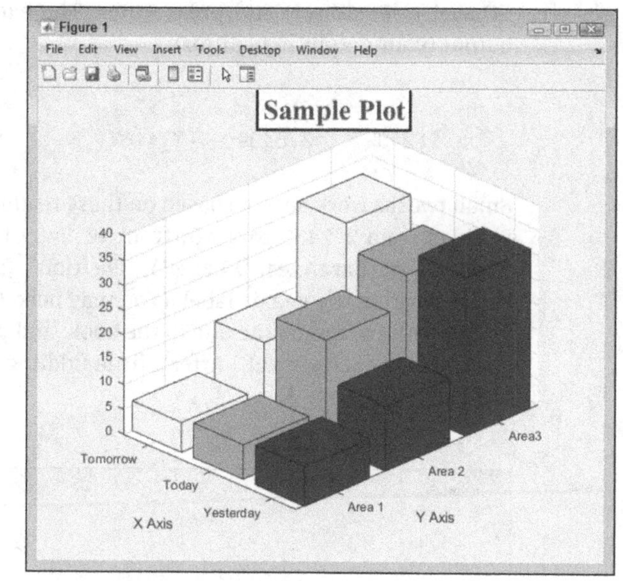

FIGURE 7-7: A title can use properties to create a pleasing appearance.

Rotating label text

In some cases, the text added to a plot just doesn't look right because it doesn't quite reflect the orientation of the plot itself. The title in Figure 7-7 looks just fine, but the x-axis and y-axis labels look slightly askew. You can modify them so that they look better.

When you review some properties using the `get()` function, you see a handle value instead of an actual value. For example, when you look at the `XLabel` value, you see a handle that lets you work more intimately with the underlying label. To see this value, you use the `get(Bar2Axes, 'XLabel')` command. If you don't want to use a variable to hold the handle, you can see the `XLabel` properties by typing **get(get(Bar2Axes, 'XLabel'))** and pressing Enter. What you're telling MATLAB to do is to get the properties that are pointed to by the `XLabel` value obtained with the `Bar2Axes` handle — essentially, a handle within a handle.

One of the properties within XLabel is Rotation, which controls the angle at which the text is displayed. To change how the plot looks, type **set(get(Bar2Axes, 'XLabel'), 'Rotation', -30)** and press Enter. The x-axis label is now aligned with the plot. You can do the same thing with the y-axis label by typing **set(get(Bar2Axes, 'YLabel'), 'Rotation', 30)** and pressing Enter.

You can also reposition the labels, although using the GUI to perform this task is probably easier. However, the Position property provides you with access to this feature. To see the starting position of the x-axis label, type **get(get(Bar2Axes, 'XLabel'), 'Position')** and press Enter. The example setup shows the following output (your output may differ):

```
ans =
    2.2433   -0.5489   -7.4167
```

Small tweaks work best, so based on these readings, you might want to change the settings from 2.2433 to 2.16, from -0.5489 to -0.4, and from -7.4167 to -7. Type **set(get(Bar2Axes, 'XLabel'), 'Position', [2.16 -0.4 -7])** and press Enter to better position the x-axis label. (You may need to fiddle with the numbers a bit to get your plot to match the one in the book, and your final result may not look precisely like the screenshot.) After a little fiddling, your X Axis label should look like the one in Figure 7-8.

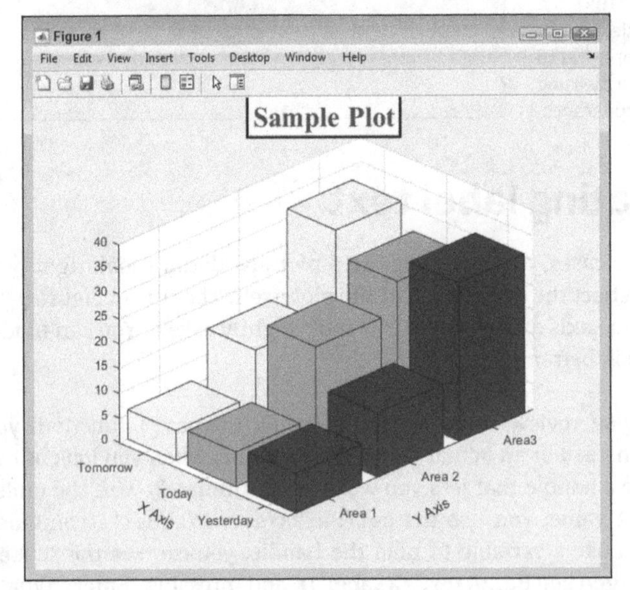

FIGURE 7-8: Any object can be rotated and repositioned as necessary.

Employing annotations

Annotations let you add additional information to a plot. For example, you might want to put a circle around a particular data point or use an arrow to point to a particular bar as part of your presentation. Of course, you may simply want to add some sort of emphasis to the plot using artistic elements. No matter how you want to work with annotations, you have access to these drawing elements:

- Line
- Arrow
- Text Arrow
- Double Arrow
- Textbox
- Rectangle
- Ellipse

To add annotations to your figure, you use the `annotation()` function. Say you want to point out that Area 3 in Figure 7-8 is the best area of the group. To add the text area, you type **TArrow = annotation('textarrow', [.7, .55], [.9, .77], 'String', 'Area 3 is the best!')** and press Enter. You see the result shown in Figure 7-9. This version of the `annotation()` function accepts the annotation type, the x location, y location, property name (`String`), and property value (`Area 3 is the best!`).

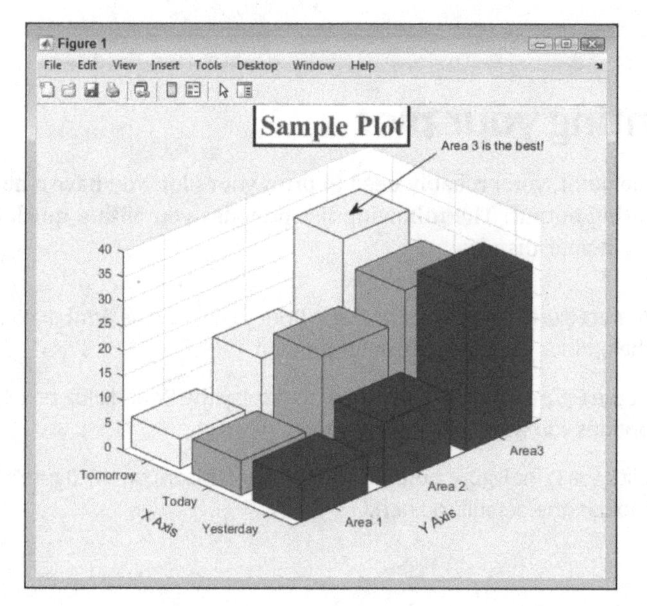

FIGURE 7-9: Add annotations to document your plot for others.

The annotations don't all use precisely the same command format. For example, when you want to add a textbox, you provide the starting location, height, and width, all within the same vector. To see this version of the `annotation()` function in action, type **TBox = annotation('textbox', [.1, .8, .11, .16], 'String', 'Areas Report', 'HorizontalAlignment', 'center', 'VerticalAlignment', 'middle')** and press Enter. In this case, you center the text within the box and place it in the upper-left corner. A textbox doesn't point to anything — it simply displays information, as shown in Figure 7-10.

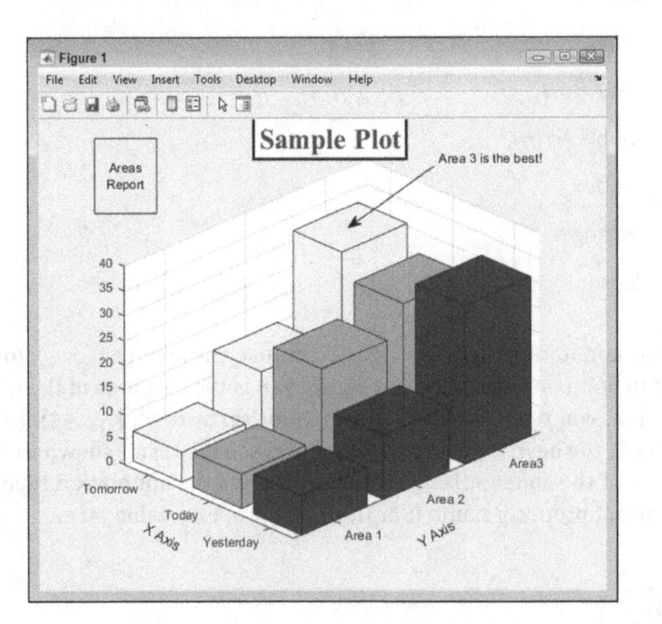

FIGURE 7-10: Annotations don't use a consistent argument setup.

Printing your plot

At some point, you probably need to print your plot. You have a number of choices in creating output. The following list provides you with a quick overview of the options at your disposal:

- At the Figure window, select File ⇨ Print to display the Print dialog box. Select the options you want to use for printing.
- At the Figure window, type Ctrl+P to display the Print dialog box. Select the options you want to use for printing.
- Click the Print Figure button (with the printer icon) on the figure's Figure toolbar (the default toolbar).

» At the Command Window, type **print()** and press Enter.
   - Using `print()` alone prints the entire figure, including any subplots.
   - Adding a handle to `print()`, such as `print(Bar2)`, prints only the object associated with the handle.

REMEMBER

In some cases, you may want to output your plot in a form that lets you print it in another location. When working in the Figure window, you select the Print to File option in the Print dialog box. MATLAB will ask you to provide a filename for printing. When working in the Command Window, you supply a filename as a second argument to the `print()` function. For example, you might use `print(Bar2, 'MyFile.prn')` as the command.

Using the Plot Extras

Chapter 6 discusses commands like `grid on`, which is used to display a grid on your plot. Earlier in this chapter, you discover essentials like adding annotation to your plot. The "Getting an axes handle" section tells you specifically about working with axes at a basic level. The following sections describe the kinds of extras you can use for emphasizing specific data. For example, the way in which you configure the grid can help make differences between data more obvious.

REMEMBER

You can always clear the current figure from the window by typing **cla** and pressing Enter. This command clears all the plot information from the figure, but doesn't close the figure window.

Creating axes dates using datetick()

You use `datetick()` to add dates to a plot axis. When using `datetick()`, you need an axis that has numbers that are in the range of the dates you need. For example, when you type **datenum('9,15,2020')** and press Enter, you get an output value of `738049`. When `datetick()` sees this value, it converts the number to a date.

REMEMBER

The `datenum()` function also accepts time as input. When you type **datenum('09/15/2020 08:00:00 AM')** and press Enter, you get `738049.333333333` as output (assuming that you first type **format longG** and press Enter). Notice that the integer portion of the value is the same as before, but the decimal portion has changed to show the time. If you don't provide a time, the output is for midnight of the day you select. You can convert a numeric date back to a string date using the `datestr()` function.

The x-axis in this example uses date values. To create an x-axis data source, type **XSource = linspace(datenum('09/15/2020'), datenum('09/19/2020'), 5);** and press Enter. This act creates a vector that contains the numeric version of dates from `09/15/2020` to `09/19/2020`. The `linspace()` function returns a vector that contains the specified number of value (5 in this case) between the two values you specify.

To create the y-axis data source, type **YSource = [1, 5, 9, 4, 3];** and press Enter. Type **Bar1 = bar(XSource, YSource)** and press Enter to create the required plot. The default tick spacing may show too many points, so type **set(gca, 'XTick', linspace (datenum('09/15/2020'), datenum('09/19/2020'), 5))** and press Enter to set the tick spacing. Notice that the x-axis doesn't use the normal numbering scheme that begins with 1 — it uses a date number instead (expressed as an exponent rather than an integer). Even though the x-axis numbers look the same, you see in the next paragraph that they aren't.

To turn the x-axis labels into dates, you now use the `datetick()` function. Type **datetick('x', 'dd mmm yy', 'keeplimits', 'keepticks')** and press Enter. Figure 7-11 shows the plot with dates in place.

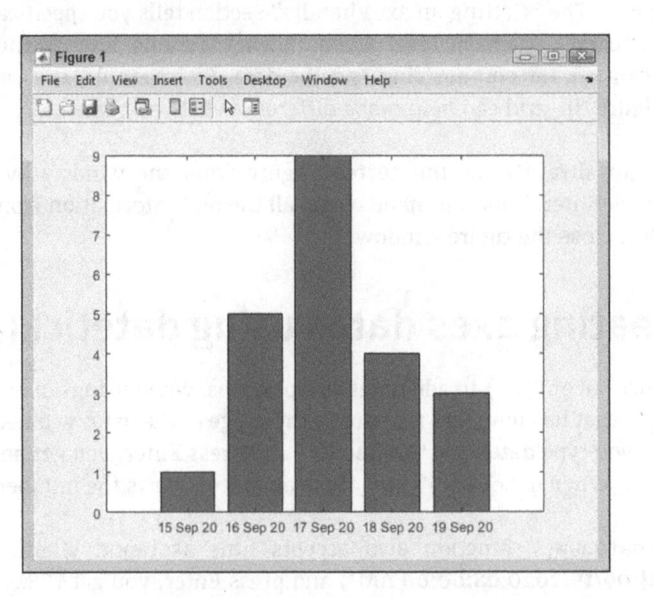

FIGURE 7-11: Creating dates in a specific format.

All the arguments used with `datetick()` are optional. When you use `datetick()` by itself, the output appears on the x-axis using a two-digit month and a two-digit day. The end points also have dates, so instead of seeing just five dates, you see seven (one each for the ends). The example uses the following arguments in this order to modify how `datetick()` normally works:

- **Axis:** Determines which axis to use. You can choose the x-, y-, or z-axis (when working with a 3D plot).
- **Date format:** Specifies how the date should appear. You can either use a string containing the format as characters or numeric values, as shown at `https://www.mathworks.com/help/matlab/ref/datetick.html#btpmlwj-1-dateFormat`. (You can also type **help datetick** and press Enter to obtain a listing of date formats.) Using characters tends to be clearer, but using numbers tends to save space and time.
- `'keeplimits'`**:** Prevents MATLAB from adding entries to either end of the axis. This means that the example plot retains five x-axis entries rather than getting seven.
- `'keepticks'`**:** Prevents MATLAB from changing the value of the ticks.

Creating plots with colorbar()

Using a color bar with your plot can help people see data values based on color rather than pure numeric value. The color bar itself can assign human-understandable values to the numeric data so that the data means something to those viewing it. The best way to work with color bars is to see them in action. The following steps help you create a color bar by using the `colorbar()` function and use it to define values in a bar chart:

1. **Type** YSource = [4, 2, 5, 6; 1, 2, 4, 3]; **and press Enter.**

   MATLAB creates a new data source for the plot.

2. **Type** Bar1 = bar3(YSource); **and press Enter.**

   You see a new bar chart. Even though the data is in graphic format, it's still pretty boring. To make the bar chart easier to work with, the next step changes the y-axis labels.

3. **Type** CB1 = colorbar('EastOutside'); **and press Enter.**

   You see a color bar appear on the right side of the plot, as shown in Figure 7-12. You can choose other places for the color bar, including inside the plot. For now, don't worry about the color bar ticks not matching those of the bar chart.

4. **Type the following code into the Command Window, pressing Enter after each line:**

```
for Element = 1:length(Bar1)
   ZData = get(Bar1(Element),'ZData');
   set(Bar1(Element), 'CData', ZData,...
      'FaceColor', 'interp')
end
```

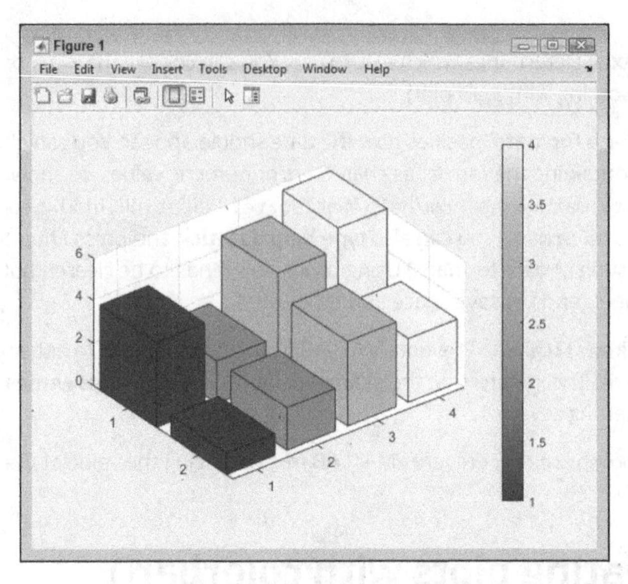

FIGURE 7-12: The color bar appears on the right side of the plot.

A number of changes take place. The bars are now colored according to their value. In addition, the ticks on the color bar now match those of the bar chart, as shown in Figure 7-13. However, the color bar just contains numbers, so it doesn't do anything more than the y-axis labels do to tell what the colors mean.

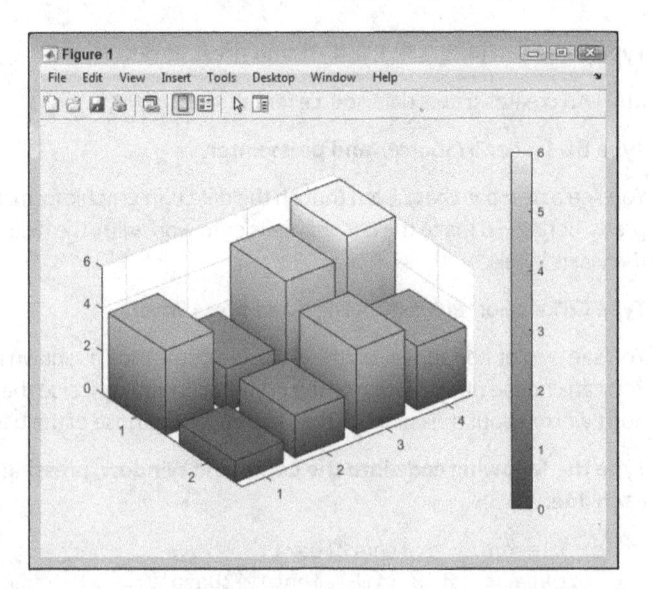

FIGURE 7-13: The bars are now colored to show their values.

5. **Type** set(CB1, 'YTickLabel', {'', 'Awful', 'OK', 'Better', 'Average', 'Great!', 'BEST'}); **and press Enter.**

   The chart now has meanings assigned to each color level, as shown in Figure 7-14.

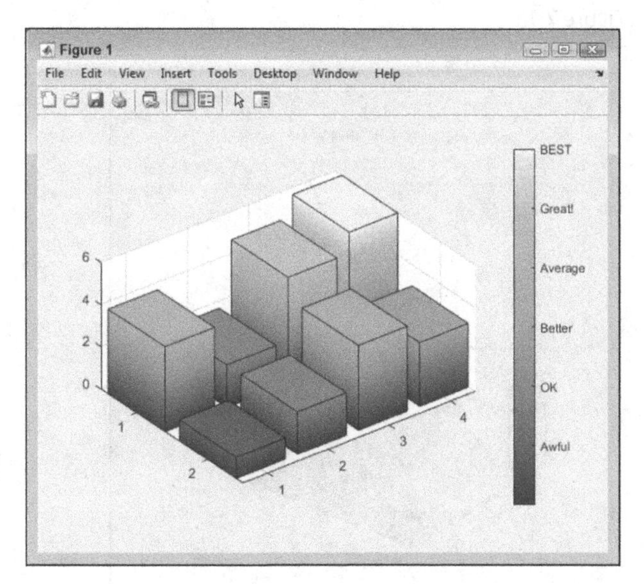

FIGURE 7-14: The color bar now conveys meaning to the bar chart.

TIP

The color scheme that MATLAB uses by default isn't the only color scheme available. The `colormap()` function lets you change the colors. For example, if you type **colormap('cool')** and press Enter, the colors change appropriately. You can also create custom color maps using a variety of techniques. To get more information, see the `colormap()` documentation at `https://www.mathworks.com/help/matlab/ref/colormap.html`.

Interacting with daspect

How the 3D effect appears onscreen depends on the data aspect ratio. The `daspect()` function lets you obtain the current aspect ratio and set a new one. The aspect ratio is a measure of how the x-, y-, and z-axis interact. For example, an aspect ratio of [1, 2, 3] would mean that for every 1 unit of the x-axis, there are two units of the y-axis and three units of the z-axis. Perform the following steps to see how this feature works:

1. **Type** YSource = [1, 3, 5; 3, 7, 9; 5, 7, 11]; **and press Enter.**

   MATLAB creates a data source for you.

2. **Type** Bar1 = bar3(YSource); **and press Enter.**

   You see a 3D bar chart appear.

3. **Type** rotate(Bar1, [0, 0, 1], 270); **and press Enter.**

   The bar chart rotates so that you can see the individual bars easier, as shown in Figure 7-15.

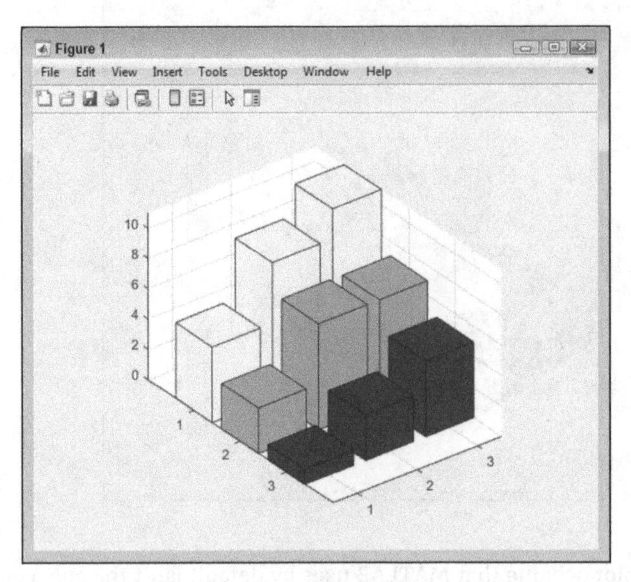

FIGURE 7-15: A 3D bar chart that you can use to work with the data aspect ratio.

4. **Type** daspect() **and press Enter.**

   The output contains three values, like this:

```
ans =
     0.3571     0.2679     2.1670
```

   So, you now know the current aspect ratio of the plot, with the first number representing the x-axis value, the second number the y-axis value, and the third number the z-axis value. Your numbers may not precisely match those shown in the book.

5. **Type** daspect([.25, 1, 1.2]); **and press Enter.**

   The data aspect ratio changes to create tall, skinny-looking bars like those shown in Figure 7-16. Compare Figures 7-15 and 7-16, and you see that the differences between the individual bars appears greater, even though nothing has changed. The data is precisely the same as before, as is the rotation, but the interpretation of the data changes.

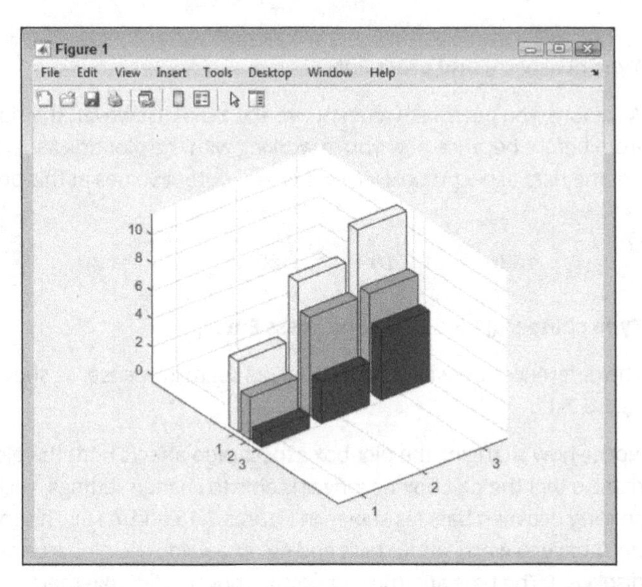

FIGURE 7-16: Modifying the aspect ratio changes how the data is perceived.

6. **Type** daspect([.65, .5, 7]); **and press Enter.**

   The impression is now that the differences between the data points are actually quite small. Again, nothing has changed in the data or the rotation. The only thing that has changed is how the data is presented.

7. **Type** daspect('auto') **and press Enter.**

   The data aspect returns to its original state.

Interacting with pbaspect

The previous section tells how to modify the data aspect ratio. In addition, you see a number of examples that show how to use rotation to modify the appearance of the data. This section discusses the plot box aspect ratio. Instead of modifying the data, the plot box aspect ratio modifies the plot box — the element that holds the plot in its entirety — as a whole. The appearance of the data still changes, but in a different way than before. The following steps get you started with this example:

1. **Type** YSource = [1, 3, 5; 3, 7, 9; 5, 7, 11]; **and press Enter.**

   MATLAB creates a data source for you.

2. **Type** Bar1 = bar3(YSource); **and press Enter.**

   You see a 3D bar chart appear.

3. **Type** rotate(Bar1, [0, 0, 1], 270); **and press Enter.**

   The bar chart rotates so that you can see the individual bars easier. (Refer to Figure 7-15.)

4. **Type** pbaspect() **and press Enter.**

   As before, you get three values: x-, y-, and z-axis. However, the numbers differ from before because now you're working with the plot box aspect ratio and not the data aspect ratio. Here are typical output values at this point:

   ```
   ans =
       1.1326    1.6180    1.0000
   ```

5. **Type** pbaspect([1.5, 1.5, 7]); **and press Enter.**

   The differences between the data points seem immense, as shown in Figure 7-17.

   Notice how changing the plot box aspect ratio affects both the plot box and the data so that the plot box no longer is able to change settings, such as the spacing between bars (as shown in Figures 7-15 and 7-16). This means that you don't have to worry about bars ending up outside the plot area and not being displayed. The bars and the plot box are now locked together.

6. **Type** pbaspect([4, 5, 1]); **and press Enter.**

   The data points now seem closer together, even though nothing has changed in the data. At this point, it helps to compare Figures 7-15 through 7-17. These figures give you a better idea of how aspect ratio affects the perception of your data in various ways.

7. **Type** pbaspect('auto'); **and press Enter.**

   The plot aspect returns to its original state.

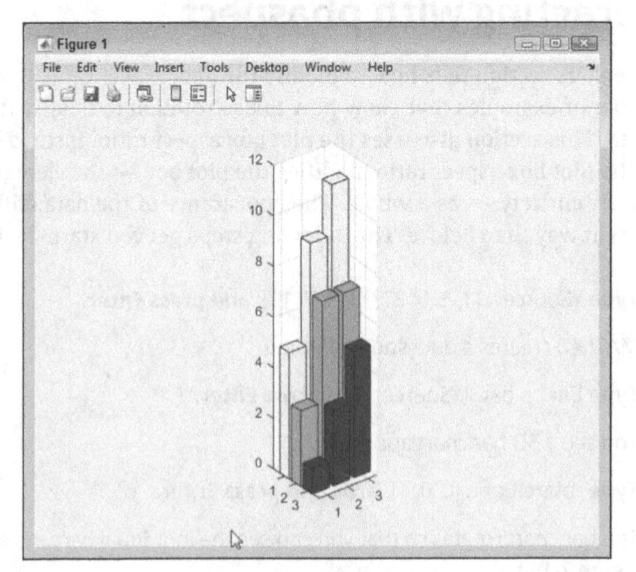

FIGURE 7-17: The data and plot box are locked together.

3 Streamlining MATLAB

IN THIS PART . . .

Making it easier to perform MATLAB tasks

Creating functions and scripts

Enhancing automation using Live Scripts and Live Functions

Working with classes

Developing apps and building projects

IN THIS CHAPTER

» Creating, modifying, and testing your script

» Using scripts in MATLAB

» Locating script errors

» Making your script run faster

Chapter 8
Automating Your Work

Getting the computer to do the work for you is probably one of the best reasons to use a computer in the first place. Anytime you can automate repetitive or mundane tasks, you free yourself to do something more interesting. MATLAB is an amazing tool for performing all sorts of creative work, but you also have a lot of mundane and repetitive tasks to perform. For example, you may need to generate the same plot every week for a report. Automating that task would free you to do something more interesting, such as discover a cure for cancer or send a rocket to Mars. The point is that you have better things to do with your time, and MATLAB is only too willing to free your time so that you can do them. That's what scripting is all about: not to make you some mad genius geek, but to automate tasks so that you can do something more interesting.

REMEMBER

Scripting is simply a matter of writing a *procedure* — writing down exactly what you want the computer to do for you, in other words. (You might compare it to a making a movie, with a writer creating the words the actors say and specifying the actions they perform.) You likely write procedures all the time for various people in your life. In fact, you may write procedures for yourself so that you remember how to perform the task later. Scripting for MATLAB is no different from any other procedure you have written in the past, except that you need to write the procedure in a manner that MATLAB understands.

This chapter helps you create basic MATLAB scripts, save them to disk so that you can access them whenever you want, and then run the scripts as needed. You also discover how to make your scripts run fast so that you don't have to wait too long for MATLAB to complete its work. This chapter helps you understand the nature of errors in scripts, and how to locate and fix them. In the final section, you use the MATLAB Profiler to verify the performance of your script and look for ways to improve it.

REMEMBER

You don't have to type the source code for this chapter manually. In fact, using the downloadable source is a lot easier. You can find the source for this chapter in the `\MATLAB2\Chapter08` folder of the downloadable source. (See the Introduction for details on how to obtain the source code.) When using the downloadable source, you may not be able to work through some of the hands-on examples because the example files will already exist on your system. In this case, you can create an alternative folder, `Chapter08a`, to hold the results of the hands-on exercises.

Understanding What Scripts Do

A script is nothing more than a means to write a procedure that MATLAB can follow to perform useful work. It's called a script and not a procedure because a script follows a specific format. MATLAB actually speaks its own English-like language that you must use to tell it what to do. The interesting thing is that you've used that language in every chapter so far by executing commands. A script doesn't do much more than link together the various commands that you have used to perform a task from one end to the other. The following sections describe what a script does in more detail.

Creating less work for yourself

The object of a script is to reduce your workload. This concept might seem straightforward now, but some people get so wrapped up in the process of creating scripts that they forget that the purpose of the script is to create less work, not more. In fact, a script should meet some (or with luck, all) of the following goals:

- Reduce the time required to perform tasks
- Reduce the effort required to perform tasks
- Allow you to pass the task along to less skilled helpers
- Make it possible to perform the tasks with fewer errors (the computer will never get bored or distracted)
- Create standardized and consistent output

» Develop a secure environment in which to perform the task (because the details are hidden from view)

REMEMBER

Notice that none of the goals in this list entails making the computer do weird things that it doesn't normally do, or wasting your time writing scripts to perform tasks that you never did in the past. The best scripts perform tasks that you already know how to do well because you have performed them so many times in the past. Yes, it's entirely possible that you could eventually create a script to perform a new task, but even in that case, the new task is likely built on tasks that you have performed many times in the past. Most people get into trouble with scripting when they try to use it for something they don't understand or haven't clearly defined.

Defining when to use a script

Scripts work well only for mundane and repetitive tasks. Sometimes writing a script is the worst possible thing you can do. In fact, many times you can find yourself in a situation in which writing a script causes real (and potentially irreparable) damage. The following list provides you with guidelines as to when to use a script:

» The task is repeated often enough that you actually save time by writing a script (the time saved more than offsets the time spent writing the script).

» The task is well defined, so you know precisely how to perform it correctly.

» There are few variables in the way in which the task is performed so that the computer doesn't have to make many decisions (and the decisions it makes are from a relatively small set of potential absolute answers).

» No creativity or unique problem-solving abilities are required to perform the task.

» All the resources required to perform the task are accessible by the host computer system.

» The computer can generally perform the task without constantly needing to obtain permissions.

» Any input required by the script is well defined so that the script and MATLAB can understand (and anticipate) the response.

Believe it or not, you likely perform regularly a huge number of tasks that fulfill all these requirements. The important thing is to weed out those tasks that you really must perform by yourself. Automation works only when used correctly to solve specific problems.

Creating a Script

Creating a script can involve nothing more than writing commands. In fact, the sections that follow show a number of ways in which you can create simple scripts without knowing anything about scripting. It may even strike you as quite odd that scripting feels much like writing commands in the Command Window. The only difference is that the commands don't execute immediately. That's the point of these following sections: Scripting doesn't have to be hard or complicated; it only needs to solve the problems you normally solve anyway.

Writing your first script

MATLAB provides many different ways to write scripts. Some of them don't actually require that you write anything at all! However, the traditional way to create a script in any application is to write it, so that's what this first section does — shows you how to write a tiny script. The most common first script in the entire world is the "Hello World" example. The following steps demonstrate how to create such a script using MATLAB:

1. **Click New Script on the Home tab of the menu.**

   You see the Editor window appear, as shown (highlighted) in Figure 8-1. This window provides the means to interact with scripts in various ways. The Editor tab shown in the figure is the one you use most often when creating new scripts.

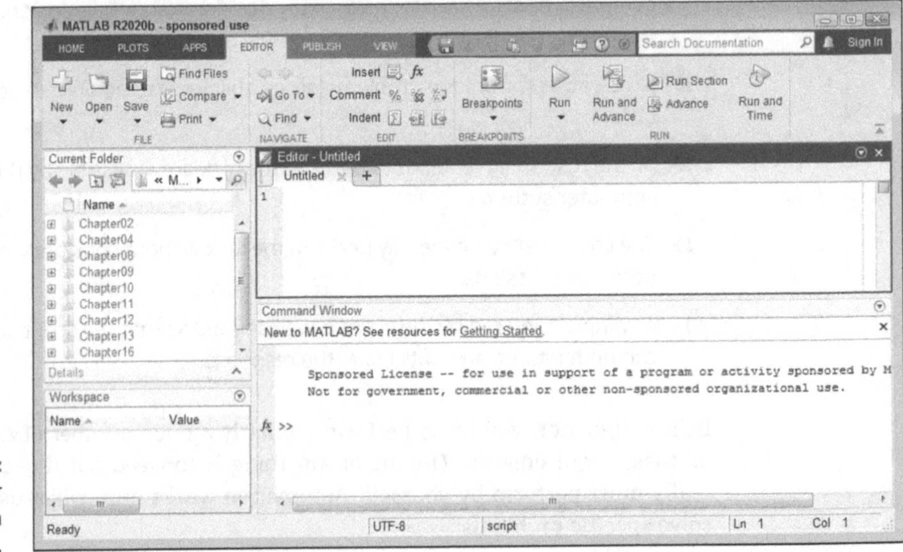

FIGURE 8-1: Use the Editor window to write a script manually.

2. **Type** disp('Hello World');.

The `disp()` function tells MATLAB that you want to display something onscreen. In this case, you display a string directly onscreen, but you can also display variables.

REMEMBER

When working with a script, you end each line with a semicolon to avoid displaying values that you don't want to display as part of the script output. You could simply type the string without a semicolon and without using the `disp()` function, but then the intent of the line of code becomes unclear. You may forget what the line of code does in the future if you leave out the details.

3. **Click Run on the Editor tab of the Editor window.**

You see a Select File for Save As dialog box, as shown in Figure 8-2. MATLAB always requests that you save your script before you run it to ensure that your script doesn't get lost or corrupted in some way should something happen when it runs.

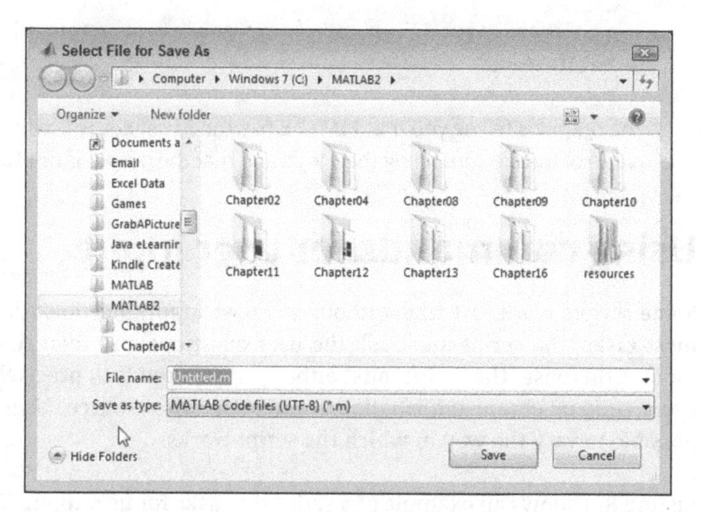

FIGURE 8-2: MATLAB always asks you to save your work before you run a script.

4. **Create or select the** `MATLAB2\Chapter08` **directory, type** FirstScript.m **in the File Name field, and click Save.**

MATLAB saves your script to disk. All your script files will have an `.m` extension.

TIP

Note that if the file isn't found in the current folder or the MATLAB path, you see another MATLAB Editor dialog box like the one shown in Figure 8-3. (As an alternative, you may see a dialog box that states that MATLAB can't run the file and provides only the Change Folder button.) You have two options to run the script:

- **Change Folder:** Change the current folder to match the saved location of the file. The Current Folder window content will change to match the save path for the script.

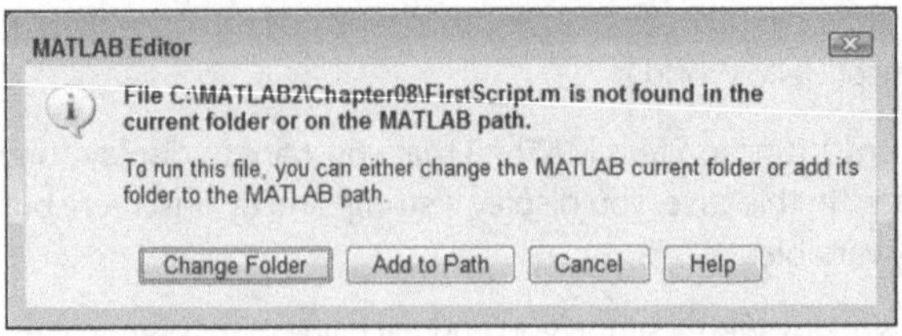

MATLAB Editor

File C:\MATLAB2\Chapter08\FirstScript.m is not found in the current folder or on the MATLAB path.

To run this file, you can either change the MATLAB current folder or add its folder to the MATLAB path.

Change Folder | Add to Path | Cancel | Help

FIGURE 8-3: The directory you use to store the script must be the current directory or in the MATLAB path.

- **Add to Path:** Add the saved location to the MATLAB path so MATLAB can find the file to run it. The added folder will appear in bold type when you view it in the Current Folder window.

Simply click the box's Change Folder button to make the dialog box disappear. This change will make all the scripts found in this chapter accessible. If you don't see this box, continue to Step 5.

5. **Select the MATLAB Command Window.**

You see the following script output:

```
>> FirstScript
Hello World
```

The output is telling you that MATLAB has run `FirstScript`, which is the name of the file containing the script, and that the output is `Hello World`.

Using commands for user input

Some scripts work just fine without any user input, but most don't. To perform most tasks, the script must ask the user questions and then react to the user's input. Otherwise, the script must either perform the task precisely the same way every time or obtain information from some other source. User input makes it possible to vary the way in which the script works.

Listing 8-1 shows an example of a script that asks for user input. You can also find this script in the `AskUser.m` file supplied with the downloadable source code.

LISTING 8-1: Asking for User Input

```
Name = input('What is your name? ', 's');
disp(['Hello ', Name]);
```

The `input()` function asks for user input. You provide a prompt that tells the user what to provide. When you want string input, as is the case in this example, you add the `'s'` argument to tell MATLAB that you want a string and not a number. When the user types a name and presses Enter, the value is placed in `Name`.

REMEMBER

The `disp()` function outputs text without assigning it to a variable. However, the `disp()` function accepts only a single input, and the example needs to output two separate strings (the `'Hello '` part and the `Name` part) as a combined whole. To fix this problem, you use the concatenation operator (`[]`). The term *concatenation* simply means to combine two strings. You separate each of the strings with a comma, as shown in the example.

When you run this example, the script asks you to type your name. Type ***your name*** and press Enter. In this case, the example uses `John` as the name, but you can use any name you choose. After you press Enter, the script outputs the result. Here is typical output from this example:

```
>> AskUser
What is your name? John
Hello John
```

Copying and pasting into a script

Experimentation is an essential part of working with MATLAB. After you get a particular command just right, you may want to add it to a script. This act involves cutting and pasting the information. When working in the Command Window, simply highlight the text you want to move into a script, right-click it, and choose Copy or Cut from the context menu. As an alternative, most platforms support speed keys for cutting and pasting, such as Ctrl+C for copy and Ctrl+X for cut.

Copying and cutting places a copy of the material on the Clipboard. Select the Editor window, right-click the location where you want to insert the material, and choose Paste from the context menu. (The pasted material is always put wherever the mouse pointer is pointing, so make sure you have the mouse cursor in the right place before you right click.) As an alternative, most platforms provide a speed key for pasting, such as Ctrl+V. In this case, you place the insertion pointer (the text pointer) where you want the new material to appear.

The Command History window succinctly stores all the commands that you type, making it easy for you to pick and choose the commands you want to place in a script. The following list provides techniques that you can use in the Command History window:

- Click a single line to use just that command.
- Ctrl+click to add additional lines to a single line selection.

» Shift+click to add all the lines between the current line and the line you clicked to a single line selection.

The result is that you end up with one or more selected lines. You can cut or copy these lines to the Clipboard and then paste them into the Editor window.

TIP

Using other sources for script material is possible, and you should use them whenever you can. For example, when you ask for help from MATLAB, the help information sometimes includes example code that you can copy and paste into your script. You can also find online sources of scripts that you can copy and paste. Checking the results of the pasting process is important in this case to ensure that you didn't inadvertently copy nonscript material. Simply delete the unwanted material before you save the script.

Converting the Command History into a script

After experimenting for a while, you might come up with a series of commands that does precisely what you'd like that series to do. Trying to cut and paste the commands from the Command Window is inconvenient. Of course, you could select the commands in the Command History window, copy them to the Clipboard, and paste them from there, but that seems like a waste of time, too.

REMEMBER

In reality, you can simply make a script out of the commands that you select in the Command History window. After you select the commands you want to use, just right-click the selected commands and choose Create Script from the context menu that appears. MATLAB opens a new Editor window with the selected commands in place (in the order they appear in the Command History window). Save the result to disk and run the script to see how it works.

Continuing long strings

Sometimes you can't get by with a short prompt — you need a longer prompt in order to obtain the information you need. When you need to create a longer string, use the continuation operator (`...`), which many people will recognize as an ellipsis. Listing 8-2 shows an example of how you can use long strings in a prompt to modify the `UserInput` example shown in Listing 8-1. You can also find this script in the `LongString.m` file supplied with the downloadable source code.

LISTING 8-2: Asking for User Input in a Specific Way

```
Prompt = [
    'Type your own name, but only if it isn''t ',...
    'Wednesday.\nType the name of the neighbor ',...
    'on your right on Wednesday.\nHowever, on ',...
    'a Wednesday with a full moon, type the ',...
    'name of\nthe neighbor on your left! '];
Name = input(Prompt, 's');
disp(['Hello ', Name]);
```

This example introduces several new features. The `Prompt` variable contains a long string with some formatting that you haven't seen before. It uses the concatenation operator to create a single string from each of the lines in the text. Each substring is self-contained and separated from the other substrings with a comma. The continuation operator lets you place the substrings on separate lines.

Notice the use of the double single quote (`isn''t`) in the text. You need to use two single quotes when you want a single quote to appear in the output as an apostrophe (isn't), rather than terminate a string. The `\n` character is new, too. This is a special character that controls how the output appears, so it is called a *control character*. In this case, the `\n` character adds a new line. When you run this example, you see output similar to that shown here:

```
LongString
Type your own name, but only if it isn't Wednesday.
Type the name of the neighbor on your right on Wednesday.
However, on a Wednesday with a full moon, type the name of
the neighbor on your left! John
Hello John
```

Everywhere a `\n` character appears in the original string, you see a new line. In addition, the word isn't contains a single quote, as expected. The following list shows the control characters that MATLAB supports, and defines how they are used.

- `''`: Single quotation mark/apostrophe
- `%%`: Percent character
- `\\`: Backslash
- `\a`: Alarm (sounds a beep or tone on the computer)
- `\b`: Backspace

- \f: Form feed
- \n: New line
- \r: Carriage return
- \t: Horizontal tab
- \v: Vertical tab
- \x*N*: Hexadecimal number, *N* (where *N* is the number of the character you want to display), where `sprintf('\x0041')` would produce a capital letter A
- *N*: Octal number, *N* (where *N* is the number of the character you want to display), where `sprintf('\0101')` would product a capital letter A

Note that control characters don't work with `disp()`. When using `disp('\x0041')`, you see an output of `\x0041`, rather than a capital letter A. However, you can use control characters with functions such as `sprint()` and `fprintf()`.

Adding comments to your script

People tend to forget things. You might know how a script works on the day you create it and possibly even for a week after that. However, six months down the road, you may find that you don't remember much about the script at all. That's where comments come into play. Using comments helps you to remember what a script does, why it does it in a certain way, and even why you created the script in the first place. The following sections describe comments in more detail.

Using the % comment

Anytime MATLAB encounters a percent sign (%), it treats the rest of the line as a comment. *Comments* are simply text that is used either to describe what is happening in a script or to *comment out* lines of code that you don't want to execute. You can comment out lines of code during the troubleshooting process to determine whether a particular line is the cause of errors in your script. The "Analyzing Scripts for Errors" section, later in this chapter, provides additional details on troubleshooting techniques. Listing 8-3 shows how comments might appear in a script. You can also find this script in the `Comments.m` file supplied with the downloadable source code.

LISTING 8-3: Using Comments to Make Code Easier to Read

```
% Tell MATLAB what to display onscreen.
Prompt = [
```

```
    'Type your own name, but only if it isn''t ',...
    'Wednesday.\nType the name of the neighbor ',...
    'on your right on Wednesday.\nHowever, on ',...
    'a Wednesday with a full moon, type the ',...
    'name of\nthe neighbor on your left! '];

% Obtain the user's name so it can
% be displayed onscreen.
Name = input(Prompt, 's');

% Output a message to make the user feel welcome.
disp(['Hello ', Name]);
```

Compare Listing 8-3 with Listing 8-2. You should see that the code is the same, but the comments make the code easier to understand. When you run this code, you see that the comments haven't changed how the script works. MATLAB also makes comments easy to see by displaying them in green letters.

Using the %% comment

MATLAB supports a double percent sign comment (%%) that supports special functionality in some cases. Here's how this comment works:

- Acts as a standard command in the Command Window.
- Allows you to execute a portion (a section) of the code when using the Run and Advance feature.
- Creates special output when using the Publish feature.

The following sections describe the special %% functionality. You won't use this functionality all the time, but it's nice to know that it's there when you do need it.

USING RUN AND ADVANCE

When you add a %% comment in the Editor window, MATLAB adds a section line above the comment (unless the comment appears at the top of the window), effectively dividing your code into discrete sections. To add a section comment, you type **%%**, a space, and the comment, as shown in Figure 8-4.

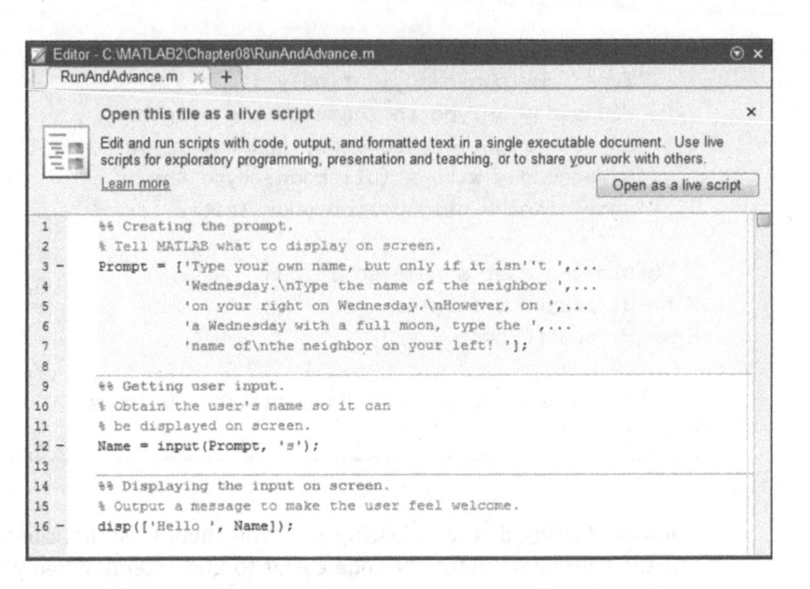

FIGURE 8-4: The %% comment adds section lines to the code.

TIP

When opening a standard script file containing a %% comment, you see a dialog at the top of the script file that asks whether you want to open the file as a Live Script. Chapter 11 tells you about working with Live Scripts, which are interactive documents managed in a single environment called the Live Editor. A Live Script can combine:

- MATLAB code
- Formatted text
- Equations
- Images

REMEMBER

As with standard comments, the %% comment appears in green type. The line above the comment is your cue that this is a special comment. In addition, the position of the text cursor (the insertion point) selects a particular section. The selected section is highlighted in a pale yellow. Only the selected section executes when you click Run and Advance. Here's how sections work:

1. **Place the cursor at the end of the** `Prompt =` **line of code and then click Run and Advance.**

   Only the first section of code executes. Notice also that the text cursor comes to rest at the beginning of the second section.

2. **Click Run and Advance.**

   The script displays a prompt asking for a name. The editor highlights the third section of code. Only the second section of code executes. You don't see the script output.

3. **Type a name and press Enter.**

   The script accepts the name provided and places it in `Name`. However, there is still no output.

4. **Place the cursor at the beginning of the second section and then click Run and Advance.**

   You are asked for your name as in Step 2, and you need to enter a name and press Enter as in Step 3. However, you don't see any output.

5. **Click Run and Advance with the text cursor at the beginning of the third `%%` comment.**

   You see the script output (the correct output, in fact) without being asked for a name.

6. **Perform Step 5 as often as desired.**

   The application displays the script output every time without asking for any further information. Using this technique lets you execute just the part of a script that you need to test rather than run the entire script every time.

TIP

You can make small changes to the code and still run a particular section. For example, change `Hello` to `Goodbye` in the code shown previously in Figure 8-4. With the third section selected, click Run and Advance. The output displays a goodbye message, rather than a hello message, without any additional input.

PUBLISHING INFORMATION

The section comments let you easily document your script. This section provides just a brief overview of the publishing functionality, but it demonstrates just how amazing this feature really is. To start with, you really do need to create useful section comments — the kind that will make sense as part of a documentation package.

When creating the setup for the script you want to publish, you need to define the output format and a few essentials. The default format is HTML, which is just fine for this example. However, if you don't make one small change, the output isn't going to appear quite as you might like it to look. On the Publish tab of the Editor window, click the down arrow under Publish and choose Edit Publishing Options. You see the Edit Configurations dialog box, shown in Figure 8-5.

The Evaluate Code option evaluates your script and outputs the result as part of the documentation. Unfortunately, MATLAB can't evaluate `input()` functions as part of publishing the documentation for a script. As a consequence, you must set Evaluate Code to false. Click Publish. MATLAB produces an HTML page like the one shown in Figure 8-6.

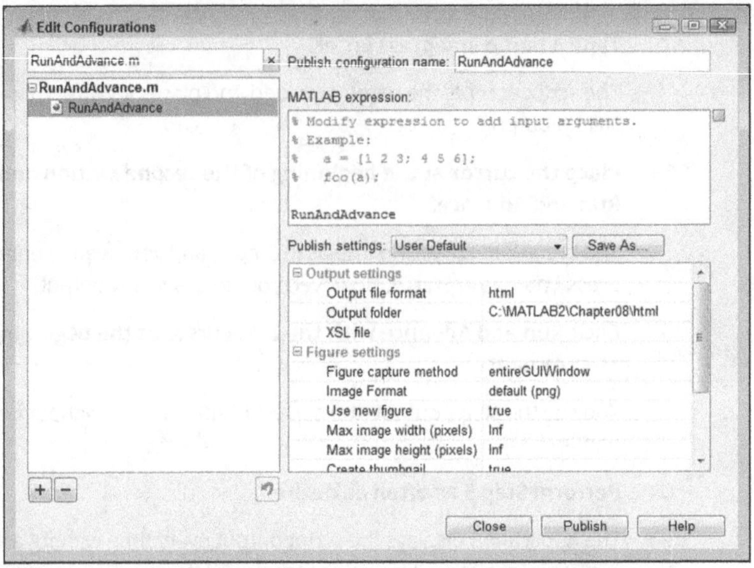

FIGURE 8-5: Modify the configuration options as needed to ensure that your script will publish correctly.

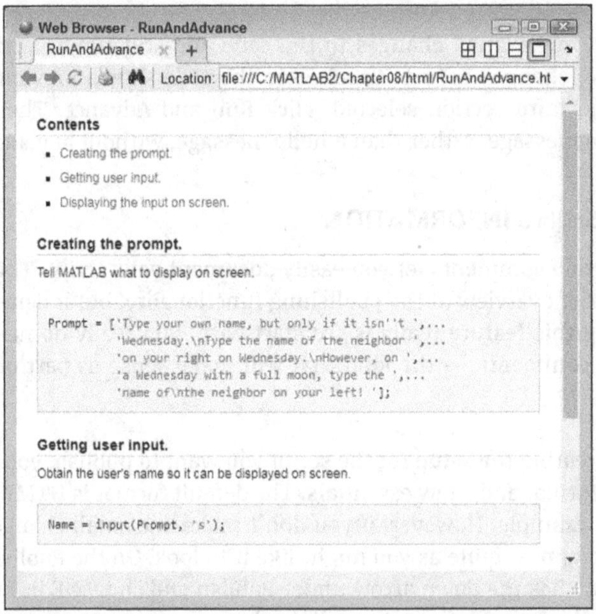

FIGURE 8-6: The published documentation looks quite nice.

Considering the little work you put into creating the documentation, it really does look quite nice. In fact, it looks professional. When working with complex scripts, documentation like this really does serve a serious need. After you're done admiring your work, close the HMTL page and the Edit Configurations dialog box.

Revising Scripts

Scripts usually aren't perfect the first time you write them. In fact, editing them quite a few times is common. Even if the script does happen to attain perfection, eventually you want to add features, which means revising the script. The point is, you commonly see your scripts in the Editor window more than once. Here are some techniques you can use to open a script file for editing:

- » Double-click the script's filename in the Current Folder window.
- » Click the down arrow on the Open option of the Home tab and select the file from the list. (The list contains every kind of file you've recently opened, not just script files.)
- » Click the down arrow on the Open option of the Editor tab and select the file from the list. (The list will include only the most recently used script files.)
- » Click Find Files in the Editor tab to display the Find Files dialog box. Enter a search criteria, such as **.m* (where the asterisk is a wild-card character for all files) and click Find. Double-click the file you want to open in the resulting list.
- » Locate the file using your platform's hard drive application (such as Windows Explorer in Windows or Finder on the Mac) and double-click the file entry.

It's a bad idea to make changes to a script and then try to use it without testing it first. Always test your changes to ensure that they work as you intend them to. Otherwise, a change that you thought would work could instead cause data damage or other problems.

Calling Scripts

Creating scripts without having some way to run them would be pointless. Fortunately, MATLAB lets you use scripts in all sorts of ways. The act of using a script — causing it to run — is known as *calling* the script. You can call scripts in these ways:

- » Right-click the script file in the Current Folder window and select Run from the context menu that appears.
- » Select the script file in the Current Folder window and press F9.
- » Type the filename on the command line and press Enter. (Adding the extension isn't necessary.)
- » Type the script filename in another script.

REMEMBER

The last method of calling a script is the most important. It enables you to create small pieces of code (scripts) and call those scripts to create larger, more powerful, and more useful pieces of code. The next step is creating functions that can send information in and out of those smaller pieces of code. (You see the topic of functions explored in Chapter 9.)

Improving Script Performance

Scripts can run only so fast. The resources offered by your system (such as memory and processor cycles), the location of data, and even the dexterity of the user all come into play. Of course, with the emphasis on "instant" in today's society, faster is always better. With this in mind, the following list provides you with some ideas on how to improve your script performance. Don't worry if you don't completely understand all these bullets; you see most of these techniques demonstrated somewhere in the book. This list serves as a reference for when you're working on creating the fastest script possible:

- Create variables once instead of multiple times.
  - Chapter 10 shows how to repeat tasks; creating variables inside these *loops* (bits of repeating code) is a bad idea.
  - An application made up of smaller files might inadvertently re-create variables, so look for this problem as you analyze your application.
- Use variables to hold just one type of data. Changing the data type of a variable takes longer than simply creating a new one.
- Make code blocks as small as possible.
  - Create several small script files rather than one large one.
  - Define small functions rather than large ones.
  - Simplify expressions and functions whenever possible.
- Use vectors whenever possible.
  - Replace multiple scalar variables with one vector.
  - Rely on vectors whenever possible to replace sparse matrices.
- Avoid running large processes in the background.

Analyzing Scripts for Errors

Ridding an application of errors is nearly impossible. As complexity grows, the chances of finding every error diminishes. Everyone makes mistakes, even professional developers. So, it shouldn't surprise you that you might make mistakes from time to time as well. Of course, the important thing is to find the errors and fix them. The process of finding errors and fixing them is called *debugging*.

Sometimes the simplest techniques for finding errors are the best. Working with your script in sections is an important asset in finding errors. The "Using the %% comment" section, earlier in this chapter, describes how to create and use sections. When you suspect that a particular section has an error in it, you can run the code in that section multiple times as you look in the Workspace window to see the condition of variables that the code creates and the Command Window to see the sort of output it creates.

Adding `disp()` statements to your code in various places lets you display the status of various objects. The information prints right in the Command Window so that you can see how your application works over time. Removing the `disp()` statements that you've added for debugging purposes is essential after the session is over. You can do this by adding a `%` in front of the `disp()` statement. This technique is called *commenting out*, and you can use it for lines of code that you suspect might contain errors as well. In addition, you can comment out debugging code that you use to troubleshoot your application in case you need it again later.

MATLAB also supports a feature called breakpoints. A *breakpoint* is a kind of stop sign in your code. It tells MATLAB to stop executing your code in a specific place so that you can see how the code is working. MATLAB supports two kinds of breakpoints:

- **Absolute:** The code stops executing every time it encounters the breakpoint. You use this kind of breakpoint when you initially start looking for errors and when you don't know what is causing the problem.
- **Conditional:** The code stops executing only when a condition is met. For example, a variable might contain a certain value that causes problems. You use this kind of breakpoint when you understand the problem but don't know precisely what causes it.

To set an absolute breakpoint, place the text cursor anywhere on the line and choose Set/Clear in the Breakpoints drop-down list on the Editor tab. When you set an absolute breakpoint, a red circle appears next to the line. To set a conditional breakpoint, place the text cursor anywhere on the line and choose Set Condition in the Breakpoints drop-down list. Type a condition, such as **x == 1**, in the dialog box that appears. Later chapters in the book demonstrate the use of breakpoints.

Creating error-handling code is also important in your application. Even though error handling doesn't fix an error, it makes the error less of a nuisance and can keep your application from damaging important data. Chapter 18 provides you with more ideas on how to locate and deal with errors in your script using error handling.

Using the MATLAB Profiler to Improve Performance

The MATLAB Profiler can help you locate the parts of your code that take too much time to execute. This allows you to spend your time efficiently in modifying the code to run faster. One way to profile your application is to click Run and Time on the Editor tab. MATLAB runs your code and displays a Profiler window like the one shown in Figure 8-7 when your code completes running.

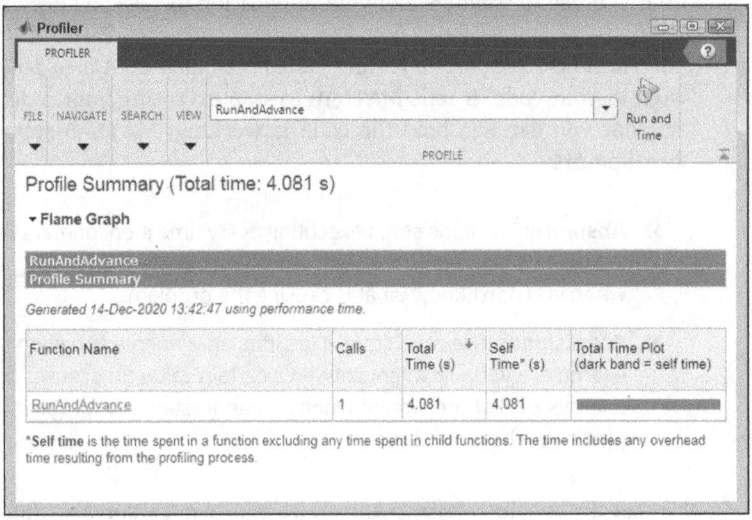

FIGURE 8-7: Viewing the run time for your scripts tells you how efficient they are.

You can also interact with the MATLAB Profiler programmatically by calling various functions. The following list provides a short overview of the functions and their purpose:

- `profile status`: Tells the current status of the Profiler, such as whether the profiler is on.
- `profile('info')`: Displays a list of Profiler data elements. You can display a particular data element by adding a period and its name to the function, such as `profile('info').FunctionHistory`, to see the function history.
- `profile on`: Turns the Profiler on. You can add the function history feature by adding the `-history` switch.
- `profile off`: Turns the Profiler off.
- `profile resume`: Restarts the Profiler without clearing its history.
- `profile clear`: Clears the Profiler statistics.
- `profile viewer`: Stops the Profiler and displays the results in the Profiler window, shown in Figure 8-7.

IN THIS CHAPTER

» Finding and using built-in functions

» Defining and using your own functions

» Understanding other function types

Chapter 9

Expanding MATLAB's Power with Functions

Simplification is an important part of creating any useful application. The better you can outline what tasks the application performs in the simplest of terms, the easier it is to define how to interact with and expand the application. Understanding how an application works is the reason you use functions. A *function* is simply a kind of box in which you put code. The function accepts certain inputs and provides outputs that reflect the input received. It isn't important to understand precisely how the function performs its task unless your task is to modify that function, but being able to visualize what task the function performs helps you understand the application as a whole. The only requirement is that you understand the inputs and resulting outputs. In short, functions simplify the coding experience.

This chapter is about three sorts of functions. If you've followed along in previous chapters, you have already used quite a few built-in functions, but simply using them may not be enough. You need to understand a little more about the inputs and outputs — the essentials of how the box works. On the other hand, you don't find out about the inner mechanisms of built-in functions in this chapter because you never need to know about those aspects.

You also get a chance to create your own functions in this chapter. The examples in previous chapters have been easy, so the need to provide simplification just isn't there. As the book progresses, you create more complex examples, so the need to simplify the code used in those examples becomes more important. Creating your own functions will make both the examples and your own code easier to understand.

MATLAB also supports some interesting alternatives to functions. They aren't functions in the traditional sense, but they make working with code simpler. These "special purpose" functions are used when the need arises to create code that is both efficient and elegant. The final part of this chapter provides a good overview of these special function types, and you see them used later in the book.

REMEMBER

You don't have to type the source code for this chapter manually. In fact, using the downloadable source is a lot easier. You can find the source for this chapter in the `\MATLAB2\Chapter09` folder of the downloadable source. When using the downloadable source, you may not be able to work through some of the hands on examples because the example files will already exist on your system. In this case, you can create an alternative folder, `Chapter09a`, to hold the results of the hands on exercises. (See the Introduction for details on how to obtain the downloadable source.)

Working with Built-in Functions

Built-in functions are those that come with MATLAB or are part of an add-on product. You typically don't have source code for built-in functions and must treat them simply as black boxes. So far, you have relied exclusively on built-in functions to perform tasks in MATLAB. For example, when you use the `input()` and `disp()` functions in Chapter 8, you're using built-in functions. The following sections tell you more about built-in functions and how you can work with them in MATLAB to achieve specific objectives.

Learning about built-in functions

There are many functions you can use to learn about built-in functions:

```
» help()
» doc()
» docsearch()
» lookfor()
» what()
```

If you already know the name of a function, one of the simplest makes use of the `help('function_name')` command, where *`function_name`* is the name of the function. Try it now. Type **help('input')** and press Enter in the Command Window. You see output similar to the output shown in Figure 9-1.

```
Command Window
New to MATLAB? See resources for Getting Started.
>> help('input')
 input  Prompt for user input.
    RESULT = input(PROMPT) displays the PROMPT string on the screen, waits
    for input from the keyboard, evaluates any expressions in the input,
    and returns the value in RESULT. To evaluate expressions, input accesses
    variables in the current workspace. If you press the return key without
    entering anything, input returns an empty matrix.

    STR = input(PROMPT,'s') returns the entered text without evaluating expressions

    To create a prompt that spans several lines, use '\n' to indicate each
    new line. To include a backslash ('\') in the prompt, use '\\'.

    Example:

       reply = input('Do you want more? Y/N [Y]:','s');
       if isempty(reply)
          reply = 'Y';
       end

    See also keyboard.

    Documentation for input

fx >>
```

FIGURE 9-1: Obtain help directly from MATLAB for built-in functions you know.

MATLAB does provide some types of category help. For example, type **help('elfun')** and press Enter to see a listing of elementary math functions at your disposal. When you type **help('specfun')** and press Enter, you see a listing of specialized math functions.

TIP

Sometimes the help information provided by the `help()` function becomes excessively long. In this case, you can use the `more()` function to present the information a page at a time. Before you use the `help()` function, type **more('on')** and press Enter to put MATLAB in paged mode. When the help information is more than a page in length, you see a

```
--more--
```

prompt at the bottom of the screen. Press the spacebar to see the next page or type **q** to end the output. If you want to see only the next line, press Enter instead. When you finish reviewing help, type **more('off')** and press Enter to turn off paged mode.

TIP

Although the `help()` function is really useful because it displays the information you need directly in the Command Window, sometimes the `doc()` function is a better choice. When using the `doc()` function, you see a nicely formatted output that includes links to example code and other information. Type **doc('input')** and press Enter, and you see the output shown in Figure 9-2. This is the option you should use when you want to get an in-depth view of a function rather than simply jog your memory as part of writing an application. In addition, when you find that the `help()` function is less helpful than you'd like, the `doc()` function generally provides more information.

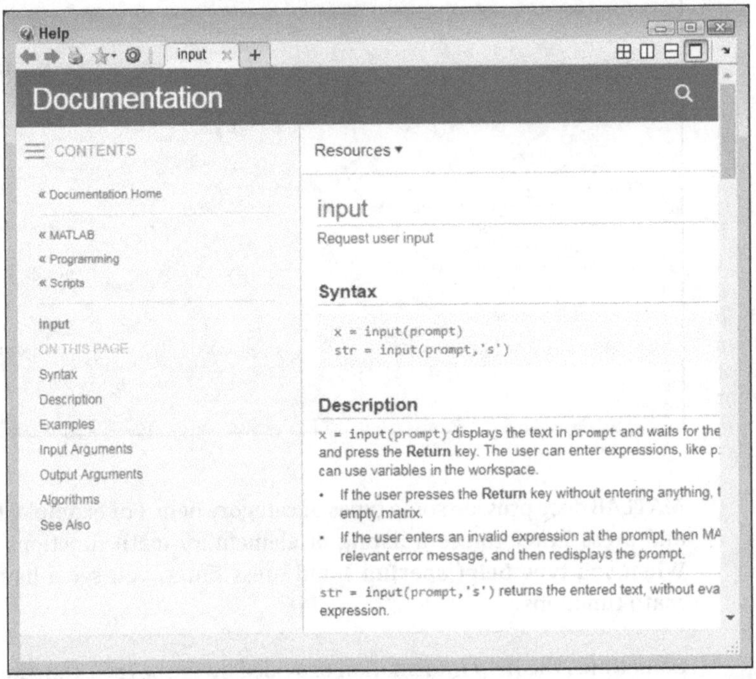

FIGURE 9-2: Use the `doc()` function when you need in-depth information about a built-in function.

Using `help()` may not always be possible because you don't know the precise name of whatever you need to find. Another useful function is `docsearch()`. You use this function when you have some idea, but not a precise one, of what you need to find. For example, type **docsearch('input')** and press Enter in the Command Window. This time you see a list of potential entries to query, as shown in Figure 9-3. Notice that the `input()` function is still the first entry in the list, but you have a number of other choices as well.

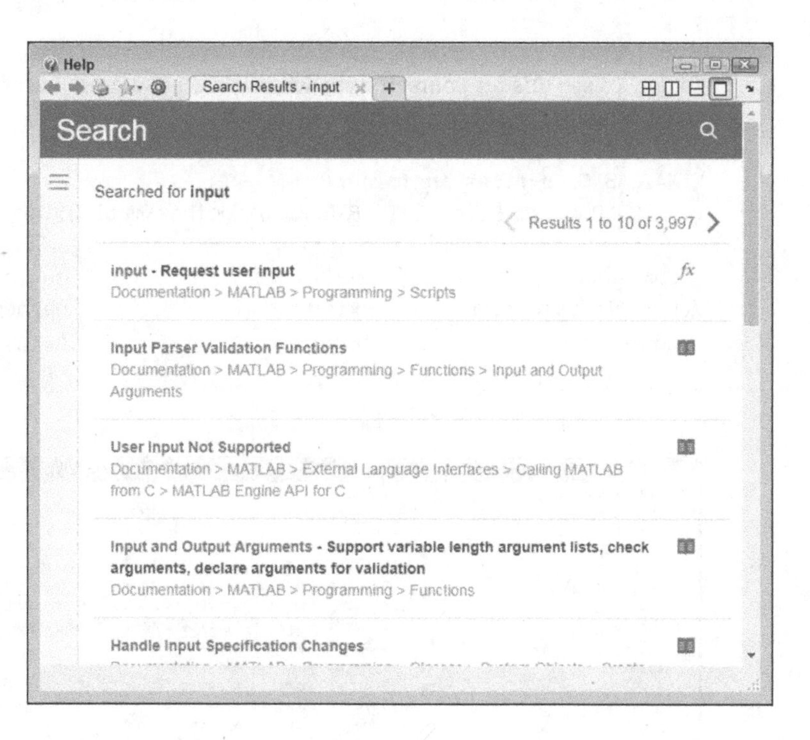

FIGURE 9-3: Search for what you need within the documentation.

One of the more interesting ways to search for built-in functions is to use the `lookfor()` function. In this case, MATLAB doesn't look in the documentation; rather, it looks in the source code files. This kind of search is important because you can sometimes see connections between functions this way and find alternatives that might not normally occur to you. To see how this kind of search works, type **lookfor('input')** and press Enter. You see the output shown in Figure 9-4. Notice that the `input()` function is in the list, but it doesn't appear at the top because the search doesn't sort the output by likely candidate.

If you really want to know more about the built-in functions from a coding perspective, start with the `which()` function, which tells you the location of the built-in function. For example, type **which('input')** and press Enter. You see the location of this built-in function on your system. On my system, I receive this output: `built-in (C:\Program Files\MATLAB\R2020b\toolbox\matlab\lang\input)`.

At this point, you know that `input()` is found in the `lang` folder. However, you really don't know what related functions might be in the same folder. Use the `what()` function to locate additional information about the content of the `lang` folder. To see this for yourself, type **what('lang')** and press Enter. You see a relatively long listing of function names like this:

```
MATLAB Code files in folder
   C:\Program Files\MATLAB\R2020b\toolbox\matlab\lang

Contents                  evalin                   munlock
ParallelException         exist                    nargchk
ans                       feval                    nargin
...
```

```
Command Window
New to MATLAB? See resources for Getting Started.
>> lookfor('input')
deal                        - Deal inputs to outputs.
isa                         - Determine if input is object of specified class.
isUnderlyingType            - Determine if input has specified underlying type
superiorfloat               - return 'double' or 'single' based on the superic
blkdiag                     - Block diagonal concatenation of matrix input arg
iscolumn                    - True if input is a column vector.
ismatrix                    - True if input is a matrix.
isrow                       - True if input is a row vector.
isscalar                    - True if input is a scalar.
isvector                    - True if input is a vector.
nargin                      - Number of input arguments to an INLINE object.
integral2ParseArgs          - Parse optional inputs to INTEGRAL2.
odefcncleanup               - This helper function incorporates any input argu
isfile                      - Determine if the input path points to a file
isfolder                    - Determine if the input path points to a folder
exportToMFile               - takes a HelpContainer object as input and genera
xmlstringinput              - Determine whether a string is a file or URL
input                       - Prompt for user input.
inputname                   - Input argument name.
iskeyword                   - Check if input is a keyword.
memoize                     - Adds memoization semantics to the input function
nargchk                     - Validate number of input arguments.
nargin                      - Number of function input arguments.
narginchk                   - Validate number of input arguments.
varargin                    - Variable length input argument list
```

FIGURE 9-4: In some cases, you need to look for associations as part of your search.

Notice that the output includes the `disp()` function that you used with the `input()` function in Chapter 8. However, you also see a number of other interesting functions in the list that could prove useful. If this listing still doesn't quench your need for more information about functions, check out the complete function list at `https://www.mathworks.com/help/matlab/referencelist.html`.

TECHNICAL STUFF

Not shown in the previous output is a listing of classes and packages found in the `lang` folder. Classes and packages are simply two other ways of packaging functionality within MATLAB. However, these two packaging methods provide more functionality than functions do in most cases, so it pays to look them up to see what sorts of things you can do with them. Using the `doc()` and `help()` functions provides you with information about the classes and packages.

Sending data in and getting data out

The essence of a function is that it presents you with a black box. In most cases, you send data in, it whirls around a bit, and then data comes back out. Managing data is an essential part of most functions.

REMEMBER

Of course, some functions require only input, some provide only output, and some perform tasks other than work directly with data. For example, the `clc()` function clears the Command Window and doesn't require any data input or produce any data output to perform the task. Every function does something; creating one that does nothing would be pointless.

The problem for many people is determining the input and output requirements for the built-in functions. The best way to discover this information is to use the `help()` or `doc()` functions. The `doc()` function is actually the easiest to use in this case. The input and output arguments appear at the bottom of the help screen. To see this for yourself, type **doc('input')** and press Enter. Scroll down to the bottom of the resulting page (refer to Figure 9-2) and you see the inputs and outputs.

In this case, you see that the input argument is a prompt and that you must provide this input as a string. The documentation explains that the prompt is there to ask the user for a specific kind of input. The output can take two forms: an array that is calculated from the input or a string that contains the precise text the user has typed.

REMEMBER

When you see a dual output for a function, it means that you need to tell the function what sort of output to provide or that there is a default. In this case, the `input()` function requires that you supply a second argument, `'s'`, to obtain the string output. The default is to provide the calculated array.

Creating a Function

Functions represent another method for packaging your code (but not the last discussed in the book; you also see classes discussed in Chapter 13). They work as an addition to scripts rather than a replacement for them. Scripts and functions each have a particular place to occupy in your MATLAB toolbox. The first section that follows explains these differences and helps you understand when you would use a script or a function. In some cases, it doesn't matter too much, but in other cases, the wrong choice can cause you a lot of frustration and wasted time.

The remainder of the sections that follow help you create custom functions of various types. You start with a simple function that doesn't require any input or output to perform a task. After that, you start to build functions with greater complexity that are also more flexible because they do accept input and produce output. Functions can be as simple or as complex as needed to perform a task, but simpler is always better (an emphasis of this chapter as a whole).

Understanding script and function differences

A *script* is a method of packaging a procedure — in other words, a series of steps that you use to perform a task. Some people have compared scripts to keyboard macros or other forms of simple step recording. On the other hand, a *function* is a method of packaging a transformation — code that is used to manage data in some manner or to perform a task that requires better data handling than a script can provide. Both types of packages contain code of a sort, but each packaging method is used differently. (Live Scripts, discussed in Chapter 11, and Live Functions, discussed in Chapter 12, are simply different ways to interact with scripts and functions.)

Scripts and functions also handle data differently. A script makes all the variables that it contains part of the workspace. As a result, after the script runs, you can easily see all the variables that the script contains as well as their ending values. A function hides its variables, and the variables become unavailable after the function runs. As a result, the actual data that the function uses internally isn't visible, and you must supply any required inputs every time you run the function.

As you see later in this section, a function also has a special header that identifies the function name, the inputs that it requires, and the outputs it provides. A function is a formal sort of coding method that's more familiar to developers. However, functions also provide greater flexibility because you can control the environment in which they perform tasks with greater ease.

The use of inputs and outputs reduces the potential for contamination by data left over from a previous run and, like Las Vegas, what happens in the function stays in the function. This feature is a big advantage: You can use the same name in a function as you would outside it without interference, and doing so avoids a lot of confusion.

Both scripts and functions reside in files that have an `.m` extension. The immediately noticeable difference between the two is that a script lacks a header. Functions always have the header that you see in the "Writing your first function" section, coming up shortly.

Understanding built-in function and custom function differences

Built-in functions (those provided with MATLAB) and custom functions (those you create yourself or that come as part of a third-party product) differ in at least one important aspect. The custom functions come with source code. You can modify this source code as needed to meet your particular needs.

The built-in `input()` function comes with MATLAB, and you can find it in the `input.m` file in the `toolbox\matlab\lang` directory used to contain part of the files for your MATLAB installation. However, if you open that file, you see documentation but no source code. The source code is truly part of MATLAB, and you can't edit it. You can modify the documentation as necessary with your own notes, but this really isn't a recommended procedure because the next MATLAB update will almost certainly overwrite your changes.

Writing your first function

Creating a function is only slightly more work than creating a script. In fact, the two processes use the same editor, so you're already familiar with what the editor can provide in the way of help. The various Editor features you'd use for creating a script all work the same way with functions, too. (You have access to the same double percent sign (`%%`) for use with sections, for example.) The following steps get you started creating your first function. You can also find this function in the `SayHello.m` file supplied with the downloadable source code.

1. **Click the arrow under the New entry on the Home tab of the MATLAB menu and select Function from the list that appears.**

   You see the Editor window, shown in Figure 9-5. Notice that the editor already has a function header in place for you, along with the inputs, outputs, and documentation comments.

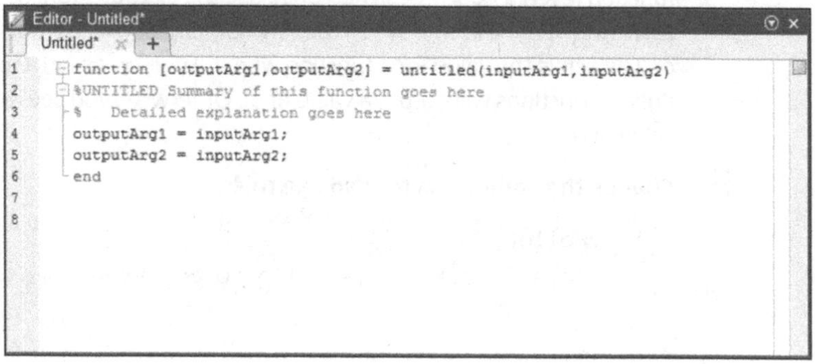

Editor - Untitled*

Untitled*

```
function [outputArg1,outputArg2] = untitled(inputArg1,inputArg2)
%UNTITLED Summary of this function goes here
%   Detailed explanation goes here
outputArg1 = inputArg1;
outputArg2 = inputArg2;
end
```

FIGURE 9-5: The Editor window helps you create new functions.

REMEMBER

Figure 9-5 may look a little complex, but that's because MATLAB includes a number of optional elements that you will see in action later in the chapter. A function has three requirements:

- A function always begins with the word `function`.
- You must include a function name.
- A function must always end with the keyword `end`.

2. **Delete** `outputArg1,outputArg2`**, but not the square brackets.**

   Functions aren't required to have output arguments. In order to keep things simple for your first function, you're not going to require any inputs or outputs.

REMEMBER

   An *argument* is simply a word for an individual data element. If you supply a number to a function, the number is considered an argument. Likewise, when you supply a string, the entire string is considered just one argument. A vector, even though it contains multiple numbers, is considered a single argument. Any single scalar or object that you provide as input or that is output from the function is considered an argument.

3. **Delete** `inputArg1,inputArg2`**, but not the parentheses.**

   Functions aren't required to have input arguments.

4. **Change the function name from** `untitled` **to** `SayHello`**.**

   Your function should have a unique name that reflects its purpose. Avoiding existing function names is essential. Before you name your function, test the name you're considering by typing **help('*NameOfYourFunction*')** and pressing Enter. If the function already exists, you see a help screen. Otherwise, MATLAB denies all knowledge of the function, and you can use the function name you have chosen.

WARNING

   Always provide help information with the functions you create. Otherwise, the `help()` function won't display any help information, and someone could think that your function doesn't exist. If you want to be certain that there is no potential conflict between a function you want to create and an existing function (even a poorly designed one), use the `exist()` function instead, such as `exist('SayHello')`. When the function exists, you see an output value of 2 (which is the value for a function with a `.m`, `.mlx`, or `.mlapp` extension; built-in functions will output a value of 5). Otherwise, you see an output value of 0.

5. **Change the comments to read like this:**

```
%SayHello()
%   This function says Hello to everyone!
```

Notice that the second line is indented. The indentation tells MATLAB that the first line is a title and the second is text that goes with the title. Formatting your comments becomes important when working with functions. Otherwise, you won't see the proper help information when you request it.

6. **Delete the existing code before the end statement; then add the following code after the comment:**

```
disp('Hello There!');
```

   The function simply displays a message onscreen.

7. **Click Save on the Editor tab.**

   You see the Select File for Save As dialog box.

8. **Select the Chapter09 directory for the source code for this book, type SayHello.m in the File Name field, and then click Save.**

   MATLAB saves your function as SayHello.m.

The filename you use to store your function must match the name of the function. MATLAB uses the filename to access the function, not the function name that appears in the file. When there is a mismatch between the function name and the filename, MATLAB displays an error message.

Using the new function

You have a shiny new function and you're just itching to use it. Before you can use the function, you must make sure that the directory containing the function file is part of the MATLAB path. You can achieve this goal in two ways:

- Double-click the directory entry in the Current Folder window.
- Right-click the directory entry in the Current Folder window and choose Add to Path ⇨ Selected Folders and Subfolders from the context menu.

You can try your new function in a number of ways. The following list contains the most common methods:

- Click Run in the Editor window, and you see the output in the Command Window. However, there is a little twist with functions that you discover in the upcoming "Passing data in" section of the chapter. You can't always click Run and get a successful outcome, even though the function will always run.
- Click Run and Advance in the Editor window. (This option runs the selected section when you have sections defined in your file.)

» Click Run and Time in the Editor window. (This option outputs profiling information — statistics about how the function performs — for the function.)

» Type the function name in the Command Window and press Enter.

Your function also has help available with it. Type **help('SayHello')** and press Enter. MATLAB displays the following help information:

```
SayHello()
    This function says Hello to everyone!
```

The output is precisely the same as it appears in the function file. The `doc()` function also works. Type **doc('SayHello')** and press Enter. You see the output shown in Figure 9-6. Notice how the title is presented in a different color and font than the text that follows. In addition, clicking the `View Code for SayHello` link displays the function's source code.

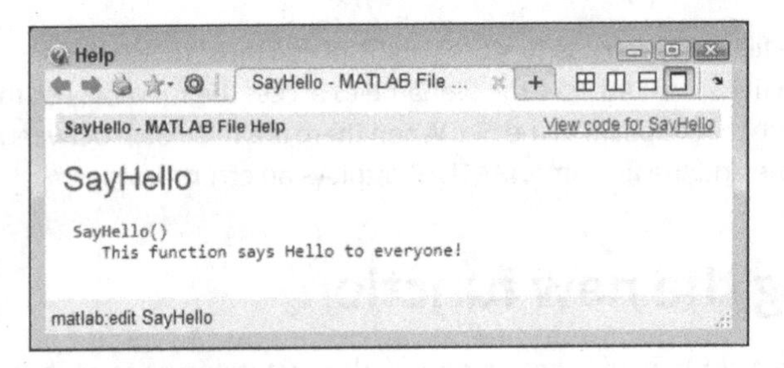

FIGURE 9-6: The help you provide is available to anyone who needs it.

Passing data in

The `SayHello()` function is a little limited. For one thing, it can't greet anyone personally. To make `SayHello()` a little more flexible, you need to pass some information to it in the form of an input argument. The following steps help you create an updated `SayHello()` that accepts input arguments. You can also find this function in the `SayHello2.m` file supplied with the downloadable source code.

1. **Click the down arrow under the Save option on the Editor tab of the Editor window and choose Save As.**

   You see the Select File for Save As dialog box.

2. **Type** `SayHello2.m` **in the File Name field and click Save.**

   MATLAB saves the function that you created earlier using a new name. Notice that the function name is now highlighted in orange. The highlight tells you that the function name no longer matches the filename.

3. **Change the function name from `SayHello` to `SayHello2`.**

   The orange highlight disappears when you place the text cursor in another location in the Editor window.

4. **Add the input argument `Name` to the function header so that the header looks like this:**

   ```
   function [ ] = SayHello2( Name )
   ```

   Notice that `Name` is now highlighted in orange to show that you haven't used it anywhere. The highlight will go away after you make the next change. The Editor window always displays an orange highlight when it detects a problem with your code. It's important to realize that the Code Analyzer feature of MATLAB detects only potential errors. (You can read more about the Code Analyzer at `https://www.mathworks.com/help/matlab/matlab_prog/matlab-code-analyzer-report.html`.) It can't absolutely tell you that a problem exists, so you need to look carefully at the highlights.

5. **Change the `disp()` function call so that it looks like this:**

   ```
   disp(['Hello There ', Name, '!']);
   ```

   The `disp()` function now requires use of the concatenation operator that was introduced in Chapter 8 to combine the text with the input argument. The output will contain a more personalized message.

6. **Click Run.**

   MATLAB displays a message telling you that the `SayHello2()` function requires input, as shown in Figure 9-7. You see this message every time you try to use Run to start the function because functions don't store argument information as scripts do.

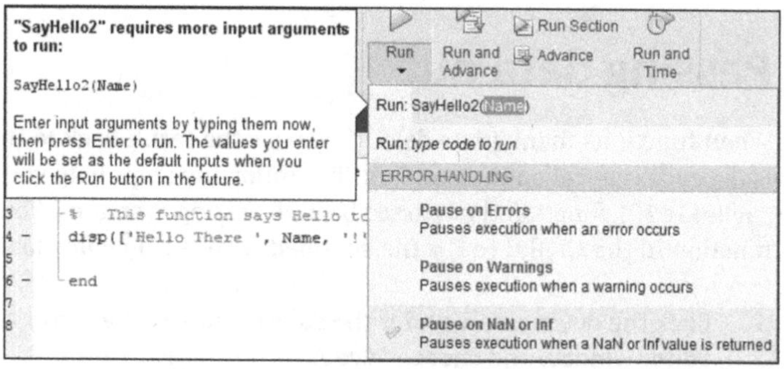

FIGURE 9-7: MATLAB knows to ask you for the input arguments as needed.

7. **Type** 'Ann' **in the Run drop-down and press Enter.**

   You see the expected output of:

   ```
   Hello There Ann!
   ```

TIP

The Run button will remember what you used for input the next time you click it. Here is how you can interact with the button's memory:

- To clear this value, click the down arrow under the Run button, right-click the Run command, and then choose Delete from the context menu that appears.
- To modify this value, click the down arrow under the Run button, right click the Run command, and then choose Edit from the context menu that appears. You will then be able to edit the existing value by typing a new value.

8. **Type** SayHello2('Josh') **in the Command Window and press Enter.**

   You see the expected output.

REMEMBER

As you modify your functions, you should also modify the comments for them. Because you add `Name` as an input argument, you should document it in the comments for this function, as shown in the following in bold:

```
%SayHello2()
%   This function says Hello to everyone!
%   It requires a string, Name, as input.
```

Adding this comment will also change the output of functions like `help()` and `doc()` so that people know why `SayHello2()` is different from `SayHello()`. If you don't take this extra step, other people could use the wrong function for a particular need.

Passing data out

When functions manipulate data, they pass the result back to the caller. The *caller* is the code that called the function. The following steps help you create an updated `SayHello2()` function that passes back the string it creates. You can also find this function in the `SayHello3.m` file supplied with the downloadable source code.

1. **Click the down arrow under the Save option on the Editor tab of the Editor window and choose Save As.**

   You see the Select File for Save As dialog box.

2. **Type** SayHello3.m **in the File Name field and click Save.**

   MATLAB saves the function you created earlier using a new name.

3. **Change the function name from** `SayHello2` **to** `SayHello3`.

   The orange highlight disappears when you place the text cursor in another location in the Editor window.

4. **Type** `HelloString` **in the square brackets before the function name so that your function header looks like this:**

   ```
   function [ HelloString ] = SayHello3( Name )
   ```

   The function now returns a value to the caller. You see the orange highlight again because `SayHello3( )` doesn't assign a value to `HelloString` yet.

5. **Modify the function code to include an assignment to** `HelloString`, **like this:**

   ```
   HelloString = ['Hello There ', Name, '!'];
   disp(HelloString);
   ```

   The function now assigns a value to `HelloString` and then uses that value as output. It also returns the output to the caller.

6. **Save the changes you've made.**

7. **Type** Output = SayHello3('Ambrose') **in the Command Window and press Enter.**

   You see the following output:

   ```
   Hello There Ambrose!
   Output =
       'Hello There Ambrose!'
   ```

8. **Type** `disp(Output)` **in the Command Window and press Enter.**

   You see the expected greeting as output.

Creating and using global variables

Functions normally use *local variables* — that is, they aren't visible to anyone but the function. Using local variables avoids confusion because each function is self-contained. In addition, using local variables makes functions more secure and reliable because only the function can access the data in the variable.

USING OPTIONAL ARGUMENTS

Dealing with optional arguments requires some MATLAB features that are covered in later chapters, so you can feel free to skip this sidebar for now and come back to it later. The current version of the SayHello2() function requires an argument. You must supply a name or the function won't run. A number of functions that you have already used, such as disp(), provide optional arguments — you can choose to use them or not. Optional arguments are important because there are times when a default argument works just fine. In fact, you can change SayHello2() so that it doesn't require an argument.

To make SayHello2() work without an input argument, you must assign a value to Name, but only if Name doesn't already have a value. In order to make the assignment, you first need to determine whether Name does have a value by checking a variable that MATLAB provides for what you called nargin (for *N*umber of *Arg*uments *In*put). If nargin equals 1, then the caller — the code that called this function — has provided an input. Otherwise, you need to supply the input to Name. Here's the updated version of SayHello2() that you can find in the downloadable source as SayHello2a.m.

```
function [ ] = SayHello2a( Name )
%SayHello2a()
%   This function says Hello to everyone!
%   It allows an optional a string, Name, as input;
%   Name has a default value of 'Good Looking'.
if nargin < 1
    Name = 'Good Looking';
end
disp(['Hello There ', Name, '!']);
end
```

The additional code states that if nargin is less than 1, the function needs to assign it a value of 'Good Looking'. Otherwise, the function uses the name provided by the caller. To test this code, type **SayHello2a** and press Enter in the Command Window. You see Hello There Good Looking! as output. Of course, the function could be broken, so type **SayHello2a('Selma')** and press Enter. You see Hello There Selma! as output, so the function works precisely as it should, and Name is now an optional argument.

WARNING

You may find that you need to make a variable visible, either because it is used by a number of functions or the caller needs to know the value of the variable. When a function makes a local variable visible to everyone, it becomes a *global variable*. Global variables can be misused because they're common to every function that wants to access them and they can present security issues because the data becomes public.

The following steps show how to create a global variable using `SayHello3` as a starting point. You can also find these functions in the `SayHello4.m` and `SayHello5.m` files supplied with the downloadable source code.

1. **Click the down arrow under the Save option on the Editor tab of the Editor window and choose Save As.**

   You see the Select File for Save As dialog box.

2. **Type** SayHello4.m **in the File Name field and click Save.**

   MATLAB saves the function you created earlier using a new name.

3. **Change the function name from** `SayHello3` **to** `SayHello4`**.**

   The orange highlight disappears when you place the text cursor in another location in the Editor window.

4. **Remove** `HelloString` **from the square brackets before the function name so that your function header looks like this:**

   ```
   function [ ] = SayHello4( Name )
   ```

   When a variable is global, you can't return it as data from a function call. The data is already available globally, so there is no point in returning it from the function.

5. **Change the** `HelloString` **assignment so that it now contains the global keyword, as shown here:**

   ```
   global HelloString;
   HelloString = ['Hello There ', Name, '!'];
   ```

6. **Save the changes you've made.**

7. **Type** SayHello4('George') **in the Command Window and press Enter.**

   You see the following output:

   ```
   Hello There George!
   ```

   At this point, there is a global variable named `HelloString` sitting in memory. Unfortunately you can't see it, so you don't really know that it exists for certain.

8. **Perform Steps 1 through 3 again to create** `SayHello5()`**.**

9. **Modify the** `SayHello5()` code **so that it looks like this:**

   ```
   function [ ] = SayHello5( )
   %SayHello5()
   %   This function says Hello to everyone using
   %   a global variable!
   ```

```
%   It requires a string, Name, as input.
%   HelloString is now a global variable.
global HelloString
disp(HelloString);
end
```

Notice that SayHello5() doesn't accept input or provide output arguments. In addition, it only declares HelloString; it doesn't actually assign a value to it, so the function should fail when it calls the disp() function.

10. **Type** SayHello5 **in the Command Window and press Enter.**

    You see Hello There George! as the output. The global variable really is accessible from another function.

Using subfunctions

A single function file can contain multiple functions. However, only one function, the *primary function* (the one that has the same name as the original file), is callable. Any other functions in the file, known as *subfunctions*, are local to that file. The primary function or other subfunctions can call on any subfunction, as long as that subfunction appears in the same file.

The main reason to use subfunctions is to simplify your code by breaking it into smaller pieces. In addition, placing common code in a subfunction means that you don't have to copy and paste it all over the place — you have to write it only once. As far as anyone else is concerned, however, the file contains only one function. The inner workings of your code are visible only to you and anyone else who can view the source code.

Listing 9-1 shows an example of how a subfunction might work. You can also find this function in the SayHello6.m file supplied with the downloadable source code.

LISTING 9-1: Creating a Subfunction

```
function [ HelloString ] = SayHello6( Name )
%SayHello6()
%   This function says Hello to everyone!
%   It requires a string, Name, as input.
%   It passes back the result as HelloString.
HelloString = [GetGreeting(), Name, '!'];
disp(HelloString);

end
```

```
function [ Greeting ] = GetGreeting ( )
Greeting = 'Hello There ';
end
```

This code is actually another version of the SayHello3 code that you worked with earlier. The only difference is that the greeting is now part of the GetGreeting() subfunction, rather than a simple string. Notice that SayHello6() can call GetGreeting(), using the same technique that it could use for any other function.

After you create this code, type **Output = SayHello6('Stan')** in the Command Window and press Enter. You see the following output:

```
Hello There Stan!
Output =
    'Hello There Stan!'
```

The output is precisely as you expect. However, now type **GetGreeting()** and press Enter in the Command Window. Instead of a greeting, you see an error message:

```
Unrecognized function or variable 'GetGreeting'.
```

The GetGreeting() subfunction isn't accessible to the outside world. As a result, you can use GetGreeting() with SayHello6() and not have to worry about outsiders using the subfunction incorrectly.

Nesting functions

You can also nest functions one inside the other in MATLAB. The *nested function* physically resides within the primary function. The difference between a primary function and a nested one is that the nested function can access all the primary function data, but the primary function can't access any of the nested function data.

In all other respects, subfunctions and nested functions behave in a similar manner (for example, outsiders can't call either subfunctions or nested functions directly). Listing 9-2 shows a typical example of a nested function. You can also find this function in the SayHello7.m file supplied with the downloadable source code.

LISTING 9-2: Creating a Nested Function

```
function [ HelloString ] = SayHello7( Name )
%SayHello7()
%   This function says Hello to everyone!
%   It requires a string, Name, as input.
%   It passes back the result as HelloString.
HelloString = [GetGreeting(), Name, '!'];
disp(HelloString);

    function [ Greeting ] = GetGreeting ( )
    Greeting = 'Hello There ';
    end
end
```

This is another permutation of the SayHello3() example, but notice how the GetGreeting() nested function now resides inside SayHello7(). After you create this code, type **Output = SayHello7('Stan')** in the Command Window and press Enter. You see the following output:

```
Hello There Stan!
Output =
    'Hello There Stan!'
```

Using Other Types of Functions

MATLAB supports a few interesting additions to the standard functions. In general, these additions are used to support complex applications that require unusual programming techniques. However, it pays to know that the functions exist for situations in which they come in handy. The following sections provide a brief overview of these additions.

Inline functions

An *inline function* is one that performs a small task and doesn't actually reside in a function file. You can create an inline function right in the Command Window if you want. The main purpose for an inline function is to make it easier to perform a calculation or manipulate data in other ways. You use an inline function as a kind of macro. Instead of typing a lot of information every time, you define the inline function once and then use the inline function to perform all the extra typing.

To see an inline function in action, type **SayHello8 = inline('[''Hello There '', Name, ''!'']')** in the Command Window and press Enter. Note that you're using two single quotes in this case, not a double quote. You see the following output:

```
SayHello8 =
     Inline function:
     SayHello8(Name) = ['Hello There ', Name, '!']
```

This function returns a combined greeting string. All you need to do is type the function name and supply the required input value. Test this inline function by typing **disp(SayHello8('Robert'))** and pressing Enter. You see the expected output:

```
Hello There Robert!
```

REMEMBER

Notice that the inline function doesn't actually include the `disp()` function call. An inline function must return a value, not perform output. If you try to include the `disp()` function call by using the `SayHello8 = inline('[disp(''Hello There '', Name, ''!'')]')` function instead and calling it with `SayHello8('Robert')`, you see the following error message:

```
Error using inlineeval (line 15)
Error in inline expression ==> disp(['Hello There ', Name, '!'])
Too many output arguments.
Error in inline/subsref (line 24)
            INLINE_OUT_ = inlineeval(INLINE_INPUTS_,
            INLINE_OBJ_.inputExpr, INLINE_OBJ_.expr);
```

Anonymous functions

An anonymous function is an even shorter version of the inline function. It can contain only a single executable statement. The single statement can accept input arguments and provide output data.

To see how an anonymous function works, type **SayHello9 = @(Name) ['Hello There ', Name, '!']** and press Enter. You see the following output:

```
SayHello9 =
    @(Name)['Hello There ',Name,'!']
```

The at (`@`) symbol identifies the code that follows as an anonymous function. Any input arguments you want to accept must appear in the parentheses that follow the `@` symbol. The code follows after the input argument declaration. In this case, you get yet another greeting as output.

To test this example, type **disp(SayHello9('Evan'))** in the Command Window and press Enter. You see the following output:

```
Hello There Evan!
```

TIP

You generally use anonymous functions for incredibly short pieces of code that you need to use repetitively. Inline functions execute more slowly than anonymous functions for a comparable piece of code. So whenever possible, use an anonymous function in place of an inline function. However, inline functions also provide the extra flexibility of allowing multiple lines of code, so you need to base your decision partly on how small you can make the code that you need to execute.

IN THIS CHAPTER

» Adding decision-making to your scripts

» Using recursive functions

» Repeating tasks

» Defining menus

Chapter 10
Adding Structure to Your Scripts

The scripts and functions you have created so far have all performed a series of tasks, in order, one at a time. However, sometimes it's important to skip steps or to perform the same step more than once. Humans make decisions about what to do with ease, but computers need a little help by using *conditional* statements that allow actions only when needed. Humans also know when to do something more than once, but computers need help in this regard as well, by using *loop* statements that define when and how many times to perform a task. This chapter helps you understand the use of both conditional and loop statements when creating scripts and functions.

As part of discovering how decisions are made and just how repetition works, you see two practical examples of how to employ your new skills. The first is a technique called *recursion*, which is simply a method of performing a task more than once, but in an elegant way. Using recursion makes solving some math problems significantly easier. The second involves the use of a menu. Most multifunction applications rely on *menus* to allow a user to select one option out of a number of possibilities.

Making Decisions

When you come to an intersection, you make a decision: go through or stop. If the light is red, you stop. However, when the light is green, you go through the intersection. The "if this condition exists, then do this" structure is something humans use almost constantly. In fact, we sometimes don't think about it consciously at all. A decision is often made without conscious thought because we have made it so many times.

A computer needs guidance in order to make a decision. It lacks a subconscious and therefore needs to work through each decision. When working with MATLAB, you might need to tell the computer that if a value is above a certain level, the computer should perform a certain set of steps. Perhaps you need to compensate in the computation of a result (some factor has affected the result beyond the normal value) or you simply need to look at the formula in a different way. The decision could be procedural — a user selects a particular option and the computer needs to perform the associated task. The thing to remember about a computer is that it follows whatever you tell it precisely, so you need to provide precise decision-making instructions.

The following sections describe two different decision-making structures that MATLAB provides: the `if` statement and the `switch` statement. Each of these statements has a specific format as well as specific times when you'd use it.

Decision-making code has a number of terms associated with it. A *statement* simply indicates what the code should do. It's the line of code that appears first in a block of tasks. A *structure* describes the statement and all the code that follows until the end keyword is reached. A *condition* specifies how to make the decision.

Using the if statement

The simplest decision to make is whether to do something — or not. However, you might need to decide between two alternatives. When a situation is true, you perform one task, but when it's false, you perform another task. There are still other times when you have multiple alternatives and must choose a course of action based on multiple scenarios and using multiple related decisions. The following sections cover all these options.

Making a simple decision

Starting simply is always best. The `if` statement makes it possible to either do something when the condition you provide is true or not do something when the condition you provide is false. The following steps show how to create a function that includes an `if` statement. You can also find this function in the `SimpleIf.m` file supplied with the downloadable source code.

1. **Click the arrow under the New entry on the Home tab of the MATLAB menu and select Function from the list that appears.**

   You see the Editor window that you used for the examples in Chapter 9.

2. **Delete** `outputArg1,outputArg2`**.**

   The example doesn't provide an output argument, but it does require an input argument.

3. **Change the function name from** `Untitled` **to** `SimpleIf`**.**

   The primary function name must match the name of the file.

4. **Change** `inputArg1,inputArg2` **to** `Value`**.**

   The function receives a single value from the caller to use in the decision-making process.

5. **Delete the existing function code.**

6. **Type the following code into the function between the comment and the** end **keyword.**

```
if Value > 5
    disp('The input value is greater than 5!');
end
```

   This code makes a simple comparison. When the input argument, `Value`, is greater than 5, the function tells you about it. Otherwise, the function doesn't provide any output at all.

7. **Create or select the Chapter10 folder, if you haven't done so already. Click Save.**

   You see the Select File for Save As dialog box. Notice that the File Name field has the correct filename entered for you. This is the advantage of changing the function name before you save the file for the first time.

8. **Click Save in the Select File for Save As dialog box.**

   The function file is saved to disk.

9. **Type** SimpleIf(6) **and press Enter in the Command Window.**

   You see the following output:

```
The input value is greater than 5!
```

10. **Type** SimpleIf(4) **and press Enter in the Command Window.**

    The function doesn't provide any output. Of course, this is the expected reaction.

Adding an alternative option

Many decisions that people make are choices between two options. For example, you might go to the beach today, or choose to stay home and play dominoes, based on whether it is sunny. When the weather is sunny, you go to the beach. MATLAB has a similar structure. The application chooses between two options based on a condition. The second option is separated from the first by an `else` clause — the application performs the first task, or else it performs the second. The following steps demonstrate how the `else` clause works. (These steps assume that you completed the `SimpleIf` example in the preceding section.) You can also find this function in the `IfElse.m` file supplied with the downloadable source code.

1. **In the Editor window, with the `SimpleIf.m` file selected, click the down arrow under Save and choose Save As from the list that appears.**

   You see the Select File for Save As dialog box.

2. **Type** IfElse.m **in the File Name field and click Save.**

   MATLAB saves the example using a new name.

3. **Replace the `SimpleIf` function name with `IfElse`.**

4. **Add the following code after the `disp()` function call:**

   ```
   else
       disp('The input value is less than 6!');
   ```

   The function can now respond even when the primary condition isn't met. When `Value` is greater than 5, you see one message; otherwise, you see the other message.

5. **Click Save.**

   The function file is saved to disk.

6. **Type** IfElse(6) **and press Enter in the Command Window.**

   You see the following output:

   ```
   The input value is greater than 5!
   ```

7. **Type** IfElse(4) **and press Enter in the Command Window.**

   You see the following output:

   ```
   The input value is less than 6!
   ```

   The example demonstrates that you can provide alternative outputs depending on what is happening within the application. Many situations arise in which you must choose an either/or type of condition.

Creating multiple alternative options

Many life decisions require more than two alternatives. For example, you're faced with a menu at a restaurant and want to choose just one of the many delicious options. Applications can encounter the same situation. A user may select only one of the many options from a menu, as an example. The following steps show one method of choosing between multiple options. (The steps assume that you completed the `IfElse` example in the preceding section.) You can also find this function in the `IfElseIf.m` file supplied with the downloadable source code.

1. **In the Editor window, with the `IfElse.m` file selected, click the down arrow under Save and choose Save As from the list that appears.**

   You see the Select File for Save As dialog box.

2. **Type** IfElseIf.m **in the File Name field and click Save.**

   MATLAB saves the example using a new name.

3. **Replace the `IfElse` function name with `IfElseIf`.**

4. **Add the following code after the first `disp()` function call:**

```
elseif Value == 5
    disp('The input value is equal to 5!');
```

   At this point, the code provides separate handling for inputs greater than, equal to, and less than 5.

TIP

   It's possible to have an `if...elseif` statement that doesn't include the `else` statement in it. In this case, every option would have a condition and there wouldn't be any default code block, as you would have with the `else` statement.

5. **Modify the third `disp()` function statement to read as follows:**

```
disp('The input value is less than 5!');
```

REMEMBER

   Many people make the mistake of not modifying everything that needs to be modified by an application change. Because you now have a way of handling inputs equal to five, you must change the message so that it makes sense to the user. Failure to modify statements often leads to odd output messages that serve only to confuse users.

6. **Click Save.**

   The function file is saved to disk.

7. **Type** IfElseIf(6) **and press Enter in the Command Window.**

   You see the following output:

   ```
   The input value is greater than 5!
   ```

8. **Type** IfElseIf(5) **and press Enter in the Command Window.**

   You see the following output:

   ```
   The input value is equal to 5!
   ```

9. **Type** IfElseIf(4) **and press Enter in the Command Window.**

   You see the following output:

   ```
   The input value is less than 5!
   ```

Using the switch statement

You can create any multiple alternative selection code needed by using the `if...elseif` statement. However, you have another good way to make selections. A `switch` statement lets you choose one of a number of options using code that is both easier to read and less time-consuming to type than using a lot of `if...elseif` statements. The result is essentially the same, but the method of obtaining the result is different. The following steps demonstrate how to use a `switch` statement. You can also find this function in the `SimpleSwitch.m` file supplied with the downloadable source code.

1. **Click the arrow under the New entry on the Home tab of the MATLAB menu and select Function from the list that appears.**

   You see the Editor window.

2. **Delete** `outputArg1,outputArg2`.

   The example doesn't provide an output argument, but it does require an input argument.

3. **Change the function name from** `Untitled` **to** `SimpleSwitch`.

   The primary function name must match the name of the file.

4. **Change** `inputArg1,inputArg2` **to** `Value`.

   The function receives a value from the caller to use in the decision-making process.

5. **Delete the existing function code.**

6. **Type the following code into the function between the comment and the** end **keyword.**

```
switch Value
    case 1
        disp('You typed 1.');
    case 2
        disp('You typed 2.');
    case 3
        disp('You typed 3.');
    otherwise
        disp('You typed an alternative value.');
end
```

This code specifically compares `Value` to the values provided. When `Value` matches a specific value, the application outputs an appropriate message.

At times, the input value doesn't match the values you expect. In such cases, the `otherwise` clause comes into play. It provides the means for doing something even if the input wasn't what you expected. If nothing else, you can use this clause to tell the user to input an appropriate value.

7. **Click Save.**

You see the Select File for Save As dialog box. Notice that the File Name field has the correct filename entered for you.

8. **Click Save.**

The function file is saved to disk.

9. **Type** SimpleSwitch(1) **and press Enter in the Command Window.**

You see the following output:

```
You typed 1.
```

10. **Type** SimpleSwitch(4) **and press Enter in the Command Window.**

You see the following output:

```
You typed an alternative value.
```

11. **Type** SimpleSwitch(-1) **and press Enter in the Command Window.**

You see the following output:

```
You typed an alternative value.
```

It's important to note that the current code checks only for values of 1, 2, or 3. Consequently, the user could provide a negative number, even if you're not expecting one. Later you discover how to check the input to ensure that it's in the correct range.

12. **Type** SimpleSwitch('Hello') **and press Enter in the Command Window.**

    You see the following output:

    ```
    You typed an alternative value.
    ```

    It's easy to become focused on the input you're expecting, rather than the input the user will provide. Not only do you have to check the range of incoming values, you must also check their data type to ensure that you're not getting something like a string.

Understanding the switch difference

A `switch` provides a short method of making specific decisions. You can't make a generalize decision, such as whether a value is greater than some amount. In order to make a match, the value must equal a specific value. The specific nature of a `switch` means that:

- The code you write is shorter than a comparable `if...elseif` structure with more than three alternatives.
- Others are better able to understand your code because it's cleaner and more precise.
- The `switch` statement tends to produce focused code that avoids the odd mixing of checks that can occur when using the `if...elseif` structure.
- MATLAB is able to optimize the application for better performance because it doesn't have to check a range of values.
- Using an `otherwise` clause can help ensure that people understand that you didn't anticipate a particular value or that the value is less important.

Deciding between if and switch

Whenever possible, use a `switch` statement when you have more than three options to choose from and you can focus attention on specific options for just one variable. Using the `switch` statement has some significant advantages, as described in the preceding section.

The `if` statement provides you with flexibility. You can use it when a range of values is acceptable or when you need to perform multiple checks before allowing a task to complete. For example, when opening a file, you might need to verify that the file actually does exist, that the user has the required rights to access the file, and that the hard drive isn't full. An `if` statement would allow you to check all these conditions in a single piece of code, making the conditions easier to see and understand.

When writing your own applications, you need to keep in mind that there isn't a single solution to any particular problem. The overlap between the `if` and `switch` statements makes it clear that you could use either statement in many situations. Both statements would produce the same output, so there isn't a problem with using them. However, `if` statements are better when flexibility is required, but `switch` statements are better when precision and speed are important. Choose the option that works best for a particular situation, rather than simply choosing an option that works.

Creating Recursive Functions

Many elegant programming techniques exist in the world, but none are quite as elegant as the recursive function. The concept is simple: You create a function that keeps calling itself until a condition is satisfied, and then the function delivers an answer based on the results of all those calls. This process of the function calling itself multiple times is known as *recursion,* and a function that implements it is a *recursive function.*

The most common recursion example is calculating factorial (n!), where *n* is a non-negative number (0! equals 1). (Calculating a factorial means multiplying the number by each number below it in the hierarchy. For example, 4! is equal to 4 * 3 * 2 * 1, or 24.)

Most examples that show how to create a recursive function don't really demonstrate how the process works. The following steps help you create a recursive function that does demonstrate how the process works. Later in the chapter, you see a less involved version of the same code that shows how the function would normally appear. You can also find this function in the `Factorial1.m` file supplied with the downloadable source code.

1. **Click the arrow under the New entry on the Home tab of the MATLAB menu and select Function from the list that appears.**

   You see the Editor window.

2. **Change** `outputArg1,outputArg2` **to** `Result`.

   The function returns a result to each preceding cycle of the call.

3. **Change the function name from** `Untitled` **to** `Factorial1`.

   The primary function name must match the name of the file.

4. **Change** `inputArg1,inputArg2` **to** `Value, Level`.

   The `Value` received is always one less than the previous caller received. The `Level` demonstrates how `Value` is changing over time.

5. **Delete the existing function code.**
6. **Type the following code into the function between the comment and the** end **keyword:**

```
if nargin < 2
   Level = 1;
end

if Value > 1
    fprintf('Value = %d Level = %d\n', Value, Level);
    Result = Factorial1(Value - 1, Level + 1) * Value;
    disp(['Result = ', num2str(Result)]);
else
    fprintf('Value = %d Level = %d\n', Value, Level);
    Result = 1;
    disp(['Result = ', num2str(Result)]);
end
```

This example makes use of an optional argument. The first time the function is called, `Level` won't have a value, so the application automatically assigns it a value of 1.

The code breaks the multiplication task into pieces. For example, when `Value` is 4, the code needs to multiply it by 3 * 2 * 1. The 3 * 2 * 1 part of the picture is defined by the call to `Factorial1(Value - 1, Level + 1)`. During the next pass, `Value` is now 3. To get the appropriate result, the code must multiply this new value by 2 * 1. So, as long as `Value` is greater than 1 (where an actual result is possible), the cycle must continue.

WARNING

A recursive function must always have an ending point — a condition under which it won't call itself again. In this case, the ending point is the `else` clause. When `Value` finally equals 1, `Result` is assigned a value of `1` and simply returns, without calling `Factorial1()` again. At this point, the calling cycle unwinds and each level returns, one at a time, until a final answer is reached.

Notice that this example uses a new function, `fprintf()`, to display information onscreen. The `fprintf()` function accepts a formatting specification as its first input. In this case, the specification says to print the string `Value =`, followed by the information found in `Value`, then `Level =`, followed by the information found in `Level`. The %d in the format specification tells `fprintf()` to print an integer value. You use `fprintf()` as a replacement for `disp()` when the output formatting starts to become more complex. Notice that `disp()` requires the use of the `num2str()` function to convert the numeric value of `Result` to a string in order to print it.

7. **Click Save.**

   You see the Select File for Save As dialog box. Notice that the File Name field has the correct filename entered for you.

8. **Click Save.**

   The function file is saved to disk.

9. **Type** Factorial1(4) **and press Enter in the Command Window.**

   You see the following output:

   ```
   Value = 4 Level = 1
   Value = 3 Level = 2
   Value = 2 Level = 3
   Value = 1 Level = 4
   Result = 1
   Result = 2
   Result = 6
   Result = 24
   ans =
       24
   ```

   The output tells you how the recursion works. Notice that all the `Value` and `Level` outputs come first. The function must keep calling itself until `Value` reaches 1. When `Value` does reach 1, you see the first `Result` output. Of course, `Result` is also 1. Notice how the recursion unwinds. The next `Result` is 2 * 1, then 3 * 2 * 1, and finally 4 * 3 * 2 * 1.

Now that you have a better idea of how the recursion works, look at the slimmed-down version in Listing 10-1. You can also find this function in the `Factorial2.m` file supplied with the downloadable source code.

LISTING 10-1: A Method for Calculating n!

```
function [ Result ] = Factorial2( Value )
%Factorial2 - Calculates the value of n!
%   Outputs the factorial value of the input number.
    if Value > 1
        Result = Factorial2(Value - 1) * Value;
    else
        Result = 1;
    end

end
```

UNDERSTANDING THE FPRINTF() FORMAT SPECIFICATION

A format specification tells a function how to display information onscreen. The `fprintf()` function accepts a regular string as input for a format specification, reading each character. When `fprintf()` encounters a percent (%) character, it looks at the next character as a definition of what kind of formatted input to provide. The following list provides an overview of the % character combinations used to format information using `fprintf()`.

- `%bo`: Floating point, double precision (Base 8)
- `%bu`: Floating point, double precision (Base 10)
- `%bx`: Floating point, double precision (Base 16, using lowercase letters for the numbers a through f)
- `%bX`: Floating point, double precision (Base 16, using uppercase letters for the numbers A through F)
- `%c`: Single character
- `%d`: Signed integer
- `%e`: Floating point, exponential notation using a lowercase e
- `%E`: Floating point, exponential notation using an uppercase E
- `%f`: Floating point, fixed point notation
- `%g`: Floating point, general notation using the more compact of `%f` or `%e` with no trailing zeros
- `%G`: Floating point, general notation using the more compact of `%f` or `%E` with no trailing zeros
- `%i`: Signed integer
- `%o`: Unsigned integer (Base 8)
- `%s`: String of characters
- `%to`: Floating point, single precision (Base 8)
- `%tu`: Floating point, single precision (Base 10)
- `%tx`: Floating point, single precision (Base 16, using lowercase letters for the numbers a through f)

- `%tX`: Floating point, single precision (Base 16, using uppercase letters for the numbers A through F)
- `%u`: Unsigned integer (Base 10)
- `%x`: Unsigned integer (Base 16, using lowercase letters for the numbers a through f)
- `%X`: Unsigned integer (Base 16, using uppercase letters for the numbers A through F)

When working with numeric input, you can also specify additional information between the `%` and the subtype, such as `f` for floating point. For example, `%-12.5f` would display a left-justified number 12 characters in width with 5 characters after the decimal point. See the full details of formatting strings at `https://www.mathworks.com/help/matlab/matlab_prog/formatting-strings.html`.

The final version is much smaller but doesn't output any helpful information to tell you how it works. Of course, this version will run a lot faster, too.

Performing Tasks Repetitively

Giving an application the capability to perform tasks repetitively is an essential part of creating an application of any complexity. Humans don't get bored performing a task once. It's when the task becomes repetitive that true boredom begins to take hold. A computer can perform the same task in precisely the same manner as many times as needed because the computer doesn't get tired. In short, the area in which computers can help humans most is performing tasks repetitively. As with decisions, you have two kinds of structures that you can use to perform tasks repetitively, as described in the sections that follow.

REMEMBER

You see a number of terms associated with repetitive code. The same terms that you see used for decision-making code also apply to repetitive code. In addition, the term *loop* is used to describe what repetitive code does. A repetitive structure keeps executing the same series of tasks until such time as the condition for the repetition is satisfied and the loop ends.

Using the for statement

The `for` statement performs a given task a specific number of times, unless you interrupt it somehow. The examples in the "Making Decisions" section, earlier in this chapter, provide steps for creating functions. Listing 10-2 shows how to use a for loop in an example. You can also find this function in the SimpleFor.m file supplied with the downloadable source code.

LISTING 10-2: Creating Repetition Using the for Statement

```
function [ ] = SimpleFor( Times )
%SimpleFor: Demonstrates the for loop
%   Tell the application how many times to say hello!

    if nargin < 1
        Times = 3;
    end
    for SayIt = 1:Times
        disp('Howdy!')
    end
end
```

In this case, SimpleFor() accepts a number as input. However, if the user doesn't provide a number, then SimpleFor() executes the statement three times by default.

Notice how the variable, SayIt, is created by assigning it an initial value of 1 (the 1 in 1:Times) and used as the starting point for the loop. The range of 1:Times tells for to keep displaying the message Howdy! the number of times specified by Times. Every time the loop is completed, the value of SayIt increases by 1, until the value of SayIt is equal to Times. At this point, the loop ends.

Using the while statement

The while statement performs a given task until a condition is satisfied, unless you stop it somehow. The examples in the "Making Decisions" section of the chapter provide steps for creating functions. Listing 10-3 shows how to use a while loop in an example. You can also find this function in the SimpleWhile.m file supplied with the downloadable source code.

LISTING 10-3: Creating Repetition Using the while Statement

```
function [ ] = SimpleWhile( Times )
%SimpleWhile: Demonstrates the while loop
%   Tell the application how many times to say hello!
```

```
    if nargin < 1
        Times = 3;
    end
    SayIt = 1;
    while SayIt <= Times
        disp('Howdy!')
        SayIt = SayIt + 1;
    end
end
```

In this example, the function can either accept an input value or execute a default number of times based on whether the user provides an input value for `Times`. The default is to say `Howdy!` three times.

Notice that the loop code actually begins by initializing `SayIt` to `1` (so the count begins at the right place). It then compares the current value of `SayIt` to `Times`. When `SayIt` is greater than `Times`, the loop ends.

REMEMBER

You must manually update the counter variable when using a `while` loop. Notice that the line that adds `1` to `SayIt` after the call to `disp()`. If this line of code is missing, the application ends up in an *endless loop* — meaning that it never wants to end. If you accidentally create an endless loop, you can stop it by pressing Ctrl+C.

Starting a new loop iteration using continue

There are times when you won't want to perform any processing should certain events occur, such as a value you don't really need. However, you don't want to stop the processing as a whole — you only want to stop this particular iteration of the loop. The `continue` clause makes it possible to end the current iteration and move on to the next loop iteration.

Listing 10-4 shows how to use the `continue` clause with a `while` loop, but you can use it precisely the same way with the `for` loop. The examples in the "Making Decisions" section, earlier in the chapter, provide steps for creating functions. You can also find this function in the `UsingContinue.m` file supplied with the downloadable source code.

LISTING 10-4: Using the continue clause

```
function [ ] = UsingContinue( Times )
%UsingContinue: Demonstrates the while loop
%   Tell the application which iterations to process.
%   Iteration five displays a special message.

    if nargin < 1
        Times = 3;
    end

    SayIt = 1;
    while SayIt <= Times
        if SayIt == 5
            disp('Welcome to iteraction 5!')
            SayIt = SayIt + 1;
            continue;
        end
        disp('Howdy!')
        SayIt = SayIt + 1;
    end
end
```

The code executes precisely the same way that the SimpleWhile example works, except that this version contains an additional `if` statement. When someone wants to execute the loop more than five times, the `if` statement takes effect on the fifth iteration. The application displays a special message, then skips the rest of the loop. Execution begins with the next loop iteration. Notice that you must update `SayIt` or the application will end up in an endless loop. To see this function in action, type **UsingContinue(8)** and press Enter. You see the first four Howdy! messages, the special message, and then three ending Howdy! messages.

Ending processing using break

A loop can ordinarily execute a certain number of times and then stop without incident. However, when certain conditions are met, the loop may have to end early. For example, some people in your organization might be ramping up the Howdy! application into overdrive. To prevent this abuse, you want the loop to stop at five — friendly, but not too verbose. The `break` clause lets you stop the loop early.

Listing 10-5 shows how to use the `break` clause with a `while` loop, but you can use it in precisely the same way with the `for` loop. The examples in the "Making Decisions" section, earlier in the chapter, provide steps for creating functions. You can also find this function in the UsingBreak.m file supplied with the downloadable source code.

LISTING 10-5: Using the break Clause

```
function [ ] = UsingBreak( Times )
%UsingBreak: Demonstrates the while loop
%   Tell the application how many times to say hello!
%   Don't exceed five times or the application will cut you off!

    if nargin < 1
        Times = 3;
    end
    SayIt = 1;
    while SayIt <= Times
        disp('Howdy!')
        SayIt = SayIt + 1;
        if SayIt > 5
            disp('Sorry, too many Howdies')
            break;
        end
    end
end
```

The code executes precisely the same way that the SimpleWhile example works except that this version contains an additional `if` statement. When someone wants to execute the loop more than five times, the `if` statement takes effect. The application displays a message telling the user that the number of Howdies has become excessive and then calls `break` to end the loop. To see this example in action, type **UsingBreak(10)** and press Enter in the Command Window.

Ending processing using return

Another way to end a loop is to call `return` instead of `break`. The basic idea is similar. See the upcoming "Differentiating between break and return" sidebar for details on how the two clauses differ.

Listing 10-6 shows how to use the `return` clause with a `while` loop, but you can use it in precisely the same way with the `for` loop. The examples in the "Making Decisions" section, earlier in the chapter, provide steps for creating functions. You can also find this function in the UsingReturn.m file supplied with the downloadable source code.

LISTING 10-6: Using the return Clause

```
function [ Result ] = UsingBreak( Times )
%UsingBreak: Demonstrates the while loop
%   Tell the application how many times to say hello!
%   Don't exceed five times or the application will cut you off!

    if nargin < 1
        Times = 3;
    end
    Result = 'Success!';
    SayIt = 1;
    while SayIt <= Times
        disp('Howdy!')
        SayIt = SayIt + 1;
        if SayIt > 5
            disp('Sorry, too many Howdies')
            Result = 'Oops!';
            return;
        end
    end
end
```

Notice that this example returns a `Result` to the caller. The value of `Result` is initially set to `'Success!'`. However, when the user gets greedy and asks for too many Howdies, the value changes to `'Oops!'`. To test this example, begin by typing **disp(UsingReturn())** and pressing Enter. You see the following output:

```
Howdy!
Howdy!
Howdy!
Success!
```

In this case, the application meets with success because the user isn't greedy. Now type **disp(UsingReturn(10))** and press Enter. This time the application complains by providing this output:

```
Howdy!
Howdy!
Howdy!
Howdy!
Howdy!
Sorry, too many Howdies
Oops!
```

DIFFERENTIATING BETWEEN BREAK AND RETURN

It may seem that `break` and `return` are almost precisely the same, but they really aren't. The UsingReturn example demonstrates one difference. You can provide a return value when using the `return` clause. In some cases, the ability to return a value is essential, and you must use `return` in place of `break`.

However, another difference isn't quite so apparent. When you nest one loop inside another, the `break` clause ends only the inner loop, not the outer loop. As a result, the outer loop continues to run as before. In some cases, this is a desirable behavior, such as when you need to interrupt the current task but still want to execute the remaining tasks. Using `return` ends all the loops. In fact, the function containing the loops ends and control returns to the caller.

Determining which loop to use

You can understand the `for` and `while` loops better by comparing Listing 10-2 and Listing 10-3. A `for` loop provides the means to execute a set of tasks a precise number of times. You use the `for` loop when you know the number of times a task should execute in advance.

A `while` loop is based on a condition. You use it when you need to execute a series of tasks until the job is finished. (For example, when two functions are expected to converge on a particular value, you can use the `while` loop to detect the convergence and end the processing.) However, you don't know when the task will end until such time as the conditions indicate that the job is done. Because `while` loops require extra code and additional monitoring by MATLAB, they tend to be slower, so you should use the `for` loop whenever possible to create a faster application.

Creating Menus

A menu is one way in which you can start to test the abilities you've gained in this chapter. Listing 10-7 shows a menu that you could use as a model for your own menu. Notice that this menu is a script. You could just as easily create a menu as a function. However, with all the emphasis on functions in this chapter, knowing that you can also use these techniques in scripts is important. You can also find this script in the `MyMenu.m` file supplied with the downloadable source code.

LISTING 10-7: A Simple Script Menu Example

```
EndIt = false;

while not(EndIt)
    clc
    disp('Choose a Fruit');
    disp('1. Orange');
    disp('2. Grape');
    disp('3. Cherry');
    disp('4. I''m Bored, Let''s Quit!');

    Select = input('Choose an option: ', 's');
    if isnan(str2double(Select))
        disp('Provide numeric input, press Enter');
        pause
        continue;
    end

    if Select == '4'
        disp('Sorry to see you go.');
        EndIt = true;
    else
        switch Select
            case '1'
                disp('You chose an orange!');
            case '2'
                disp('You chose a grape!');
            case '3'
                disp('You chose a cherry!');
            otherwise
                disp('You''re confused, quitting!');
                break;
        end
        pause(2)
    end
end
```

The example begins by declaring a variable, EndIt, to end the while loop. A while loop is the perfect choice in this case because you don't know how long the user will want to use the menu.

The example clears the Command Window and then displays the options. After the user enters a selection, the application checks to determine when it should end. If it should, it displays a goodbye message and sets EndIt to true.

TIP

This example is the first one to provide some sort of input checking. It uses str2double() to convert Select from a string into a double value. If the input isn't numeric, str2double outputs the value NaN, or not a number. You can check for the NaN value using isnan(). When isnan() is true, the code tells the user to input only numeric values, and the user must press Enter to continue. The continue clause starts with the next iteration of the loop to give the user another chance.

When the user chooses some other numeric option, the code relies on a switch to provide a response. In this case, the response is a simple message, but your production application would perform some sort of task. When a user provides a useless response, the application detects it and ends. Notice the use of the break clause.

The pause(2) function is new. Because the Command Window is cleared after each iteration, the pause(2) function provides a way to display the response for two seconds. The user can also choose to press Enter to return to the menu early.

IN THIS CHAPTER

» Understanding Live Script

» Creating a Live Script

» Running a Live Script

» Performing diagnostics

Chapter 11
Working with Live Scripts

In the past, writing an application was all about writing code in a step-by-step, static manner, which works fine for a great many needs. Also, running regular code makes the results appear separately from the code, resulting in a disconnect between the two.

However, many people prefer to work with code in a dynamic manner so that they can see the result of a line of thought immediately after pursuing it. A Live Script is a technique of writing code that lets you work with your code dynamically and see the result alongside the code. This approach provides a less abstract method of creating MATLAB scripts. The first part of this chapter helps you understand how Live Script compares with a regular MATLAB script, allowing you to take what you've learned in previous chapters and build on it.

You use a special kind of editor, the Live Editor, to create a Live Script. This editor has a number of new features that you haven't seen before now, but some features work the same as they do in the standard Editor. The next part of the chapter gets you started with the Live Editor.

Using sections, as described in the "Using the %% comment" section of Chapter 8, becomes more important when working with a Live Script because the idea is that you update only what is needed. The next section of the chapter discusses Live Script sections and explains how to use them when running your code.

As with anything else, you eventually find errors in your Live Scripts. Because of how you interact with Live Scripts, the debugging techniques are a little different from those you use when working with a standard script.

REMEMBER

You don't have to type the source code for this chapter manually. In fact, using the downloadable source is a lot easier. You can find the source for this chapter in the `\MATLAB2\Chapter11` folder of the downloadable source. When using the downloadable source, you may not be able to work through some of the hands on examples because the example files will already exist on your system. In this case, you can create an alternative folder, `Chapter11a`, to hold the results of the hands on exercises. See the Introduction for details on how to obtain the downloadable source.

Comparing a Live Script to a Regular Script

No magic is associated with a Live Script. You need to think in terms of extended functionality. Regular scripts are stored as code in a text file with an `.m` extension. On the other hand, Live Script relies on an `.mlx` file, which is an extension of the `.zip` file format. This `.mlx` file contains three types of content:

- Code
- Formatted content
- Output

REMEMBER

The code and formatted content are combined into an XML document using the Office Open XML (ECMA-376) format. Using the `.mlx` file format has these advantages:

- **Interoperability across locales:** If you store your script using letters from one language and open the script back up using the editor for another language, the script will automatically appear with the correct lettering. So, this means that anyone should be able to read your code. However, MATLAB doesn't magically translate the language, so comments and other textual elements still appear in the original language.
- **Extensible:** You can extend the formatting capability of your script to use all the formatting options found in Microsoft Word. You can implement some custom formatting options as well, but that topic is well beyond the scope of this book.

» **Forward compatibility:** The file format allows future proofing by implementing the ECMA-376 standard's forward-compatibility strategy.

» **Backward compatible:** Future versions of MATLAB will support the features provided by previous versions of MATLAB.

As previously mentioned, the use of the `.mlx` file format allows other functionality as well. A standard script displays its output in the Command Window, and you must run it every time you start MATLAB. A Live Script stores its output in the same window as the code, and you save the output when you save the `.mlx` file. Using a Live Script saves considerable time and effort when you have to create similar reports each day, because you update only the parts that have changed.

To present nicely formatted output with a standard script, you must rely upon *publishing markup*, which is a method of telling the Live Editor how to format the text. For example, placing asterisks on either side of a word or phrase like this, `*This is bold!*`, will make the text appear in bold type. When working with a Live Script, the experience is more akin to working with a word processor. You use various controls to format the output directly so that you can see it immediately without having the run the code again.

Working with the Live Editor

The Live Editor has a different appearance and some different functionality from the standard Editor. However, many things are the same. For example, you use the same commands and functions. MATLAB has some features that the Live Editor doesn't support, such as classes (see Chapter 13). In addition, because of how it works, Live Editor doesn't support some MATLAB preferences, including custom keyboard shortcuts and Emacs-style keyboard shortcuts (see `https://www.mathworks.com/help/matlab/matlab_env/about-editor-debugger-preferences.html` for a more complete list). Working with nonstandard shortcuts is outside the scope of this book, but you can learn more about custom keyboard shortcuts at `https://www.mathworks.com/help/matlab/matlab_env/keyboard-shortcuts.html`, and about the Emacs-style shortcuts at `https://blogs.mathworks.com/community/2007/05/11/setting-up-keybindings-for-the-command-window-and-editor/`. With these differences in mind, the following sections get you started using Live Editor. You can also find this Live Script in the `SimpleLiveScript.mlx` file supplied with the downloadable source code.

Opening the Live Editor

You use Live Editor to edit Live Scripts and Live Functions. To open the Live Editor, choose any of these options:

- » Click New Live Script on the Home tab.
- » Click New ➪ Live Script on the Home tab.
- » Click New ➪ Live Function on the Home tab.

The appearance of the Live Editor for Live Functions differs from Live Scripts, and you see these differences discussed in Chapter 12. Figure 11-1 shows how the Live Editor appears for Live Scripts. Note that you can click the x in the upper-right corner of the Welcome to the Live Editor pane to free up some space.

FIGURE 11-1: The Live Editor shows an Editor pane (left) and an Output pane (right).

Working with the Output pane

The Output pane is where you see the results of any code that you write and run. In addition, it shows formatted text you create and any graphics you include. The plots you develop also appear in the Output pane. What you end up with is a report made up of everything you've done. Use the following steps to see how all this works (the steps assume you've opened the Live Editor using one of the options in the previous section).

1. **Click Text in the Text section of the Live Editor tab.**

   You see the current line (the one with the insertion pointer) change to a text line (notice there is no longer a number on the left side of the display).

2. **Type** This is a Sample Section **and then select Heading 1 from the style drop-down list in the Text section of the Live Editor tab.**

   The formatting of the text you typed changes.

3. **Press Enter to move to the next line; then click Code in the Code section of the Live Editor tab.**

   The line changes to a code line, where you can type code.

4. **Type the following code in the code area of the document:**

   ```
   x = 1 + 2;
   fprintf('X = %d', x);

   plot(1:10, sin(1:10));
   ```

5. **Press Enter and then click Section Break in the Section section of the Live Editor tab.**

   The Live Editor adds a blue line showing the end of the section.

6. **Place the cursor anywhere within the current section; then click Run Section in the Section section of the Live Editor tab.**

   You see the result shown in Figure 11-2.

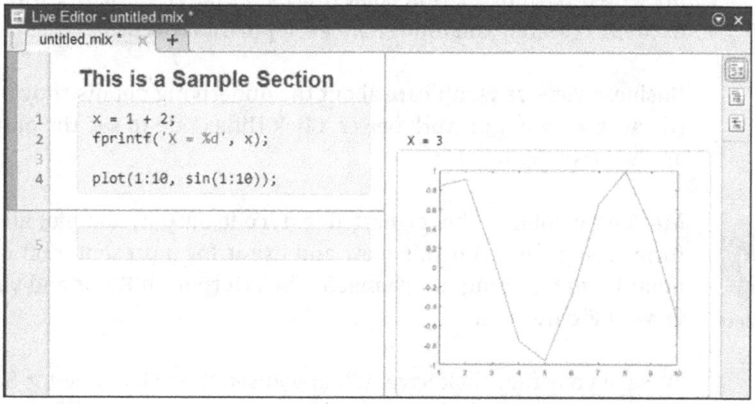

FIGURE 11-2: The Live Editor window displays the results from the formatted text, code, and plot.

The default setting for the Output pane is to show the output to the right. It's the best view to use as you work through the coding process so that you can keep code and output separate. However, you have two other ways of displaying the Output pane by using the three buttons on the right side of the Output pane. Click the Output Inline button, and the display in Figure 11-2 changes to look like Figure 11-3.

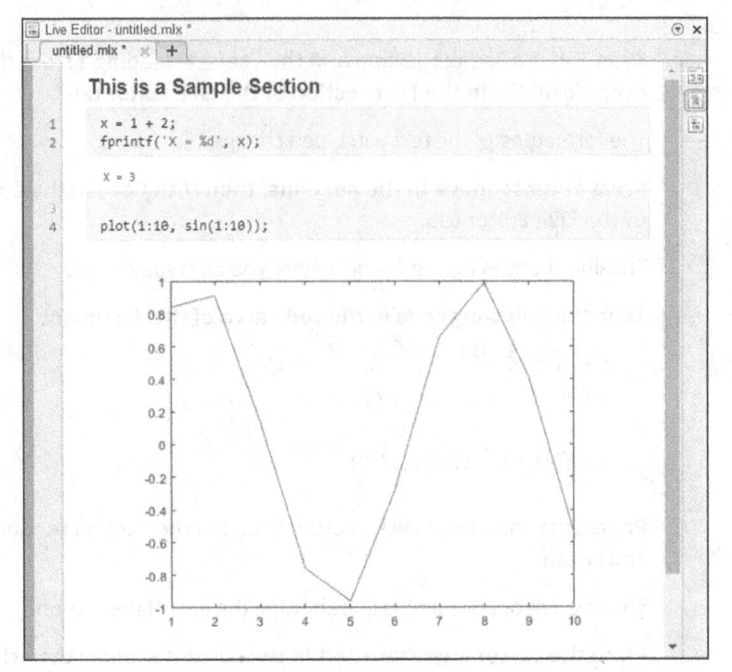

FIGURE 11-3: Using the Output Inline view lets you see code and output together.

This is a good view to use when working with colleagues who need to know the details in a nicely presented manner. In many respects, the output looks like that presented by Jupyter Notebook (`https://jupyter.org/`), a product extensively used for research and other nondevelopmental needs.

Business viewers won't care about the underlying details shown in Figure 11-3, so the next view might work better. Click Hide Code to see the output in a new way, as shown in Figure 11-4.

REMEMBER

MATLAB combines the formatted text, code output, and plot all in one easily read form. You could print this view and use it for a presentation or other need. The point is that nothing has changed. Click Output on Right and you see the original view in Figure 11-2.

To save your file, click Save. When you see the Select File for Save As dialog box, type **SimpleLiveScript** in the File Name field and then click Save.

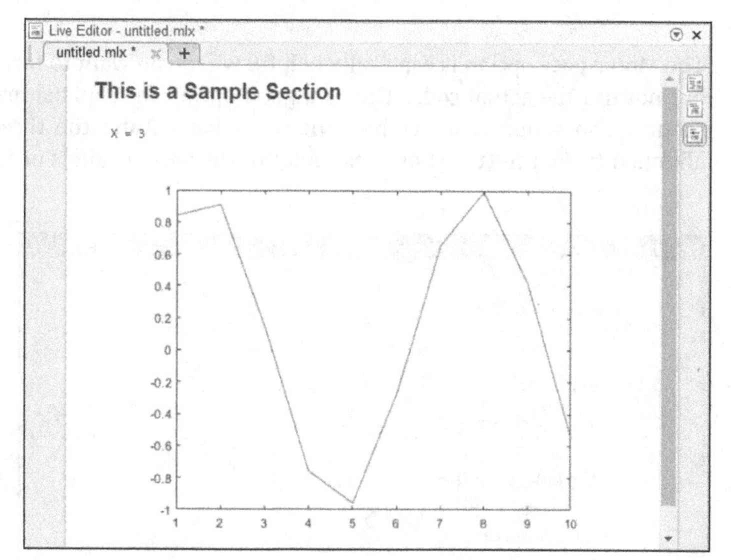

FIGURE 11-4: Using the Hide Code view creates a presentation ready report.

Adding formatted text

The previous section shows how to add a heading to a section. Of course, headings can appear anywhere needed. Live Editor supports five different paragraph styles:

- Normal
- Heading 1
- Heading 2
- Heading 3
- Title

Each of the paragraph styles offers presets for the text you need to include in your document. Below the style drop-down list are buttons for

- Bold
- Italics
- Underline
- Monospace

The Monospace option is especially helpful when you want to format text as code, but not use the actual code. This is a good option for explanatory text. The Text section also supports bullet lists, numbered lists, and the three standard text alignments. Figure 11-5 shows examples of the various kinds of formatting.

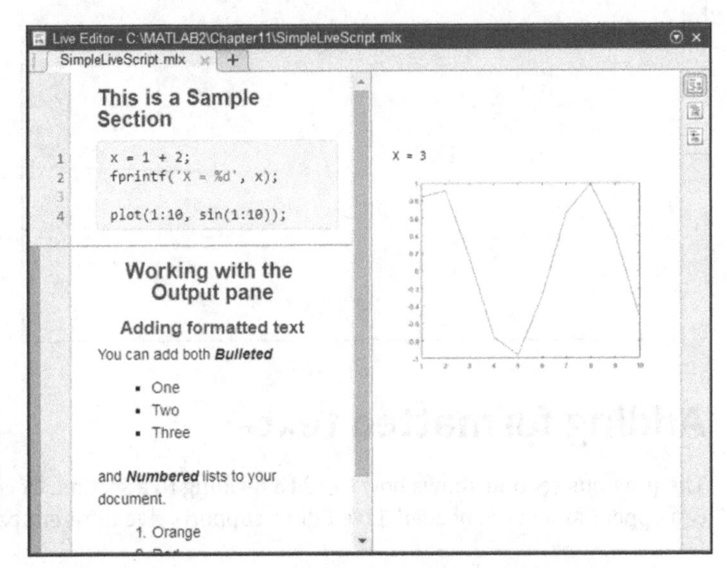

FIGURE 11-5: Examples of various text formatting features in Live Editor.

Using plots within the coded area

As demonstrated in the "Working with the Output pane" section, all you need to do to include a plot in your document is to provide the required code. However, you may find that the code becomes involved and you don't want to have a lot of hard to read code in your report. Here's a quick example:

```
x=linspace(0,2*pi,30);
y=sin(x);
z=cos(x);
plot(x,y,x,z);
```

You type this code into the Command Window to plot it. After plotting the data, you now save the resulting figure to a file named `SinCos.fig`. Simply click the Save Figure button in the Figure window. You can now add a simple piece of code, `openfig('SinCos.fig');`, to your Live Script document to use the figure. This technique works well if the figure changes regularly and you don't want to include the code for recreating it in your Live Script, but do want to display the figure as part of your report. Figure 11-6 shows the output from this example.

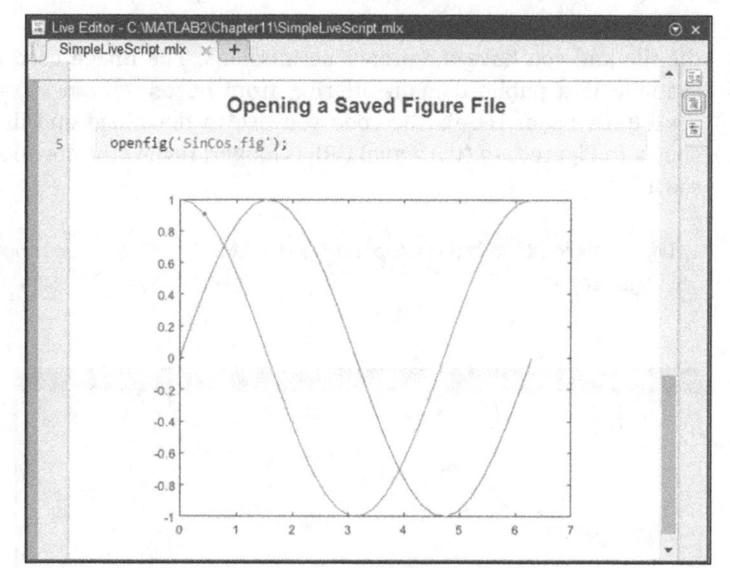

FIGURE 11-6: Use a saved figure when necessary in your document.

Incorporating graphics

You can use graphics in a number of ways within MATLAB, such as when you perform image analysis (see Chapter 20 for an overview of performing analysis tasks). However, graphics can also add eye candy to a presentation to make your point clearer and the presentation more enjoyable. Anything you can do to maintain your audience's interest will make your presentations more successful. Fortunately, you can use any of these graphics formats as part of a Live Script document:

- BMP (Microsoft Windows Bitmap)
- GIF (Graphics Interchange Files)
- HDF (Hierarchical Data Format)
- JPEG (Joint Photographic Experts Group)
- PCX (Paintbrush)
- PNG (Portable Network Graphics)
- TIFF (Tagged Image File Format)
- XWD (X Window Dump)

You usually used a two-step process to work with graphics in MATLAB: Read the graphic into a variable, and then display the content of the variable onscreen. The location of the graphic usually doesn't matter as long as you have read access to the file and you have the proper permissions. The image used in the following example is a public domain offering from https://commons.wikimedia.org/wiki/Main_Page. Here's the code you use to download and display the image shown in Figure 11-7 (the actual URL is longer; review the downloadable source to see it):

```
Img = imread('https://upload.wikimedia...Dog_face.png');
image(Img)
```

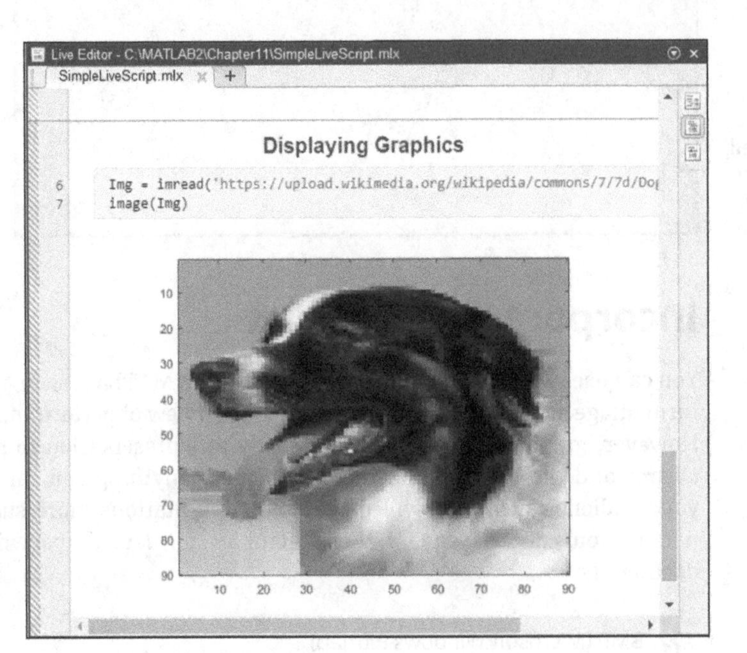

FIGURE 11-7: Add pizzazz to your document using graphics from any location.

TECHNICAL STUFF

Your graphics need not be static in MATLAB. For example, you can create an Acquire Webcam Image task, use it to connect to a webcam, and then perform as-needed tasks, as described at https://www.mathworks.com/help/supportpkg/usbwebcams/ug/acquire-images-from-webcams.html. To use this feature, you must install the MATLAB Support Package for USB Webcams.

Incorporating controls

Controls can make your documents interactive. You can do everything from very simple things, like displaying a Help button, to providing controlled inputs for your algorithm. You can even create input form documents or design a test. Using controls gives your application more of a full-fledged application feel without requiring you to get a computer science degree. The experience of using controls isn't free of some development tricks, but using the MATLAB controls is easier than working with controls in a full-fledged development environment. MATLAB supports these controls:

- **Numeric Slider:** A numeric slider allows control over a value within a specific range and prevents users from entering non-numeric values. For example, you might want to allow the viewer to choose the number of graphs displayed at any time as part of the plot. Moving the Numeric Slider control's thumb (the square box that controls its value) to the desired number of graphs will change the appearance of the plot. The advantage of using a numeric slider is that you control the correct input values and don't need to consider values outside the desired range, reducing application errors.
- **Drop-Down List:** A drop-down list box contains a set number of input values. The user selects the desired input value from the list, meaning that incorrect input values due to typos and other forms of user error are a thing of the past. Relying on drop-down list boxes instead of requiring the user to type values also makes data entry significantly faster.
- **Check Box:** A check box is a logical input, and you obtain one of two values from it. The feature is either on or off; the answer is either yes or no; and the user either wants to opt in or opt out.
- **Edit Field:** An Edit Field control provides a convenient means of allowing freeform input from the user. Avoid using the Edit Field control when you can, because it creates a potential security breach unless you perform a lot of background checks. The Edit Field control is the most flexible of all the controls because you really can perform free-form input.
- **Button:** A Button control enables you to define the precise moment when an action should occur. For example, after providing all the input required to create a plot, the user clicks the Plot button in your application and sees the resulting plot onscreen. To use the Button control effectively, you must set the Run field for any controls in the document to `Nothing`.

Controls require a certain amount of configuration before you use them. The following steps help you create a simple control example in which the user selects a value from a Numeric Slider control, and clicking a Button displays that value.

1. **Choose Numeric Slider from the Control drop-down list in the Code section of the Live Editor tab.**

   You see the slider, along with its fields as shown in Figure 11-8. Modifying the settings immediately will ensure that you don't forget to do so later. However, you can change the configuration later by right-clicking the control and choosing Configure Control from the context menu.

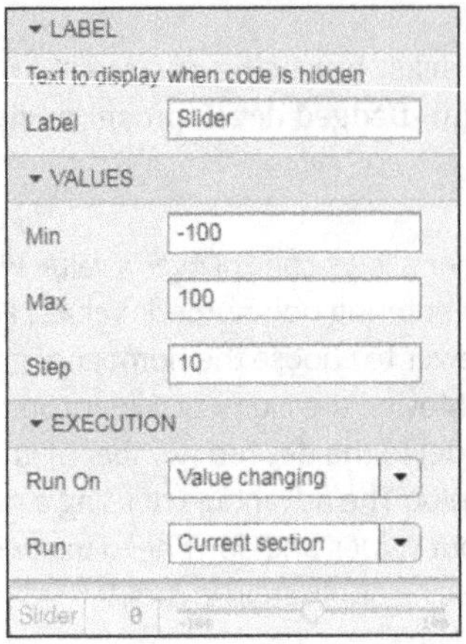

FIGURE 11-8: Controls provide access to settings through a configuration dialog box.

2. **Type** A Value **in the Label field.**

   The next three settings control the minimum Numeric Slider value, the maximum value, and the increments between each change as the user moves the thumb.

3. **Type** 0 **in the Min field.**
4. **Type** 5 **in the Max field.**
5. **Type** 1 **in the Step field.**
6. **Choose Nothing in the Run field.**

   This last setting is especially important because it allows you to control interactions with the Numeric Slider control using the Button control.

7. **Go to the next line of the document and choose Button from the Control drop-down list in the Code section of the Live Editor tab.**

   You see a control configuration dialog box.

8. **Type** See Value **in the Label field.**

   The Run field determines how much of the document the Button controls. The default setting of Current Section is sufficient for this example, but you should review the other options as well.

9. **Place the cursor anywhere within the section; then click Run Section in the Section section of the Live Editor tab.**

   The application displays the current value, which is 5 (see Figure 11-8), as shown in Figure 11-9.

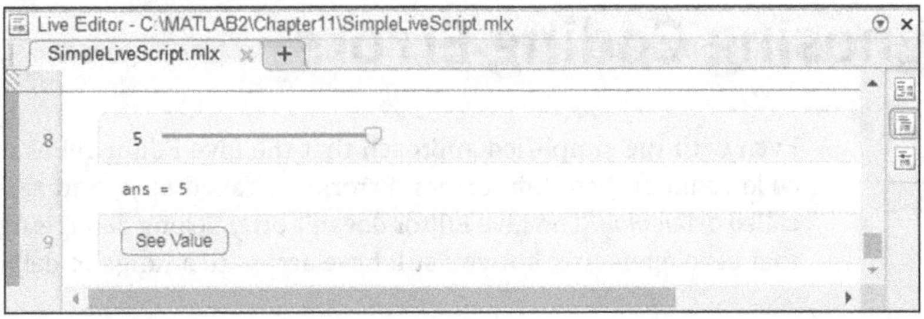

FIGURE 11-9: The controls all have default values when you run the section.

10. **Move the slider to a value of 3.**

    Notice that the output doesn't change.

11. **Click See Value.**

    The output changes to match the new value.

Running Live Script Sections

Knowing how to run a Live Script can save a lot of time. This chapter runs every example using the Run Section option. If you're building an application or performing updates, using Run Section ensures that you execute only the current section. When the rest of your code works, there is no reason to keep running it repeatedly. It's not that your computer will wear out or get annoyed — it's more a matter of preserving your time. Adding extra code to each execution will slow you down.

Of course, if you make changes to those previous sections, you need to run the code starting from the changed section to your current location. When the current location is the end of the file, you simply place the cursor where the change occurred and then choose Run to End. This button saves you from having to click Run Section repeatedly.

Another option for running Live Script sections is to rely on Run and Advance. The Run Section option keeps the cursor in the currently selected section. However, Run and Advance runs the code in the current section and then moves to the next section. Using this approach allows you to run the entire document one section at a time.

Clicking Run will execute the entire document. The cursor doesn't change positions. You use this option when you need to update the entire document and are sure that the update will succeed. This is the fastest way to run your Live Script code.

Diagnosing Coding Errors

Even with the simplified approach that the Live Editor relies on for coding, your code could still contain errors. Errors are called *bugs*, and removing the bugs is called *debugging*. The Live Editor doesn't offer strong debugging features like ones that developers use, but you still have access to a range of debugging options.

The editor is quite determined to save you from yourself. For example, when you type something wrong in the editor and MATLAB recognizes it as an error, you see a red line under the errant text with a message explaining what it thinks has gone wrong, as shown in Figure 11-10. In fact, examining your document for red underlines could save you a lot of work.

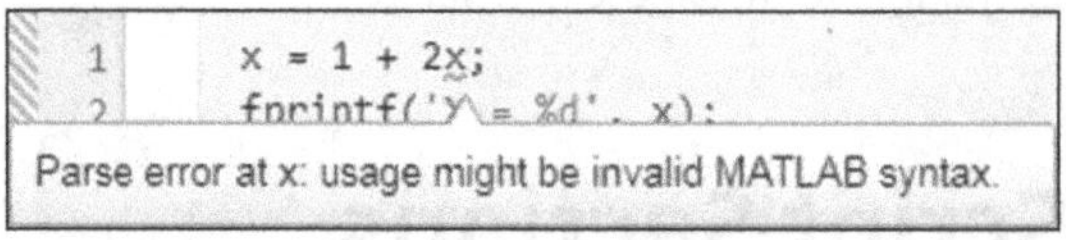

```
1    x = 1 + 2x;
2    fprintf('Y = %d', x);
Parse error at x: usage might be invalid MATLAB syntax.
```

FIGURE 11-10: Look for red lines under your code because they're errors.

Removing the semicolons from the ends of your lines of code can also help you locate errors by showing the output from suspect code. You don't have to remove all the semicolons, which can take a lot of time; just remove them for calls that you suspect. Run the section again to determine whether you made a wrong assumption somewhere.

The debug type features that the Live Editor provides allow you detailed running of code lines. The Step drop-down list in the Run section of the Live Editor tab contains options to

- **Step:** Runs a single line of code so that you can see how the function call performs.
- **Step In:** When calling your own custom functions, you step into those functions from the current thread of execution by clicking Step In. This feature lets you take a detailed look at the code to see whether it has any underlying problems.
- **Step Out:** After examining a particular function, clicking Step Out takes you one level up so that you're back to where the function was called in your script. Otherwise, you'd need to single-step through the entire function.
- **Run to Cursor:** If you know where a problem occurs, you can place the cursor there and run the code from the current location to the problem area. Sometimes, focusing on a specific area of the code helps you locate the source of errors.

Clicking Step places the Live Editor into a debugging mode, and the editor's controls change, as shown in Figure 11-11. The current line of code is highlighted in light green, so you know where you're executing the code.

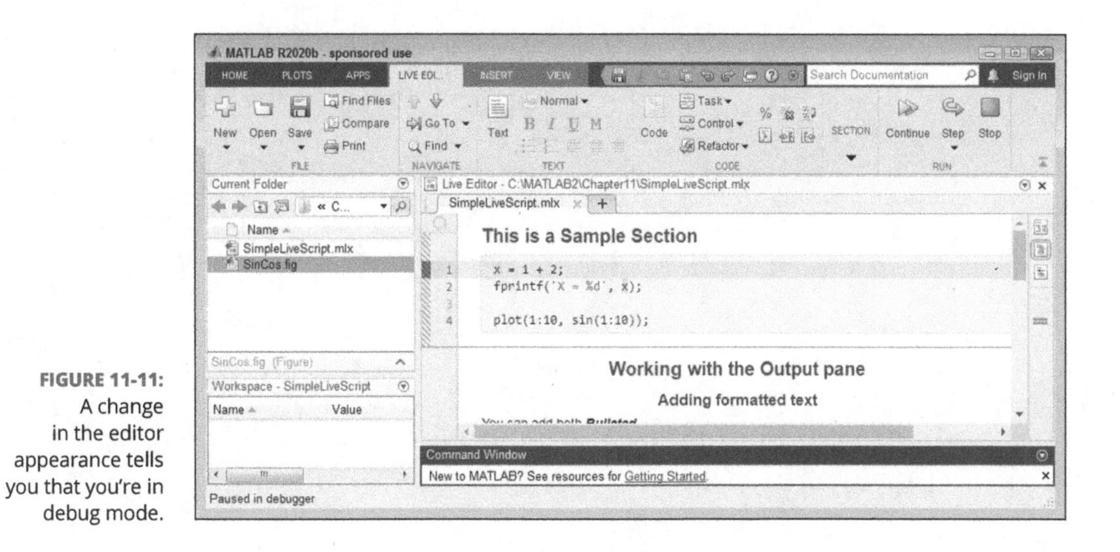

FIGURE 11-11: A change in the editor appearance tells you that you're in debug mode.

The Step drop-down list enables the options that are useful to the current situation. In this case, you'd click Step to see the next line because you aren't calling a function. If you want to continue the execution of the current section, you click Continue. Likewise, if you think you've found the error, you click Stop, fix it, and try again.

TIP

After you execute the code to create variables, you can see each variable's value by hovering the mouse over it. This technique makes it very fast to determine whether a variable contains what you think it should contain. Other debugging techniques are available, but they're common to all the editors, so Chapter 18 covers those.

IN THIS CHAPTER

» **Understanding the Live Function difference**

» **Using functions to start Live Functions**

» **Combining Live Functions with Live Scripts**

Chapter 12

Working with Live Functions

Chapter 11 discusses Live Scripts, which are the counterpart to scripts in MATLAB. Live Functions represent the counterpart to functions in MATLAB as well, and they embody many of the same characteristics as Live Scripts do. For example, you use the Live Editor when working with Live Functions (see the "Working with the Live Editor" section of Chapter 11 for details). However, there are significant differences between Live Scripts and Live Functions, even in the manner of creating them in the Live Editor. This chapter helps you understand where Live Functions fit in with functions and Live Scripts as part of your MATLAB development strategy.

One of the things that sets Live Functions apart from their plain function alternative is the flexibility they provide in creating reusable code. This chapter helps you understand the flexibility that Live Functions provide and demonstrates how you can move your current functions to Live Functions. More important, this chapter helps you understand whether to use functions or Live Functions as your development strategy. Sometimes using the simpler function can be better if you don't actually need the special features that Live Functions can provide.

Live Functions can stand alone, just as functions do. However, you gain the most power from combining Live Functions with Live Scripts as needed. In fact, when performing complex analysis, the combination of Live Scripts and Live Functions can be unbeatable in creating a solution that's ultimately simpler than any other means at your disposal.

REMEMBER

You don't have to type the source code for this chapter manually. In fact, using the downloadable source is a lot easier. You can find the source for this chapter in the \MATLAB2\Chapter12 folder of the downloadable source. When using the downloadable source, you may not be able to work through some of the hands-on examples because the example files will already exist on your system. In this case, you can create an alternative folder, Chapter12a, to hold the results of the hands-on exercises. See the Introduction for details on how to obtain the downloadable source.

Comparing a Live Function to a Regular Function

A regular MATLAB function exists in a text file that uses an .m file extension. Chapter 9 tells you about various kinds of functions and demonstrates how to create them. A Live Function exists within an .mlx file, just as a Live Script does, with all the added functionality described in the "Comparing a Live Script to a Regular Script" section of Chapter 11. You can also add formatted text, images, hyperlinks, and equations to Live Functions — something that isn't available with a regular function.

The most important difference between Live Functions and regular functions is that it's easier to partition a Live Function using sections. The benefits of working with sections become especially noticeable when using the following (as explained in Chapter 9):

- Subfunctions (also called local functions)
- Nested functions
- Inline functions
- Anonymous functions

Sections enable you to separate these elements so that they're easier to see, test, and debug. Suddenly, instead of one huge bulky piece of text, you work with small, individual elements that are easier to understand. So, the actual function may not differ much between Live Functions and regular functions, but your perception of them will, which can reduce the work required to interact with code you didn't write, or wrote a long time ago.

REMEMBER

You call both functions and Live Functions from outside the function file. The easiest form of access is from the Command Window. However, you can also call them from scripts and Live Scripts. In contrast to functions, you can *merge* (create a single entity) from a Live Script and a Live Function, as described in the "Merging Live Functions and Live Scripts," near the end of this chapter.

Understanding Live Function Flexibility Differences

The easiest way to understand the flexibility differences between a Live Function and a regular function is to create and interact with a Live Function. The following sections assume that you've already worked with the Live Editor, as explained in the "Working with the Live Editor" section of Chapter 11. These sections focus on differences between these Live Functions and regular functions, but they also help you understand differences between Live Functions and Live Scripts.

Creating a Live Function

The following steps help get you started creating your first Live Function. You can also find this Live Function in the `DoAdd.mlx` file supplied with the downloadable source code.

1. **Choose New ⇨ Live Function in the File section of the Home tab.**

   You see the new Live Function shown in Figure 12-1. Comparing this figure with Figure 11-1 shows that the beginning of a Live Script looks different from that of a Live Function. A Live Function starts with a text section for documentation purposes and a code section for the actual function code.

TIP

   The Live Editor may display a warning message for the newly created function. This is because you haven't given the function a name and haven't saved the file, so you can safely ignore the warning for now.

FIGURE 12-1: A Live Function begins with two sections: one text and one code.

2. **Change the first text line to read:** Add Two Numbers.

3. **Change the second text line to read:** Provide the x and y input values to receive the summed output value.

4. **Change the untitled entry in the first code line to read** DoAdd.

5. **Click Save in the File section of the Live Editor tab.**

   You see the Select File for Save As dialog box, shown in Figure 12-2. Notice that the dialog box automatically suggests `DoAdd.mlx` as the filename.

6. **Click Save.**

   The Live Function is added to the current directory.

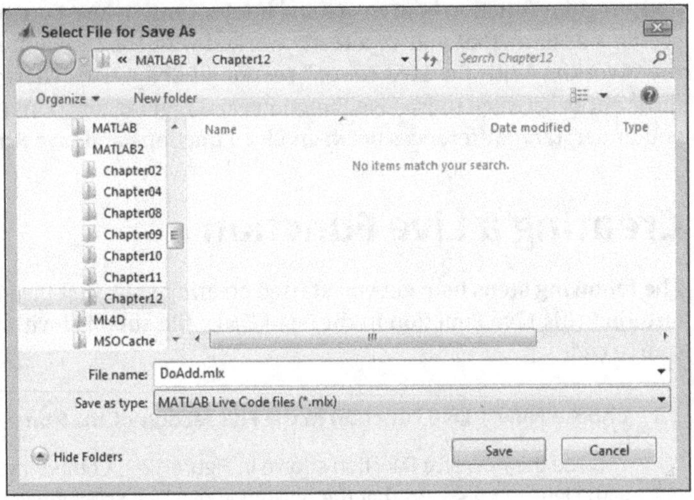

FIGURE 12-2: Save the file using the name of the function you created.

Running a Live Function

One of the first things you notice when working with a Live Function is that you can't click any of the run options in the Section or Run sections of the Live Editor tab. Instead, you run a Live Function just as you would a regular function. To try the function shown in the previous section, type **DoAdd(1, 2)** and press Enter in the Command Window. You see:

```
ans =
     3
```

However, there is a difference when obtaining help for a Live Function. Type **help('DoAdd')** and press Enter. The help text looks like this:

```
DoAdd   Add Two Numbers
   z = DoAdd(x, y)

    Provide the x and y input values to receive the summed
          output value.

    Documentation for DoAdd
```

When you click the `Documentation for DoAdd` link, you see a Help dialog like the one shown in Figure 12-3. `DoAdd()` is a simple function, so there isn't much documentation for it. However, as you add formatted text, images, hyperlinks, and equations to the documentation, the output of the Help dialog box also changes. Everything you add to the function also appears as part of help, making your function significantly easier to use.

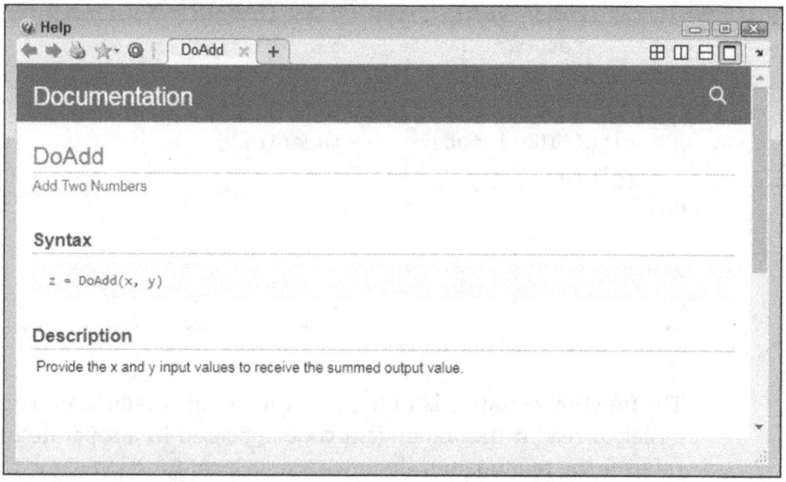

FIGURE 12-3: Help works differently for Live Functions; you get more, with less effort.

Refactoring a Live Function

As you work on your functions, you might be amazed at just how quickly they grow, often becoming unmanageable. The problem is that you've already invested a lot of time and effort in creating the function, so rewriting it is unappealing. *Refactoring,* the process of moving pieces of code around without changing its functionality, is often the right answer for this situation. For example, consider the code in Listing 12-1.

LISTING 12-1: An Integer Version of DoAdd()

```
function z = DoAddInt(x, y)
try
    if isreal(x) && rem(x, 1) == 0
        x = int64(x);
    else
        disp('Value x isn''t an integer.')
        return;
    end
catch
    disp('The input is non-numeric.')
    return;
end

try
    if isreal(y) && rem(y, 1) == 0
        y = int64(y);
    else
        disp('Value y isn''t an integer.')
        return;
    end
catch
    disp('The input is non-numeric.')
    return;
end

z = x + y;
end
```

The function certainly isn't huge, but it contains redundant code, and it would be easier to read if that redundant code appeared in a separate function. You could refactor the redundant code into a single local function using these steps.

1. **Remove the first two `return;` statements so that the first block of redundant code looks like this:**

```
try
    if isreal(x) && rem(x, 1) == 0
        x = int64(x);
    else
        disp('Value x isn''t an integer.')
    end
```

```
catch
    disp('The input is non-numeric.')
end
```

You can't refactor code that contains `return;` statements. MATLAB displays an error message if you try.

2. **Highlight the code block shown in the previous step and then choose Refactor ⇨ Convert to Local Function in the Code section of the Live Editor tab.**

   MATLAB refactors the code so that it appears like the code shown in Figure 12-4.

   The result in Figure 12-4 still isn't quite right, but it's close.

```
Live Editor - C:\MATLAB2\Chapter12\DoAddInt.mlx
DoAddInt.mlx

1   function z = DoAddInt(x, y)
2   x = untitled(x);
3
4   try
5       if isreal(y) && rem(y, 1) == 0
6           y = int64(y);
7       else
8           disp('Value y isn''t an integer.')
9           return;
10      end
11  catch
12      disp('The input is non-numeric.')
13      return;
14  end
15
16  z = x + y;
17  end
18

19  function x = untitled(x)
20  try
21      if isreal(x) && rem(x, 1) == 0
22          x = int64(x);
23      else
24          disp('Value x isn''t an integer.')
25      end
26  catch
27      disp('The input is non-numeric.')
28  end
29  end
```

FIGURE 12-4: MATLAB refactors the code by creating a local function.

3. **Type** CheckInt **to provide a valid function name.**

   Notice that the name of the new function changes in both places.

REMEMBER

You need some method of determining whether `x` is actually an integer, so the next step adds a special `x` output to the two error outputs.

4. **Add return value statements to the local function, as shown in bold:**

```
function x = CheckInt(x)
try
    if isreal(x) && rem(x, 1) == 0
        x = int64(x);
    else
        disp('Value isn''t an integer.')
        x = nan;
    end
catch
    disp('The input is non-numeric.')
    x = nan;
end
end
```

You can now check for nan (not a number) as part of the main function. In addition, you can use the same approach for the value y.

5. **Modify the DoAddInt() function so that it looks like this:**

```
function z = DoAddInt(x, y)
x = CheckInt(x);
if isnan(x)
    return;
end

y = CheckInt(y);
if isnan(y)
    return;
end

z = x + y;
end
```

Compare this version of DoAddInt() with the version found in Listing 12-1 and you see that this version is simpler.

TIP

Refactoring to a local function is a good idea when the refactored code is unique to the Live Function or Live Script that uses it. However, you might decide that you want to use a particular piece of code in a number of your Live Scripts and Live Functions. If this is the case, you highlight the section of code you want to refactor and reuse; then you choose Refactor⇨Convert to Function in the Code section of the Live Editor tab. MATLAB will create a new file that contains the code in question. You make the same sorts of modifications as shown earlier in this section, except that you make them in the separate file.

Using the specialized coding buttons

Specialized features can make your coding experience easier and possibly save time as well. MATLAB provides a set of six specialized coding buttons in the Code section of the Live Editor tab, as shown in Figure 12-5.

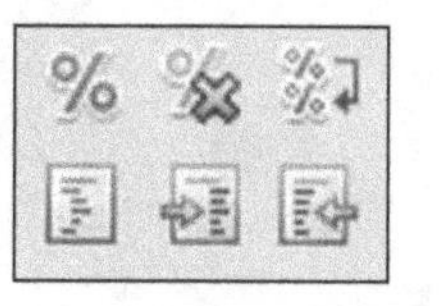

FIGURE 12-5: Use these specialized buttons to save time and effort.

Table 12-1 provides you with a quick summary of how you use each button to perform tasks in the Live Editor.

TABLE 12-1 Live Editor Coding Buttons

Button Name	Speed Key	Purpose
Comment	Ctrl+R	Adds a % in front of each selected line in a group, commenting the code out for testing or debugging.
Uncomment	Ctrl+T	Removes the first % in front of each selected line in a group, making the code active again. If the group contains a comment line, only the first % is removed, which means that the comment stays a comment.
Wrap Comments	Ctrl+J	Combines lines with a similar indent to make a comment block smaller. Works only with lines that are already commented out.
Smart Indent	Ctrl+I	Adds indentation as needed to make the code more readable. MATLAB uses natural code boundaries, such as `if...else` statements, to make the indentation.
Increase Indent	Ctrl+] or Tab	Adds a single level of indentation to the selected lines of code.
Decrease Indent	Ctrl+[or Shift+Tab	Removes a single level of indentation from the selected lines of code. If the selected code is already at the left margin, the button has no effect on that line.

The Smart Indent button requires a little more explanation. When you click this button, MATLAB uses the default indentation scheme, which appears in Listing 12-1. However, some people prefer to have function code indented within the function block. Consequently, you must select the function code (without the function declaration and final `end`) and use Increase Indent to obtain the desired result. Here's the modified version of `DoAddInt()` with function level indentation added:

```
function z = DoAddInt(x, y)
    x = CheckInt(x);
    if isnan(x)
        return;
    end

    y = CheckInt(y);
    if isnan(y)
        return;
    end

    z = x + y;
end
```

TIP

Many people find this form of indentation easier to read because the function boundaries stick out more. However, the use of sections in Live Editor tend to make the function boundaries more obvious as well. So, whether you indent function code or not is a matter of personal taste.

Going to a specific function

If your function file contains a number of functions, locating a particular function can be difficult. To make things easier, you can use the Go To drop-down list in the Navigate section of the Live Editor tab. When you click the down arrow, you see a list of function names in the currently selected file similar to those shown in Figure 12-6. Simply click the function you want to work with.

FIGURE 12-6: Use the Go To drop-down list to access functions quickly.

Converting a Function to a Live Function

You don't have to start from scratch to use a Live Function. MATLAB provides the means to use existing code, such as `SayHello3.m` from Chapter 9. Use these steps to copy `SayHello3.m` from the `Chapter09` folder to the `Chapter12` folder used for this chapter.

1. **Right-click `SayHello3.m` in the `Chapter09` in the Current Folder window and choose Copy from the context menu.**

   MATLAB places a copy of the file on the Clipboard.

2. **Right-click in the `Chapter12` folder and choose Paste from the context menu.**

   You see a copy of `SayHello3.m` placed in the `Chapter12` folder.

At this point, you can open `SayHello3.m` as a Live Function by right-clicking the file and choosing Open as Live Function from the context menu. MATLAB performs the required conversion for you automatically, as shown in Figure 12-7.

Live Editor - SayHello3.mlx *

SayHello3.mlx *

SayHello()

This function says Hello to everyone! It requires a string, Name, as input. It passes back the result as HelloString.

```
1  function [ HelloString ] = SayHello3( Name )
2  HelloString = ['Hello There ', Name, '!'];
3  disp(HelloString);
4
5  end
```

FIGURE 12-7: MATLAB automatically converts the format of your function file.

REMEMBER

Notice that the file no longer uses the document format used for functions:

```
%SayHello()
%   This function says Hello to everyone!
%   It requires a string, Name, as input.
%   It passes back the result as HelloString.
```

The documentation automatically appears in the form used for a Live Function. When you save the function, it receives the name `SayHello3.mlx`, which means you now have two functions with the same name. Change the word `Hello` to `Goodbye` in the `HelloString = ['Hello There ', Name, '!'];` line of the function and save the Live Function to disk. Now type **SayHello3('John')** and press Enter. You see this output showing that MATLAB will execute the Live Function before it executes the standard function.

```
Goodbye There John!
ans =
    'Goodbye There John!'
```

WARNING

When there are two files, one Live Function and one normal function, with the same function name, MATLAB always executes the Live Function. This means that if you have a naming conflict, you may obtain an unexpected result from your code. Always ensure that the standard functions you plan to use have a different name from the Live Functions in the same folder.

Sharing Live Functions and Live Scripts

Eventually, you'll want to show off your work to impress colleagues or as a means of teaching students. The "Publishing information" section of Chapter 8 describes how to publish your script, which isn't quite the same as sharing a Live Function or Live Script. You have these options when sharing a Live Function or Live Script:

- Interactive document
- Full screen presentation
- Plain text
- Static document

The following sections provide an overview of these methods of sharing your code with others. Each technique has benefits, so you should use the approach that best meets your specific sharing need.

Using an interactive document

You can simply send the `.mlx` file to others. It's recommended that you hide the code for any Live Script you send out this way so that recipients focus on the results rather than the underlying code. This is the option to use for peer review. Anyone receiving the file will have full access to your code because MATLAB doesn't provide any form of document protection such as that found in applications like Microsoft Word. So, the downside of this approach is that you won't be able to hide any great ideas from view — everything is open to everyone.

Employing a full screen presentation

When providing a presentation in public, you can place MATLAB in full-screen mode by choosing Full Screen in the Display section of the View tab. What you see is a help-type display that works exceptionally well for teaching needs.

A Live Function display will show the function documentation and code. When working with a Live Script, you can show the output to the right, in line with the code, or hide the code from view. To change the view, move the cursor to the top of the screen to display a Toolstrip containing these tabs:

- Live Editor
- Insert
- View

To leave full-screen mode, press Ctrl+F11 or Full Screen in the Display section of the View tab. Note that MATLAB places the full-screen display on the primary screen of a two-or-more-screen computer, so you can have MATLAB on a secondary screen and display the primary screen for your audience.

Working with plain text

Even though it's a big step backward, you can save your Live Function or Live Script as a plain function or plain script for those who have older versions of MATLAB. To perform this task, choose Save⇨Save As in the File section of the Live Editor tab. You see a Select File for Save As dialog box like the one shown previously in Figure 12-2. Choose MATLAB Code Files (UTF-8)(*.m) in the Save as Type field; then click Save.

The resulting file converts any special Live Function or Live Script features to comments. Figure 12-8 shows an example of the `SimpleLiveScript.mlx` file from Chapter 8. It's definitely less informative than using a Live Script (see the figures in Chapter 11 as examples), but the result does provide the essential code and the details as comments.

Creating a static document

Sharing your Live Function or Live Script as a static document makes it easier for people who lack a copy of MATLAB to interact with you. In addition, this method of sharing is better when you want to focus on Live Script output rather than code, with a greatly reduced chance of anyone's seeing the code at all. This is also a great method for producing handouts for a class or other presentation venue.

Editor - C:\MATLAB2\Chapter12\SimpleLiveScript.m

SimpleLiveScript.mlx | SimpleLiveScript.m

```
1      %% This is a Sample Section
2
3  -   x = 1 + 2;
4  -   fprintf('X = %d', x);
5
6  -   plot(1:10, sin(1:10));
7      %% Working with the Output pane
8      % Adding formatted text
9      % You can add both _*Bulleted*_
10     %%
11     % * One
12     % * Two
13     % * Three
14     %%
15     % and _*Numbered*_ lists to your document.
16     %%
17     % # Orange
```

FIGURE 12-8: Live Function and Live Script features are converted to comments when using plain text.

To use this technique, make sure you have the document you want to share selected in the editor. Choose one of the Export options on the Save drop-down list found in the File Section of the Live Editor tab. You have access to these output formats:

- PDF
- Word
- HTML
- LaTeX

This option also supports a batch mode. Choose Save⇨Export Folder, and you see the Export Files in Folder dialog box, shown in Figure 12-9. You can export the files to any of the four supported file formats. In addition, you can choose the source and target folders, require MATLAB to ask before overwriting any existing output files, and request that MATLAB also copy supporting files, such as graphics. Just configure the dialog box options and click Export to make the process happen.

Export Files in Folder

Export to:
HTML
Folder to export:
C:\MATLAB2\Chapter12\
Location for exported files :
C:\MATLAB2\Chapter12\Export\
Ask before overwriting
Copy supporting files
Export
Cancel

FIGURE 12-9: Use batch mode when converting a large number of code files.

Performing Comparisons and Merges

Managing code files can become cumbersome when you start to create multiple versions of the same file. In addition, you may want to move useful features from one file to another without having to cut and paste the changes. The act of checking two files for differences is *comparison*, while the act of moving differences from one file to another is *merging*. The following sections explain both activities.

Comparing Live Functions and Live Scripts

Sometimes you need to know how one file differs from another file, which isn't hard if the two files are short. However, performing a manual comparison of two larger files is problematic because even the most astute viewer will likely make errors. To avoid this problem, you can ask MATLAB to perform the comparison for you. For example, consider the differences between `DoAdd.mlx` and `DoAddInt.mlx`. Use these steps to perform a comparison:

1. **Right-click the source file or folder and choose Compare Against ⇨ Choose from the context menu.**

   The example uses the `DoAdd.mlx` file. You see the Select Files or Folders for comparison dialog box, shown in Figure 12-10. Note that you can perform either individual file comparisons or complete folder comparisons.

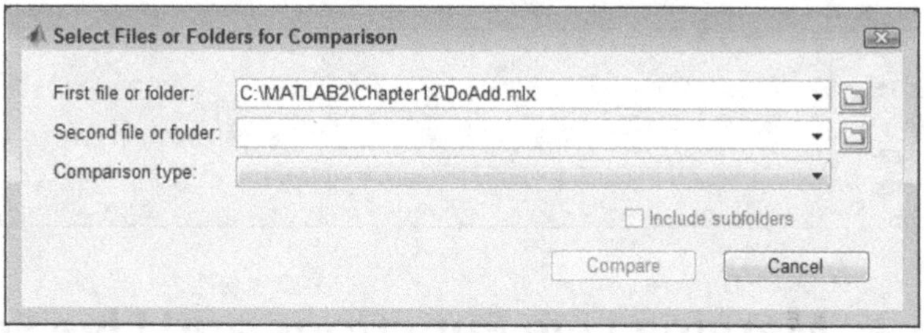

FIGURE 12-10: Choose the source and target file or folder, plus a comparison type.

2. **Add a file or folder to use for comparison to the Second File or Folder field.**

   You can use the browse button to locate the file or folder as needed. The example uses the `DoAddInt.mlx` file.

3. **Choose a Comparison Type drop-down list box entry.**

   The entries you see depend on the kind of comparison you perform. When working with Live Function or Live Script files, you can choose from Live Code

Comparison (the kind used for this example) and Binary Comparison (which provides a very low-level view of file differences).

4. **(Optional) Select Include Subfolders.**

   This option is available only when comparing folders that contain subfolders.

5. **Click Compare.**

   You see the output of the comparison. Figure 12-11 shows the comparison for this example. Even though you can't see it in the print book, MATLAB color-codes the comparison so that you can see modifications, additions, and deletions.

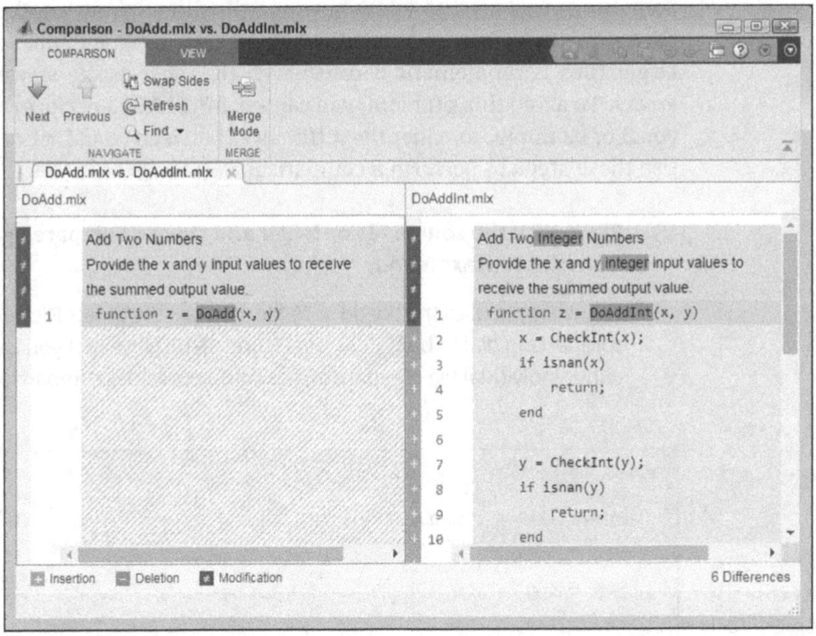

FIGURE 12-11: The comparison shows modifications, additions, and deletions.

Merging Live Functions and Live Scripts

The previous section of the chapter tells you how to see differences between two files. You can also use this approach for seeing differences between file folders. Whether the changes appear in files or folders, you can use a special Merge mode in MATLAB to move differences from one file to another. The process begins in the Comparison dialog box (refer to Figure 12-11). In that dialog box, click the Merge Mode button in the Merge section of the Comparison tab to show the Merge tab shown in Figure 12-12.

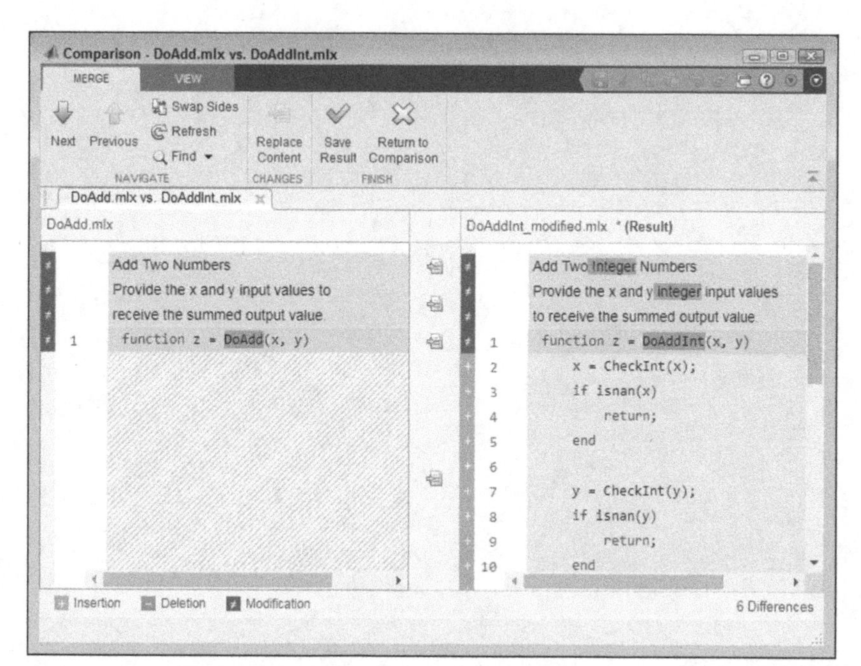

FIGURE 12-12: Use Merge Mode to move differences from one file to another.

REMEMBER

Notice that there are now document icons with right-pointing arrows next to each change. To move a change from one document to another, simply click the button. The movement always occurs between the source document on the left to the target document on the right. Consequently, opening the files in the correct order is essential. Clicking Swap Sides only changes the view — movement still occurs from the original source document to the target document.

After you complete the required moves, click Save Result in the Finish section of the Merge tab. MATLAB won't save the changes to the original target document. Instead, it recommends an alternative filename in the Select File for Save As dialog box. Click Return to Comparison to perform more document comparison tasks.

IN THIS CHAPTER

» Defining Object Oriented Programming (OOP) principles

» Considering MATLAB OOP differences

» Creating a class

» Using OOP in MATLAB

Chapter 13

Designing and Using Classes

Objects are things that everyone uses every day. You grab a doorknob and turn it to open a door. Both the doorknob and the door are objects. Each has its own characteristics, and you wouldn't confuse a doorknob for a door. The doorknob turns; the door opens. Each object has its own behaviors. In short, the use of objects is intuitive.

MATLAB supports the use of objects in the form of Object-Oriented Programming (OOP). OOP may sound complex, but like the doorknob and the door, using objects in MATLAB is actually quite intuitive, as described in the first part of the chapter.

Many programming languages support OOP, but they each have nuances. If you have used OOP with another language, you will find that working with OOP in MATLAB makes use of that knowledge. The second part of the chapter looks at differences between MATLAB OOP support and that provided in some common languages. You also discover how to use objects in MATLAB.

Creating your own MATLAB class, which is the code used to build an object, is the topic of the third part of the chapter. MATLAB tries to keep things simple. Consequently, the classes you create in MATLAB may not be as powerful as those you create in other languages, but the MATLAB classes are easy to understand, and they do everything you need to do with objects in MATLAB.

The fourth part of the chapter helps you use the class you designed to create objects and employ them to do something productive. This part of the chapter also helps you understand why certain class designs work better than others do. Creating truly useful classes means that you'll ultimately do less work in MATLAB to produce some truly amazing results.

REMEMBER

You don't have to type the source code for this chapter manually. In fact, using the downloadable source is a lot easier. You can find the source for this chapter in the `\MATLAB2\Chapter13` folder of the downloadable source. When using the downloadable source, you may not be able to work through some of the hands on examples because the example files will already exist on your system. In this case, you can create an alternative folder, `Chapter13a`, to hold the results of the hands on exercises. See the Introduction for details on how to obtain the downloadable source.

A Brief Overview of Object-Oriented Programming (OOP)

OOP is a formal method of defining how to interact with data in a safe manner using *properties,* which define the characteristics of the data, and *methods,* which define how to interact with the data. Take, for example, a doorknob made of brass. The brass is a property of the doorknob because it affects what you know about it; it's the doorknob's data. When turning a doorknob, you don't change any of the doorknob's characteristics; rather, you interact with the doorknob using the data defined for it. For example, some doorknobs may open only when turned clockwise, so the method of interacting with the doorknob is to turn it clockwise. Seeing objects in MATLAB helps you interact with abstract data in a more realistic manner. That's what the following sections are about — helping you to see data in a new way.

Defining an object

An *object* in MATLAB is a real-world implementation of an underlying mathematical abstraction. This definition differs somewhat from the definition of an object in general-purpose languages. To put it into perspective, consider that you might create a complex algorithm in MATLAB that you want others to manipulate. Perhaps this algorithm calculates the amount of actual sunlight hitting the plants in a greenhouse based on the sunrise and sunset times, as well as the amount of cloud cover. The algorithm is abstract but produces a real-world result that someone can easily understand and use. The properties in this case would include:

- Sunrise time
- Sunset time
- Amount of cloud cover over time

These properties affect just the amount of sunshine, however. They don't account for the needs of the plants. So, you might also need these properties:

- Sensitivity of plants to sunshine
- Amount of daily sunshine required by plants

Now you need some way of interacting with the data. The methods would include:

- Current amount of sunlight hitting the plants
- Adjustments needed for the greenhouse cover

REMEMBER

Underlying these real-world concerns is a complex algorithm. The algorithm may be too abstract and complicated for most users to understand, but the terms used for the object make things simpler by relating them to the real world. Getting into the mindset that objects make abstractions real will help you create better classes, which in turn help more people use your truly amazing algorithms.

Considering how properties define an object

Your algorithm uses variables to define how it works — that is, what sort of output it produces. These variables, usually created as separate entities, are the actual data within the object. As a real-world example, an apple has properties like color, taste, and sweetness. Each of these properties is data about the apple. However, you don't want anyone to modify the data in a way that will produce incorrect results from your algorithm or simply cause the entire application to crash. Properties allow you to manage how someone interacts with the data used by your algorithm. In fact, you should consider using properties to manage data in these ways:

- The data type of the input, such as an integer
- The range of the input, such as accepting only values greater than

- The value of the input based on the values of other data elements, such as accepting one value when another data element is true and other values when another data element is false
- The number of items required to perform analysis, such as a certain size vector
- Whether the property is read-only, such as when the property is an output value of the algorithm
- A default input value to use when the user doesn't supply one

REMEMBER

A class uses specific techniques to perform all these tasks. You see how to create a set of simple properties in the "Adding properties" section, later in this chapter. The important thing to remember is that properties always control how someone interacts with object data when created correctly.

Using methods to interact with an object

Methods are a kind of function. (See both Chapters 9 and 12 for more on functions.) The differences between functions and methods is that methods are part of an object and are designed to work with the data managed by that object. Because of these differences, methods work somewhat differently from functions, as you'll see in the "Specifying methods" section, later in this chapter. For example, a special variable provides you with access to the underlying object. You often use methods to interact with data in these ways:

- Determine the value of a property
- Change the value of a property
- Perform tasks with object properties, such as creating a plot
- Verify that object properties will work together properly, such as by determining whether sensor inputs are within acceptable ranges
- Reset the object data to a known good state
- Define object behaviors, such as performing an action when inputs meet certain criteria
- Implement other object behaviors, such as raising an event (see the next section for details) when needed

REMEMBER

There are many ways to use methods within an object to affect object data or use object data to perform specific tasks. However, methods always relate to the object in some way.

A special kind of method, called a *constructor*, creates an object based on the class you define. MATLAB classes can include one or more constructor methods as needed to help create the objects in a special way (such as by using arguments to automatically define property values). You always define a constructor for MATLAB classes, but MATLAB provides a default constructor when you create a new class. Creating an object from a class definition (which acts as a blueprint for the object) is called creating an *instance* of the class.

Listening to an object using events

An *event* is something that occurs because of a change. For example, when you turn the doorknob, the door opens. Likewise, having too much sun hitting the plants causes the greenhouse cover to close. When working with data, reaching a particular sensor threshold could cause the application to adjust the system associated with the sensor. In fact, events occur all the time in real life — we're inundated with them and ignore many of them (or possibly respond automatically without thinking).

Just as in the real world, a MATLAB object can act as a source for events. It can also listen for events by registering itself as a *listener*. The events, even if no one is listening, are generated (*triggered*) as needed when certain things happen (such as reaching a threshold). There is a four-step process for events:

1. Define the event.
2. Add listeners for the event.
3. Trigger the event when the conditions are right.
4. React to the event as needed.

Unlike mandatory code, in which a function performs a task in MATLAB every time, events are far more flexible. An external piece of code must register as a listener to receive an event. So, not every piece of external code receives the event, and if there are no listeners, nothing receives the event at all. Even after receiving the event, a piece of code may choose not to act upon it depending on event *attributes* (values that indicate event conditions). Creating and using events is outside the scope of this chapter, but the list of articles at `https://www.mathworks.com/help/matlab/events-sending-and-responding-to-messages.html` offers helpful insights. The article at `https://www.mathworks.com/help/matlab/matlab_oop/events-and-listeners-syntax-and-techniques.html` provides you with some sample code examples.

Understanding the need for privacy

Properties provide access to data within an object. They rely on *getters* to obtain the data and *setters* to change the data. However, you might not always want properties to allow complete access. Perhaps you don't want outsiders to see the property at all, which mean that it's a *nonpublic property*. Protecting the data from harm is one of the tenets of OOP. However, sometimes it can be difficult to figure out just how much protection the data needs. The following list shows keywords used with properties to define a level of protection for properties. (You see the protection in action later in the chapter as part of creating a class, and then using it.)

- **Constant:** Allows someone to view the data but not modify it, even by internal methods. For example, the value of pi is something that you would allow an outsider to see, but you wouldn't allow the outsider to modify because the value of pi never changes.
- **GetAccess=private:** Reduces property access to just the methods within the object so that it's possible to create *getters* (methods that get the values) and *setters* (methods that set the values). Using getters and setters reduces risk of data corruption by allowing your code to validate any access or modification before it occurs.
- **SetAccess=private:** Makes a property read-only. Outside code can read the value but can't modify it. Internal methods, however, can change the value of the property as needed.
- **Dependent:** Calculates the value of a property only when external code requests it using its dependent attribute. A dependent property has no value of its own. It doesn't store a value, so you must calculate it each time.

REMEMBER

Using getters and setters, disallowing free access to data, and protecting the methods used to manage data accomplish *encapsulation*, which is making the object a black box. The data is input and managed, but external code can't determine how these processes take place. All that matters is that given a certain input, the caller can rely on a certain output. Using this approach leaves you free to change implementation details as needed.

Understanding OOP in MATLAB

When working in certain environments, you use objects all the time because that's the preferred way of doing things. For example, when working with C++ or Java, you really do depend heavily on OOP. These languages support the OOP *paradigm* (pattern of solving problems) to the exclusion of every other paradigm.

TIP

MATLAB is more like Python because it supports multiple paradigms (see the article at `https://blog.newrelic.com/engineering/python-programming-styles/` for a discussion about Python programming paradigms), only one of which is OOP. The following sections help you understand how OOP differs in MATLAB from other languages. You can easily skip the following sections if you aren't coming from another language background and really don't care to know OOP in any more detail than is needed to work with MATLAB.

Comparing MATLAB OOP to other languages

One of the overriding considerations of use of OOP in MATLAB is to make things simple. With this in mind, think about MATLAB as being less formal than many languages out there. For example, when you define a variable in a language such as C++ or Java, the variable is called a *field,* and you can access it directly without anything interfering with that access. In MATLAB, when you define a public property, you can also define a set and a get method for that property, which controls access to it. In this way, MATLAB makes it easier to ensure that your property is fully protected, even when you make the property public.

MATLAB classes deal with values most of the time, rather than references. When working with values, any change you make to the object only occurs in the current context. When working with references, changes occur to the underlying object and extend outside the current context. The one exception to this rule is if you create a class that derives from the `handle` class (see `https://www.mathworks.com/help/matlab/ref/handle-class.html` for details). In this case, you can use the object's handle to work with it as a reference. If this whole issue of values versus references seems a little too difficult, the example code sections later in the chapter will help you.

WARNING

There is no such thing as a static property in MATLAB. A static property is one that you access from the class, rather than from the object. You can use it to track class-specific variables, such as the number of existing objects in use. Because of how MATLAB works, using static properties would cause collisions between objects. You might actually find code examples purporting to provide a workaround for the static property limitation online, but the discussion at `https://stackoverflow.com/questions/6450204/how-to-obtain-static-member-variables-in-matlab-classes` makes it clear that using static properties is a recipe for disaster.

In some cases, it's handy to create a copy constructor for your objects. A *copy constructor* is a special method that creates an object using a class as a blueprint and then fills the properties with values from another object. However, MATLAB doesn't provide support for a copy constructor. One way to get around this problem is to define the original class as a handle class and then implement a copy method, as described at https://www.mathworks.com/help/matlab/matlab_oop/custom-copy-behavior.html. Using a copy method isn't quite the same as a copy constructor because this approach requires multiple steps, but it should work for most needs.

TECHNICAL STUFF

OOP can become pretty complicated, and covering it in detail would require another book. Even covering the differences between MATLAB and languages like Java and C++ would consume a considerable amount of space. If you need additional details on how MATLAB compares to other languages, read the article at https://www.mathworks.com/help/matlab/matlab_oop/matlab-vs-other-oo-languages.html. This article provides significant detail, but the learning curve is also a little steep. It's also helpful to review the code comparison at https://www.mathworks.com/matlabcentral/fileexchange/18972-comparison-of-c-java-python-ruby-and-matlab-oop-example for an understanding of how you'd implement a class in MATLAB when you have experience with other languages.

Uses of classes and objects in MATLAB

The essential reason to use classes and objects in MATLAB is to reduce complexity when your application gets large. You use OOP to break up code into smaller, more easily understood pieces. However, you must also consider that you have other options, such as using Live Script (see Chapter 11) to perform tasks in a different manner. In addition, you can combine methods to achieve specific results. For example, you could create all your plots as objects but then access them in a Live Script. With these various approaches in mind, here is how you generally use classes and objects in MATLAB:

- To hide the complexity of an underlying task
- To ensure that the focus remains on an outcome rather than the process for deriving that outcome
- When performing a task multiple times and a function or Live Function won't reasonably achieve the desired result (such as processing a large number of input sensors)
- When working with other researchers, scientists, or professionals in other disciplines that may need a simple interface for complex tasks
- To keep data safe and uncorrupted

Performing tasks with objects in MATLAB

Later sections of the chapter show how to create and use classes from scratch. This section offers a brief overview of operations that you perform with objects. Of course, the most basic task is to create an object from a class definition:

```
object = MyClass;
```

In this case, `object` is an instance of `MyClass`. If you provide one or more constructors for `MyClass`, you can follow `MyClass` with parentheses and the required input values. Say that `MyClass` contains a property called `Property1`. You could then set this property in `object`, like this:

```
object.Property1 = 2;
```

Now `Property1` contains the value `2`, but only for the instance of `MyClass` called `object`. If you create another instance of `MyClass`, named `object2`, you could set `Property1` to another value without affecting the value of `Property1` in `object`. Each object you create is separate.

Unlike many other programming languages, MATLAB uses a different approach to working with methods. You specify the method, followed by the object to exercise the method in parentheses. For example, if an instance of `MyClass` named `object` has a method named `DoAdd()`, you use:

```
DoAdd(object);
```

At some point, you need to interact with objects to determine their characteristics. Even if you created the class, you might forget about the presence of a particular property, method, or event. The following list provides functions that you can use to interact with objects:

- `class()`: Returns the class name of the object.
- `enumeration()`: Displays any class enumeration members and names. The class must actually contain an enumeration to allow use of this function.
- `events()`: Provides a list of event names defined by the class.
- `help()`: Displays help information as it would for any script or function.
- `methods()`: Provides a list of methods implemented by the class.
- `methodsview()`: Creates a separate window to list the methods implemented by the class.
- `properties()`: Provides a list of class property names. You can also obtain a list of property names by typing the object name and pressing Enter.

You may also need to compare two objects or an object and a class. The following functions help you perform these tasks.

- `isa()`: Determines whether an object is a member of a specific class.
- `isequal()`: Determines whether two objects are equal with regard to characteristics. Equality means that both objects are of the same class and size and their corresponding property values are equal.
- `a == b (eq)`: Determines whether the handle variable `a` refers to the same object as handle variable `b`. In essence, it determines whether the two objects occupy the same memory — changes to `a` will affect `b`.
- `isobject()`: Outputs a logical true when the variable is a MATLAB object.

Creating a Basic MATLAB Class

It's finally time to work through the class creation process. The following sections help you create a basic class that you can use for experimentation purposes.

Starting the class

The first thing you need to do is create the class file. The following steps get you started.

1. **Choose New ⇨ Class in the File section of the Home tab.**

   You see a class definition (`classdef`) file like the one shown in Figure 13-1.

   The class template supplies you with an outline of what your class might look like. You see a `properties` section and a `methods` section with a constructor and basic method. Each class you create will require some basic changes.

2. **Change the `classdef` section so that it looks like this:**

   ```
   classdef MyClass
       %MYCLASS provides an example of class creation.
       % It shows how to define properties and methods.
   ```

   These changes give your class a specific name and provide help information that you can obtain using the `help()` function.

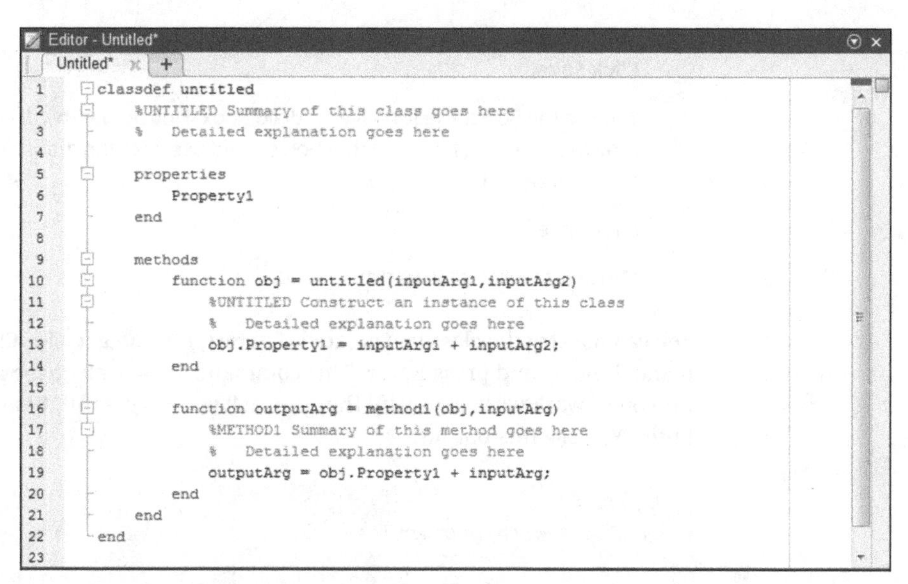

Editor - Untitled*
Untitled*
classdef untitled
%UNTITLED Summary of this class goes here
% Detailed explanation goes here
properties
Property1
end
methods
function obj = untitled(inputArg1,inputArg2)
%UNTITLED Construct an instance of this class
% Detailed explanation goes here
obj.Property1 = inputArg1 + inputArg2;
end
function outputArg = method1(obj,inputArg)
%METHOD1 Summary of this method goes here
% Detailed explanation goes here
outputArg = obj.Property1 + inputArg;
end
end
end

FIGURE 13-1: MATLAB gets you started with a basic class outline.

3. **Change the constructor method (the first one in the `methods` section) to look like this:**

```
function obj = MyClass()
   %MYCLASS Construct an instance of this class
   %   This is the default constructor.
   obj.Property1 = 0;
end
```

This change ensures that you can create an instance of your class without supplying any arguments. It also sets `Property1` to a default value. The pointer to this class instance appears in `obj`, which you use for other object-specific tasks within the class. In many languages, this kind of constructor is called a *default constructor*.

4. **Change the `method1()` method to look like this:**

```
function outputArg = method1(obj)
    %METHOD1 Returns the Property1 value
    %   Just a starting point for methods.
    outputArg = obj.Property1;
end
```

Normally, you replace `method1()` (and `Property1`, for that matter) with specific code. However, this section helps you get up and running with a very basic class quickly so that you get an idea of how things work.

5. **Click Save.**

   You see the Select File for Save As dialog box. Notice that the dialog box provides a default filename that you should use. Like functions and scripts, this one uses an `.m` file extension.

6. **Click Save.**

   MATLAB saves the file to disk for you.

You can now try the class you created. To start, type **object = MyClass;** in the Command Window and press Enter. This command creates an `object` of type `MyClass` for you to work with. To verify that the task was successful, type **object** and press Enter. You see this output:

```
object =
  MyClass with properties:

    Property1: 0
```

One of the things you should try is help. Type **help('object')** and press Enter; then click the Documentation for MyClass link. You see the window shown in Figure 13-2.

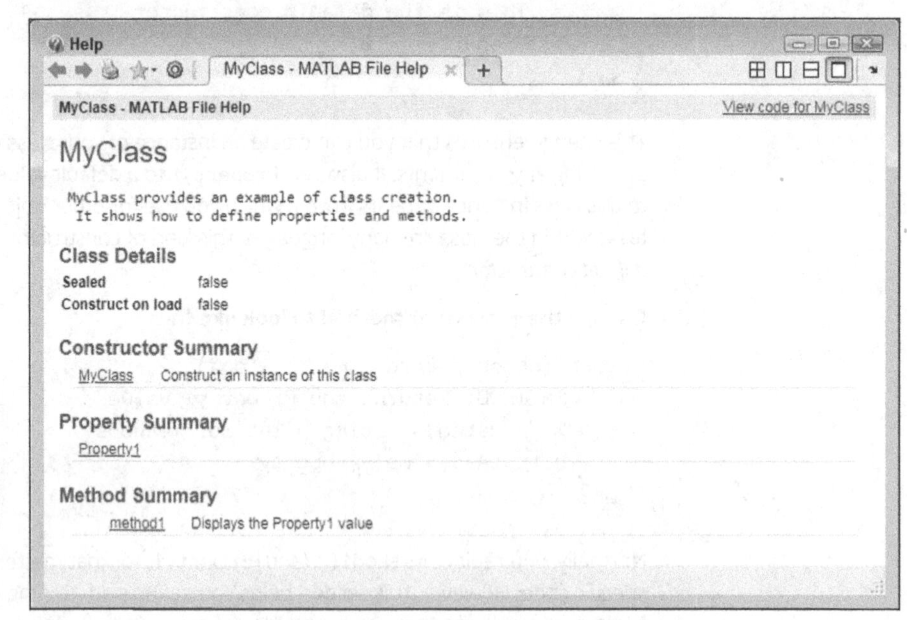

FIGURE 13-2: Class help screens give you detailed information about class characteristics.

Each of the items listed in the help window provides links for additional information. For example, if you click the method1 link, you see the help display shown in Figure 13-3. The interesting thing about help is that you can ask for specifics using a command as well. Type **help('object.method1')** and press Enter to see the details about `method1()` in the Command Window.

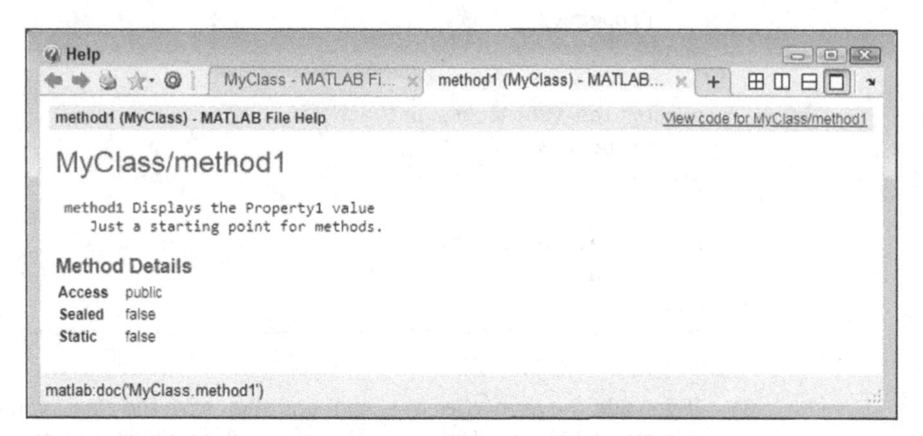

FIGURE 13-3: Drill down into the help display to get additional details.

To change the value of `Property1`, type **object.Property1 = 5;** and press Enter. You can verify the change by typing **method1(object)** and pressing Enter. MATLAB displays the following output:

```
ans =
     5
```

Adding properties

Properties come in a number of forms in MATLAB, each with their specific attributes. If a property has no attributes, it's a public property, and external code can access it freely. However, even with public properties, you can add get and set methods. The "Understanding the need for privacy" section, earlier in this chapter, describes the various nonpublic properties. Listing 13-1 shows these kinds of properties added to `MyClass`.

LISTING 13-1: Adding Properties to MyClass

```
properties
    Property1
end
```

(continued)

LISTING 13-1: ***(continued)***

```
properties (Constant)
    pi = 3.141592653589793238;
end

properties (GetAccess=private)
    Property2
end

properties (SetAccess=private)
    Property3
end

properties (Dependent)
    Property4
end
```

If you add the properties as shown and then save the class, you can see how the various attributes affect property access. Type **object** and press Enter. You see output similar to this (Property1 might have a value of 5 instead):

```
object =
  MyClass with properties:

    Property1: 0
           pi: 3.1416
    Property3: []
```

TIP

Notice that any changes you make to the class definition automatically affect existing objects. The public property, `Property1`, works just as it did before. Providing a custom setter will ensure that `Property1` contains values in the correct range. For example, you might use a custom setter like this (placed in the `methods` section of the code):

```
function obj = set.Property1(obj, value)
    if isnumeric(value)
        obj.Property1 = value;
    else
        error('Input must be numeric!');
    end
end
```

Notice that the `set.Property1()` function starts with `set`, followed by a period, followed by `Property1`. This is the pattern used for get and set functions. The function code checks to see whether `value` is numeric. If so, it gives `Property1` a new value. Otherwise, it displays an error onscreen. You must return a new object when working with a setter, as shown by `function obj` in the preceding code. To try this new function, type **object.Property1 = 2** and press Enter; the value of `Property1` changes. However, when you type **object.Property1 = 'Hello'** and press Enter, you see an error message instead.

You can also see the constant property, `pi`, but MATLAB truncates its value in the output to match the current default format. The `GetAccess=private` attribute keeps you from seeing `Property2`. However, `SetAccess=private` lets you see `Property3`, which hasn't been initialized, so you see an empty vector.

Type **object.Property4** and press Enter. You see this error message:

```
In class 'MyClass', no get method is defined for
Dependent property 'Property4'. A Dependent property
needs a get method to access its value.
```

A property that has the `Dependent` attribute set stores no value, so requesting `Property4` without defining a get method causes an error message. Adding the following function to the `methods` section of `MyClass` will fix the problem:

```
function value = get.Property4(obj)
    value = class(obj.Property1);
end
```

The `get.Property4()` function code lacks any mention of `Property4`. This is because `Property4` contains no value — it's a calculated property. If you type **object.Property4** and press Enter, you see the `Property1` class, which is `double`. Likewise, if you type **object** and press Enter now, you also see the `Property4` output.

REMEMBER

MATLAB acts as if `Property2` doesn't exist because it has the `GetAccess=private` attribute set. If you type **object.Property2** and press Enter, you see this error message.

```
No public property 'Property2' for class 'MyClass'.
```

Likewise, if you attempt to set the Property2 value by typing **object.Property2 = 2** and pressing Enter, MATLAB acts as if you typed `object` by itself. In other words, `Property2` has no outside access. To use `Property2`, you need to create a method that relies on it for a calculation or outputs it as a value.

You can't set the value of Property3 because it has the SetAccess=private attribute. If you type **object.Property3 = 2** and press Enter, you see this error message:

```
You cannot set the read-only property 'Property3' of
        MyClass.
```

Specifying methods

You see a basic method defined in the "Starting the class" section of the chapter, along with a simple constructor. The preceding section, "Adding properties," shows how to create methods associated with various kinds of properties. The MyClass example has a number of properties associated with it whose values can't change directly while using the instantiated object. Consequently, it would be nice if you could create a constructor to set those properties. Fortunately, constructors can accept a variable number of inputs, as shown here (MATLAB classes can have only one constructor, so you need to delete or comment out the previous constructor before adding this one):

```
function obj = MyClass(Prop1, Prop2, Prop3)
    %MYCLASS Construct an instance of this class
    %   This constructor will take a variable
    %   number of arguments. The arguments appear
    %   in order for Property1, Property2,
    %   and Property3.

    if nargin >= 1
        obj.Property1 = Prop1;
    else
        obj.Property1 = 0;
    end

    if nargin >= 2
        obj.Property2 = Prop2;
    else
        obj.Property2 = 0;
    end

    if nargin == 3
        obj.Property3 = Prop3;
    else
        obj.Property3 = 0;
    end

end
```

Using this constructor will allow you to provide a variable number of values as input. For example, type **obj = MyClass(1, 2, 3)** and press Enter to define values for Property1, Property2, and Property3.

Methods can have attributes, just as properties can. You can see a list of method attributes at https://www.mathworks.com/help/matlab/matlab_oop/method-attributes.html. One of the more interesting method attributes is the Static attribute. You use this attribute to create methods that you can execute as part of the class, rather than as part of an instance. Here's an example of a Static method:

```
methods (Static)
    function name = ShowName()
        name = 'This is MyClass.';
    end
end
```

Notice that you define a new methods section with the Static attribute, just as you do with property attributes. Defining the function ShowName() works like any other function, except that you don't work with an object variable. In fact, you can't access any of the property values, because they're defined as part of an instance. To use this method, type **MyClass.ShowName()** and press Enter. You see the output shown here:

```
ans =
    'This is MyClass.'
```

IN THIS CHAPTER

» Considering the app design approach

» Creating a basic app

» Testing the app

» Packaging the app

Chapter 14
Creating MATLAB Apps

Conceptually, a MATLAB app is about the same as any other app, except that it reflects the origin of its creation. In other words, you won't find the next sophisticated gaming program as a MATLAB app. Likewise, you won't build a word processor or a web app using MATLAB. A MATLAB app puts a nice interface on the various pieces of code you've built, and it enables you to view that code as an application. So if you've created scripts and functions to interact with sensors, you can now build an application to see the data without worrying about the underlying code. The first section of this chapter gives you additional app details, including how to work with the App Designer.

The next three sections of the chapter tell you how to create, test, and package a MATLAB app. When you're done, you'll have a packaged app that you can send to someone or share online with other MATLAB users. Even if you don't share your app with anyone else, just having all your code neatly packaged will make it a lot easier to use. Note that you need to know about MATLAB classes, as discussed in Chapter 13, to build apps.

You don't have to type the source code for this chapter manually. In fact, using the downloadable source is a lot easier. You can find the source for this chapter in the `\MATLAB2\Chapter14` folder of the downloadable source. When using the downloadable source, you may not be able to work through some of the hands-on examples because the example files will already exist on your system. In this case, you can create an alternative folder, `Chapter14a`, to hold the results of the hands on exercises. See the Introduction for details on how to obtain the downloadable source.

Working with the App Designer

The App Designer is the starting point for creating a MATLAB app. However, before you start the App Designer, you want to know a little more about apps, such as what apps can do and where you can find example apps to use as templates for your own app. The following sections help you become acquainted with apps and help you get started with your first app by starting the App Designer.

Understanding apps

An app helps you automate tasks in a way that scripts, functions, Live Scripts, and Live Functions really can't. The difference is the inclusion of a Graphical User Interface (GUI), where app users configure the characteristics of a calculation just as they would when using any other custom app. You create a GUI that helps guide the user into the correct sequence of events, and the user doesn't worry about working with code at all.

Apps use a special file format. The "Understanding the MATLAB files and what they do" section of Chapter 4 describes the `.mlapp` and `.mlappinstall` files associated with apps. When creating an app, you begin with the `.mlapp` file, which stores things like the GUI you create for your app. You see how this process works as the chapter proceeds, so just know for now that you'll be working with a different file type than you've used in the past.

Of course, you might wonder about all the extra work required to create an app. After all, you could simply provide a set of written instructions for the existing code you've created. There are good reasons for using apps, such as these:

- Reducing human errors due to a misunderstanding of required input, values, or processes
- Increasing the number of people who can work with your code by reducing the complexity of the process
- Ensuring that people will actually work with your code by making the process easier
- Improving output results because everyone will use the same consistent process
- Removing ambiguity from the data analysis
- Creating a single package for the entire app so that you don't have to worry about some pieces getting lost
- Defining an environment in which all the code elements are designed to work together so that you don't encounter version differences

Getting apps

You should play with a few apps before you start designing your own. For one thing, playing with other people's apps will help you see what works and what doesn't in the way of design. You also gain insights into what you need to include as part of your app so that it works properly. There are currently three sources for apps:

- **Toolboxes:** Some toolboxes come with apps to make the toolbox easier to use. You can find some of these toolboxes at `https://www.mathworks.com/help/referencelist.html?type=app`.
- **File Exchange:** Many people have already developed apps for MATLAB. In fact, you might find the app you need and not need to develop an app of your own. You can find a list of community-supported apps at `https://www.mathworks.com/matlabcentral/fileexchange/?term=type%3A%22App%22`.
- **Other users:** The people who work in your community might actually be the best source of apps. They likely encounter the same problems that you do, so they may have created an app to solve the problem.

MATLAB makes it easy to locate new apps to try. Click Get More Apps in the File section of the Apps tab to display an Add-On Explorer window similar to the one shown in Figure 14-1. The left side of the window provides filters that you can use to narrow your search. The right side of the display shows the apps that match any filtering you use. Notice especially the Using MATLAB/App Building category, which you can use to locate examples specific to app building.

One of the more interesting examples is regexpBuilder. You can find it in the Using MATLAB/Language Fundamentals category (you may have to select the Data Types subcategory). When you select this app, you see a page similar to the one shown in Figure 14-2. To install regexpBuilder, follow these steps:

1. **Click Add.**

   Add-On Explorer asks for your email address and password for verification. You then see a license agreement dialog box, where you click I Accept. The app downloads and becomes available for use. Unfortunately, you don't see it in the Apps list.

2. **Click Get More Apps in the File section of the Apps tab.**

   You see the Add-On Explorer display, shown in Figure 14-1.

3. **Locate the regexpBuilder app again.**

   This time you notice an Open Folder button.

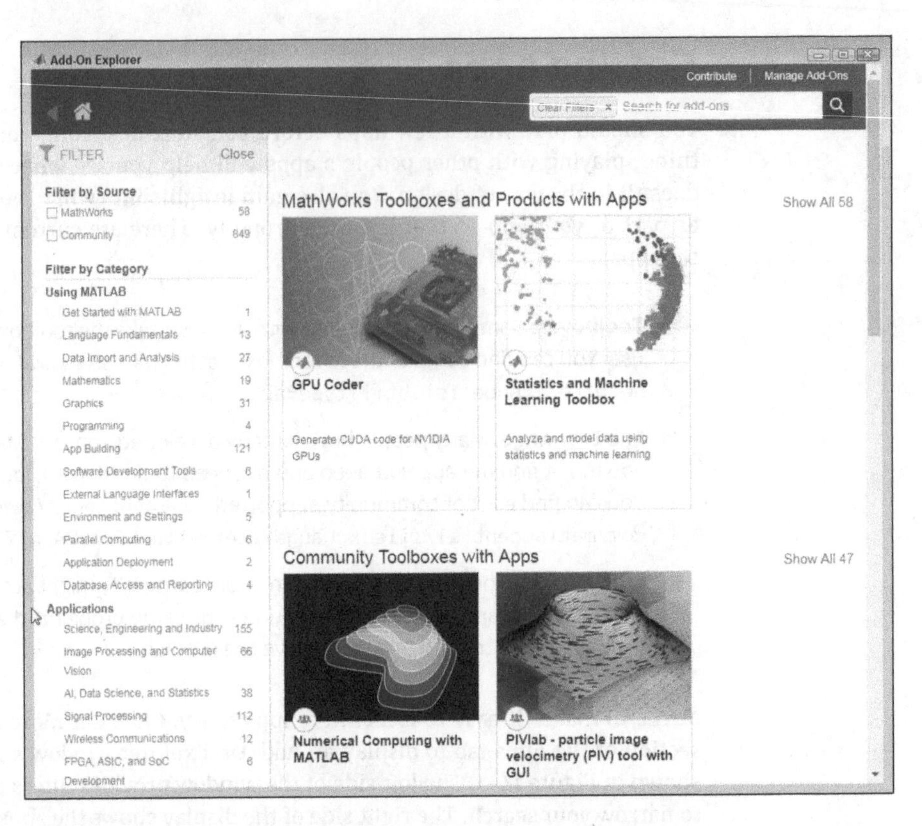

FIGURE 14-1: Use the Add-On Explorer to locate potential apps to try.

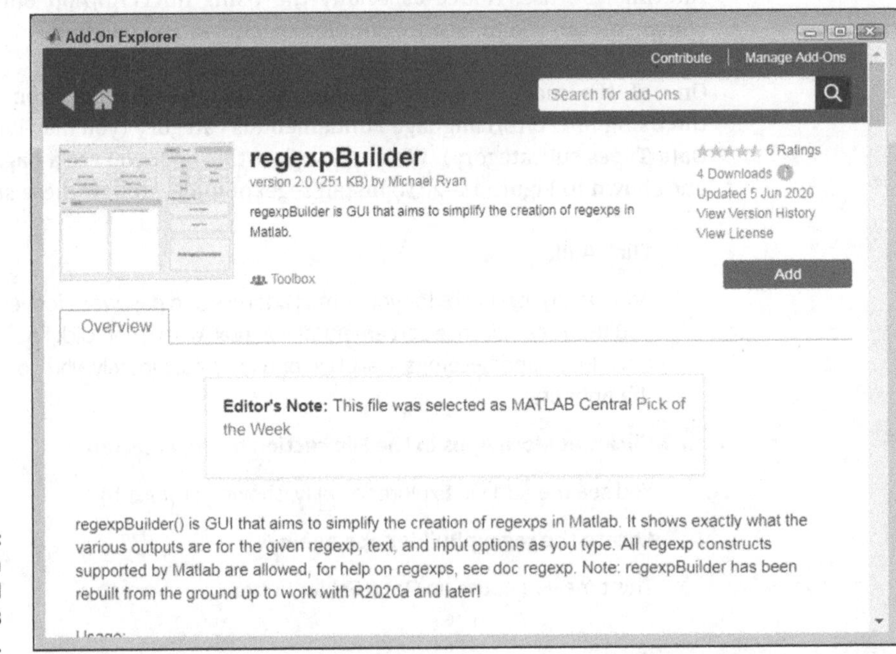

FIGURE 14-2: When you find an app you like, add it to your MATLAB setup.

4. **Click the Open Folder button.**

   You see the `regexpBuilder.mlappinstall` file in the Current Folder window, as shown in Figure 14-3.

FIGURE 14-3: Locate the required installation file.

5. **Right-click the `regexpBuilder.mlappinstall` file and choose Install from the context menu.**

   You see the Install dialog box shown in Figure 14-4, which asks where you want to install the app (you may not be able to select any tab other than Apps). In general, you want to keep your apps in the Apps tab to make them easy to find.

6. **Click Install.**

   You see regexpBuilder appear in the Apps section of the Apps tab. The app is ready to use.

FIGURE 14-4: Choose an installation location for the app.

TIP

Remember that you can click the Back button in the Current Folder window to move from the `regexpBuilder` folder to your code folder again. You could also select the folder from the drop-down list box showing the previous folder history.

At this point, you can give regexpBuilder a try. Click the regexpBuilder entry in the Apps section of the Apps tab to display the window shown in Figure 14-5. Now that you have the app, take some time to play with it using the examples found at `https://www.mathworks.com/help/matlab/ref/regexp.html`. The results are comprehensive.

FIGURE 14-5: regexpBuilder helps you build regular expressions for use in MATLAB.

Starting the App Designer

It's time to look at the App Designer. You use it to create your app or to view other people's apps, assuming that they send you the `.mlapp` file. Click Design App in the File section of the Apps tab to display the App Designer Start Page, shown in Figure 14-6.

TIP

The banner at the top of the window is important because it provides access to a number of interesting bits of information (simply click the right- or left-pointing arrows on each side of the banner as needed). The first of these banner entries tells you about a three-minute tutorial that is worth trying if you want to get a little additional hands-on time.

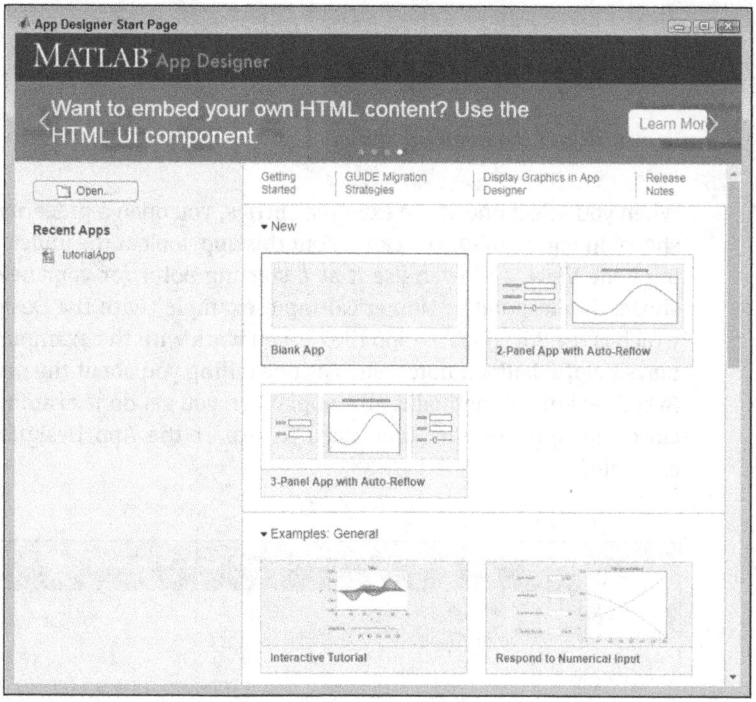

FIGURE 14-6: Building an app starts with the selection of a template.

Immediately below the banner are template sections. The first of these sections simply says New, and this is where you choose a template for your app. Below the New category are various categories that cover important areas of app development:

- **»** Examples: General
  - Interactive Tutorial
  - Respond to Numerical Input
  - Respond to User Selections
  - Embed HTML Content
  - Lay Out Controls in a Grid
- **»** Examples: Programming Tasks
  - Link Data to a Tree
  - Analyze an Image
  - Configure a Timer
  - Display Specialized Axes

* Create a Table
* Query Website Data
* Pass Data Between Apps

When you select one of the example entries, you open a predefined app like the one shown in Figure 14-7. You can try out this app, look at the underlying code by clicking Code View, and even use it as a starting point for your new app. Figure 14-7 shows the Respond to Numerical Input example (with the Designer tab selected), which is the `Mortgage.mlapp` file. As you work with the example file, MATLAB displays helpful balloon notes, such as one telling you about the need to add callbacks (a special kind of method) to the app. When you get done examining an app, simply close the App Designer (you need to restart the App Designer to view another example).

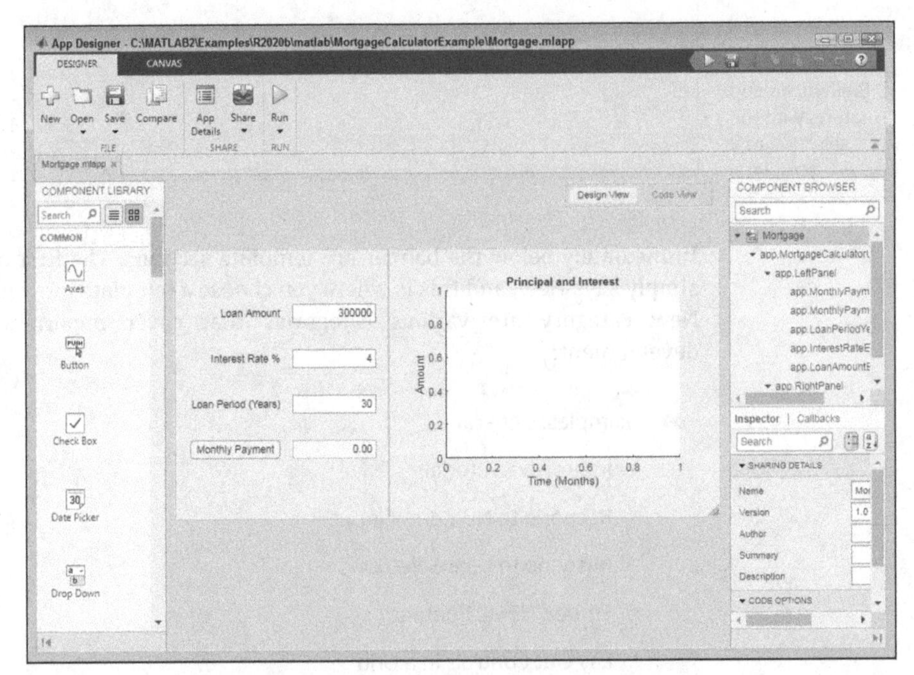

FIGURE 14-7: Looking at the examples can provide insights into building your own app.

The Respond to Numerical Input example relies on the 2-Panel with Auto-Reflow template. In fact, all the examples rely on one of three templates supplied as part of the App Designer (all of which appear in the New section). For this chapter, you need to create a new app using the Blank App template. To create the new blank app, open the App Designer Start Page (refer to Figure 14-6) and click Blank App in the New category. You see the blank App Designer display, shown in Figure 14-8 (with the Canvas tab selected).

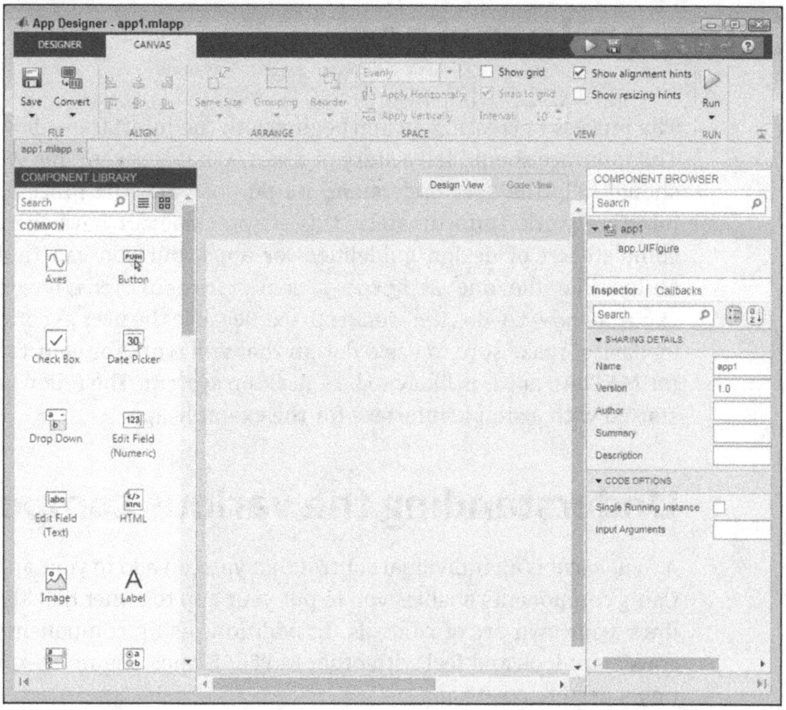

FIGURE 14-8: The example in this chapter begins with a blank design area.

The initial view shows you the Component Library, which contains the components you use to build an interface, on the left side of the display. In the center is the canvas used to draw your interface. The right pane is the Component Browser, which you use to configure the components on the canvas.

The canvas provides two views. The first is Design View, which provides a graphical view of your app. The second is Code View, which shows the underlying code used to interact with the app components.

At the top of the display are two tabs. Designer gives you the controls needed to interact with the app as a whole. Canvas contains the controls needed to help you create a better interface. For example, there are controls used to align the app components, arrange them in specific ways, or space them evenly across the canvas. You can also choose to use various design aids, such as a grid to see how components will align better. The sections that follow offer more insights into how everything works.

Defining an Interface

The process of creating an app begins with the interface — how the app will look. The interface should tell a story of a sort. Whatever your app will do, the interface should help the user understand its purpose and the process used to make the interface work. Unfortunately, this chapter doesn't have the space to provide a complete set of design guidelines for apps, but you can find many guidelines online like the one at `https://docs.microsoft.com/en-us/windows/win32/uxguide/how-to-design-desktop-ux`. Because the user experience tends to differ by device, make sure to use a design that will work for your target device, which, for MATLAB apps, is likely to be a desktop system. The following sections get you started with a simple interface for the example app.

Understanding the various components

A *component* is an individual control that you can add to your app, such as a button. Using components enables you to put your app together quickly without having to draw your own set of controls. In addition, using components gives your app a consistent look and feel with other MATLAB apps. Figure 14-9 shows the components used to create apps.

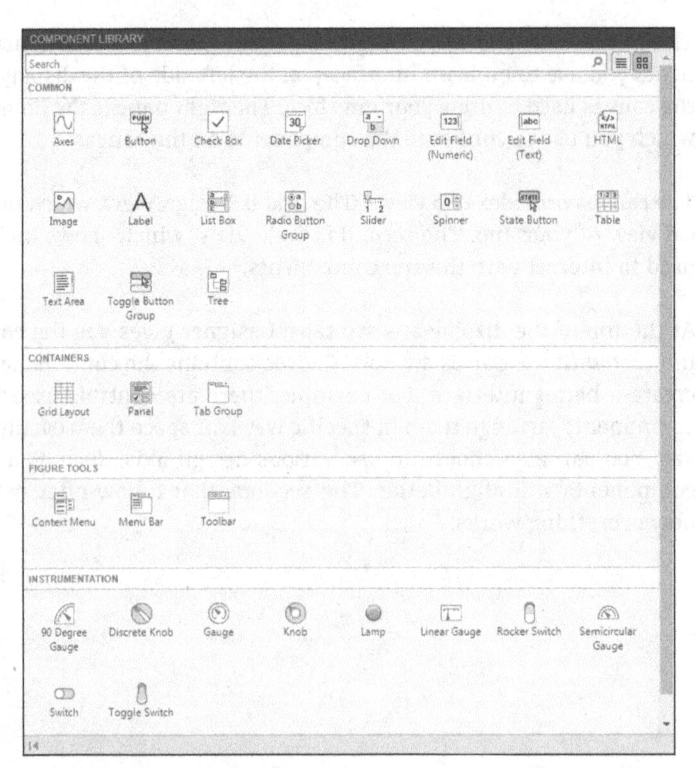

FIGURE 14-9: Using these components will make app design faster and easier.

The list of components that MATLAB provides is somewhat limited compared to other products and languages, but the number of components is sufficient for the kinds of apps that most people create. The example app will show a customized message with a selected piece of data in a particular format. The purpose of this app is to help you become acquainted with a variety of components. To begin, it's helpful to select Show Grid in the View section of the Canvas tab so that you can more easily align the components. Selecting this option automatically selects Snap to Grid, in most cases as well (because it's selected by default).

The top of the application should have a title so that everyone knows the app name. You need an `Edit Field (Text)`, as contrasted to an `Edit Field (Numeric)`, to provide a place for the user to type a name. Below the name entry, you add a `Date Picker` with the current date selected, and a `Drop Down` to select the date format. The `Drop Down` will have a first entry of `Default Date` to allow selection of the default date format. It's important to wait for the user to make selections before doing anything, so below the date selections, you provide a `Button` for the user to request the formatted message. A Label will receive the formatted message, so you need to resize it to receive the information. Figure 14-10 shows the interface for the example application.

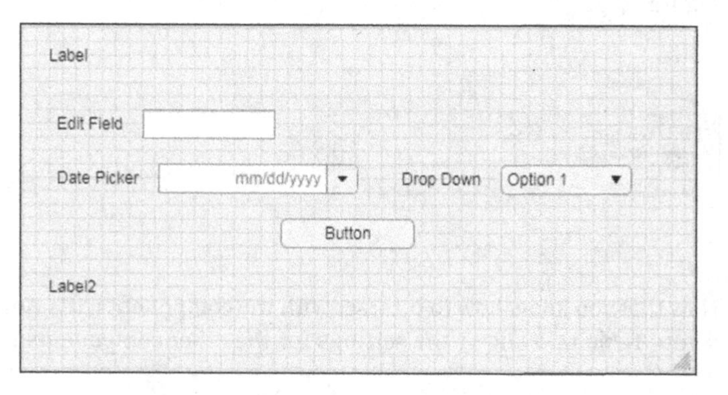

FIGURE 14-10: The example app uses a top-to-bottom design.

After you finish adding the components, saving your app is a good idea. Click Save in the File section of the Designer or Canvas tab, and you see the Save File dialog box. MATLAB doesn't automatically suggest a filename for you, so type **DateDisplay.mlapp** in the File Name field and click Save.

Even though the list of components in the Component Library is fixed, you can create custom components. The process requires advanced programming skills, and component implementation is outside the scope of this book, but you can discover details at `https://www.mathworks.com/help/matlab/creating_guis/develop-classes-of-ui-component-objects.html`. Starting with the R2020b MATLAB release, you can create a class implementation for your UI components by defining a subclass of the `ComponentContainer` base class instead of using a script or function. None of your custom components will appear in the Component Library list, so you simply need to know that they exist to use them.

Changing the component properties

The layout shown in Figure 14-10 will probably work fine, but the app itself won't work well because no one will know what to type in each of the fields. To make this app work, you need to define component properties. Begin with the first `Label`. You want to provide a welcoming message to people using your app, so select the label and then type **Welcome to the Date Display App!** in the `Text` property of the Component Browser. The message isn't very exciting yet. Select `Center` for the `HorizontalAlignment` property value, type **16** in `FontSize`, and select `B` in `FontWeight`. The label looks better.

Note that some components contain two items, so you must select the item you want to work with. For example, an `Edit Field (Text)` consists of a label and a text box. To change the label, you must specifically select it, as shown in Figure 14-11.

FIGURE 14-11: Make sure to select the element you want to work with in a component.

Now that you know how to interact with the components, it's time to complete the setup. Table 14-1 shows the settings for the other components for this app.

After you make the changes shown in Table 14-1, you may need to rearrange the components to provide a more pleasing appearance. Figure 14-12 shows the final appearance of the app.

TIP

You can resize the canvas so that it better matches your components. You can use a sizing arrow in the bottom-right corner for the purpose. Making your canvas fit the components gives your app a more finished feel.

TABLE 14-1 **Configuring the Date Display App components**

Component	Property	Setting
Label portion of `Edit Field (Text)`	`Text`	Your Name
Label portion of `Date Picker`	`Text`	Selected Date
Date picker portion of `Date Picker`	`Value`	Current date
Label portion of `Drop Down`	`Text`	Date Format
Drop-down portion of `Drop Down`	`Items` (each on their own line)	Default yyyy-mmm-dd yyyy-mm-dd dd/mmm/yyyy dd mmm yyyy
`Button`	`Text`	Show Messsage
`Label2`	`BackgroundColor`	White (you must select this value from the drop-down list, rather than simply typing it in the field)
`Label2`	`Text`	Blank

FIGURE 14-12: The app looks nicer now and provides useful prompts.

Making the Interface Functional

If you were to run your app right now, you'd see an interesting interface, but nothing more. You need to add code to make the app do something. You have two approaches to adding code for your app. The easiest method is to allow MATLAB to manage most tasks for you. However, no matter what you do, you eventually end up in Code View, where you can type at least some minimalistic code. The following sections help you understand the underlying techniques for making your app functional.

Working with Code View

When you click Code View for your app, what you see is a specialized class, as shown in Figure 14-13. (See Chapter 13 for help in seeing how MATLAB classes work.) MATLAB automatically creates this class as you add components to the canvas and configure them. Every detail that you add to the app is part of the class code.

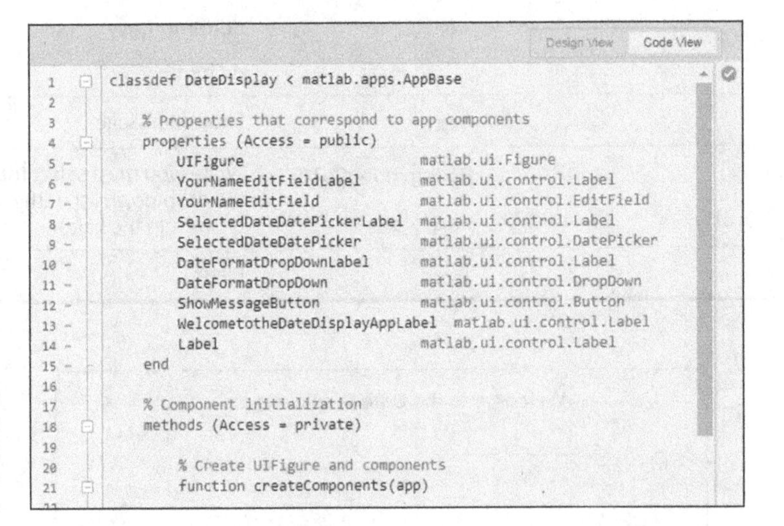

FIGURE 14-13: MATLAB wrote code while you created the app interface.

Notice that the `createComponents()` function declaration at the bottom of Figure 14-13 requires one argument, `app`. You use `app` much like you used `obj` in Chapter 13. It provides you with access to the application. Consequently, when you view the `Edit Field (Text)` component for entering a name, you see this code:

```
% Create YourNameEditFieldLabel
app.YourNameEditFieldLabel = uilabel(app.UIFigure);
app.YourNameEditFieldLabel.HorizontalAlignment = ...
    'right';
app.YourNameEditFieldLabel.Position = [21 142 65 22];
app.YourNameEditFieldLabel.Text = 'Your Name';
```

Every entry begins with app, which signifies that this piece of code interacts with the application. You don't need to know what this code does unless you plan to modify it later. It's actually a lot easier to use the Component Browser to change these settings and let MATLAB modify the underlying code for you. For now, all you really need to know is that there is a class in the background and that MATLAB manages it for you.

Creating a callback function

A *callback function* is one that registers itself to monitor and then respond to an event. For example, when the user clicks Show Message, a callback function will do something about the click. Every control in an app can have callback functions associated with it, but you need to provide callback functions only for essential tasks. The example application uses two callback functions: one for Date Format changes and another for Show Message clicks. Follow these steps to create a callback function:

1. **Right-click the control that will need the callback, and choose the Callbacks menu from the context menu.**

   You see a list of callbacks associated with the control, as shown in Figure 14-14. In this case, you see the callbacks for Show Message. There is only one: `ButtonPushedFcn`. The Date Format control has two callbacks, but the example will use only the `ValueChangedFcn` callback.

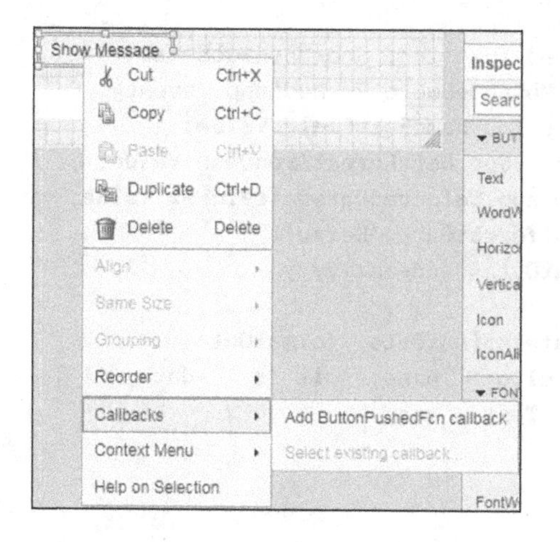

FIGURE 14-14: Choose the callback function you want to add to the app.

2. **Select the Add *Function Name* Callback entry you need.**

   You see a new callback function added to the code, as shown here for the ButtonPushedFcn, which is ShowMessageButtonPushed() for this specific instance.

```
% Callbacks that handle component events
methods (Access = private)

    % Button pushed function: ShowMessageButton
    function ShowMessageButtonPushed(app, event)

    end
end
```

3. **Type the code needed to make the callback work.**

   Remember that you access application features using the app object.

As previously mentioned, this example uses two callbacks. Listing 14-1 shows the code you need to add to the callbacks after you create them in bold.

LISTING 14-1: **Creating Callbacks for the Example App**

```
% Callbacks that handle component events
methods (Access = private)

    % Button pushed function: ShowMessageButton
    function ShowMessageButtonPushed(app, event)
        name = app.YourNameEditField.Value;
        formatOut = app.DateFormatDropDown.Value;
        selDate = app.SelectedDateDatePicker.Value;
        if strcmp(formatOut, 'Default')
            formatOut = 'dd-mmm-yyyy'
        end
        date = datestr(selDate, formatOut);
        msg = ['Hello ', name, ' it''s ', date, '!'];
        app.Label.Text = msg;
    end
```

```
    % Value changed function: DateFormatDropDown
    function DateFormatDropDownValueChanged(app, event)
        value = app.DateFormatDropDown.Value;
        if strcmp(value, 'Default')
            value = 'dd-mmm-yyyy'
        end
        app.SelectedDateDatePicker.DisplayFormat = value;
    end
end
```

The `ShowMessageButtonPushed()` code begins by obtaining all the required values from the various controls. It then ensures that `formatOut` contains a usable value in the form of a date format string. The next step is to convert the date into a date string using `datestr()`. Finally, the code creates a message, `msg`, and outputs it to `app.Label.Text`.

The `DateFormatDropDownValueChanged()` code has two purposes. First, it ensures that the selected date value is in a useful date format string form. It then assigns this `value` to `app.SelectedDateDatePicker.DisplayFormat` to modify the presentation on the form so that the user knows how the final date will look.

TIP

To remove a callback that you no longer need, select <no callback> from the drop-down list for the event, such as `ValueChangedFcn` for a Date Picker, in the Callbacks tab of the Component Browser. You can then delete the associated code in Code View. Make sure to delete the entire function.

Running the App

You need to be sure to save your app before you run it. Even though MATLAB rarely freezes, you never know when a crashed system will ruin your day. Unfortunately, MATLAB won't tell you that the app hasn't been saved. It's a good habit to click Save before you click Run. Saving your app will take a lot less time than recreating it.

To test your app, click Run. Type a name in the Your Name field, ensure that the Selected Date and Date Format fields are correct, and then click Show Message. Figure 14-15 shows a typical example of what the application might look like.

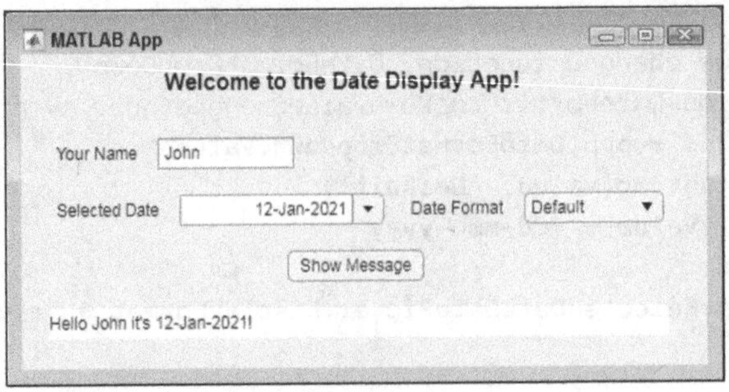

FIGURE 14-15: The final app provides a simple interface for displaying the date.

Packaging Your App

Packaging your app is a matter of using a wizard to answer some basic questions and then generating the `.mlappinst` file. The following steps give you a short overview of the process.

1. **Click Package App in the File section of the Apps tab.**

   You see the wizard, shown in Figure 14-16.

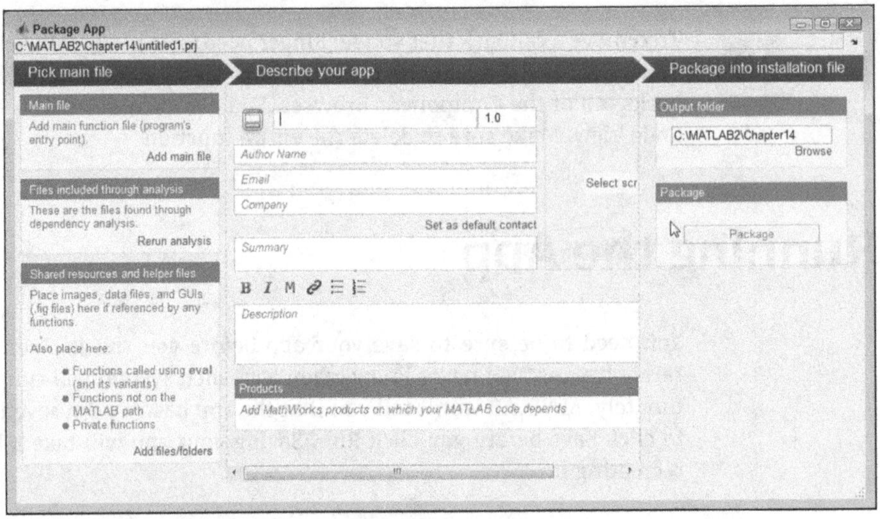

FIGURE 14-16: Start the Package App wizard.

2. **Click Add Main File.**

   You see an Add Files dialog box, where you select the `DateDisplay.mlapp` file.

3. **Click Open.**

   You see the file added to the Main File list.

4. **Fill in the details in the center section of the wizard.**

   As a minimum, you should provide a name for your app and your contact information, along with a summary of the app's purpose.

5. **Choose a folder in the Output Folder section.**
6. **Click Package.**

   After a few seconds, if there are no errors, you see a message: Packaging Complete. You can then click Open Output Folder to see the package for yourself.

REMEMBER

As part of creating the package, MATLAB also creates a `Date Display App.prj` file. Open this file if you want to make changes to how you package your app.

IN THIS CHAPTER

» Defining the project and seeing how to use it

» Creating a project structure

» Performing checks on your project setup

Chapter 15
Building Projects

The term *project* gets used in all sorts of places. In fact, most people have a to-do list right there on their desk that's packed with projects they may never accomplish. People think about them, though and then push them further down on the list as new projects (that also won't get done) take their place. A *MATLAB project* is documentation for a task that you contemplate, devise, or plan. Your MATLAB project, as described in the early part of this chapter, is a specific kind of documentation for tasks that you perform in MATLAB so that you can obtain repeatable results.

Projects have a specific format. You use a particular way of describing how to do something so that anyone who views the project also understands the process used to accomplish the task. The next part of the chapter helps you understand the project structure. When creating a MATLAB project, you can choose a blank project or one created from a source, such as a folder, Git, or Subversion (SVN). You also create projects as part of packaging an app (as described in the "Packaging Your App" section of Chapter 14).

Because a project is a kind of process, you must also ensure that your project will produce the desired result. Checking your project for flaws will reduce the risk of someone producing apps or performing other MATLAB tasks inconsistently. The last section of the chapter helps you create consistent and useful projects.

REMEMBER

You don't have to type the source code for this chapter manually. In fact, using the downloadable source is a lot easier. You can find the source for this chapter in the `\MATLAB2\Chapter15` folder of the downloadable source. When using the downloadable source, you may not be able to work through some of the hands-on examples because the example files will already exist on your system. In this case, you can create an alternative folder, `Chapter15a`, to hold the results of the hands on exercises. See the Introduction for details on how to obtain the downloadable source.

Considering the Need for Projects

If you work by yourself or in a small group on a limited research task, you likely don't need a project. However, research and other MATLAB-related tasks tend to grow and become a lot larger than anticipated. You might have hundreds of people all working together toward a common goal; the task itself might span years; or the task could become so complicated that you need organizational aids to understand it. With these sorts of issues in mind, you may need help with

- Automating startup and shutdown tasks
- Devising easy access to frequently performed tasks
- Creating projects based on current conditions
- Managing projects based on environmental changes
- Reducing the complexity of large projects
- Sharing projects with others
- Upgrading or updating an existing project
- Considering project dependencies
- Developing a method of performing tasks that aren't performed often enough to memorize the process, but are performed often enough to pay back the time for developing the project
- Using source control to manage the project

REMEMBER

The overriding goal of a project is to think about what you want to accomplish and how to accomplish it. Although some people just plunge in and start working toward a goal, using this approach is often error prone and counterproductive because you waste a lot of time pursuing false leads. Using projects is an organized method of interacting with MATLAB on a large scale.

Creating a New Project

Before you can use a project to manage anything, you must create one. MATLAB provides a number of methods to create projects, as described in the following sections. The example relies on creating a blank project so that you can see how the various features work without becoming involved in third-party aspects, such as source control.

TECHNICAL STUFF

You need to perform additional setups to use MATLAB with a source-control manager. The two integrated support mechanisms are Git (see `https://www.mathworks.com/help/matlab/matlab_prog/set-up-git-source-control.html` for setup requirements) and SVN (see `https://www.mathworks.com/help/matlab/matlab_prog/set-up-svn-source-control.html`). It's theoretically possible to use other source-control managers, but you must perform a lot of manual configuration to get the job done. The articles at `https://www.mathworks.com/help/matlab/matlab_prog/about-mathworks-source-control-integration.html` and `https://www.mathworks.com/help/matlab/matlab_prog/customize-external-source-control-to-use-matlab-for-comparison-and-merge.html` provide a starting point for this task.

Choosing a project type

You need to choose a project type as a first step to creating a project. MATLAB provides menu options to create the following four project types:

- **Blank Project:** You start the project from scratch.
- **From Folder:** The project begins with the files found in a specific folder of your choosing. Using this project type saves time when you already have the files you want to use for the project in one place.
- **From Git:** The project begins with the files found in a Git repository. MATLAB also performs some configuration steps required to use Git as a repository and source-code manager.
- **From SVN:** The project begins with the files found in a SVN repository. MATLAB also performs some configuration steps required to use SVN as a repository and source code manager.

This example uses a blank project to enable you to see the various steps involved. The following steps help you create the project:

1. **Choose New ⇨ Project ⇨ Blank Project in the File section of the Home tab.**

   You see a New Project dialog box, like the one shown in Figure 15-1. This dialog box has example entries already configured in it.

FIGURE 15-1: Provide a name and location for your project files.

2. **Type a name in the Project Name field and select a location in the Project Folder field; then click Create.**

   You see a Welcome to Your Project dialog box like the one shown in Figure 15-2. To create a project, you perform all three tasks: adding files, setting paths, and determining whether to run your project at startup, before you move on. You can click Learn More to see help files for each of the tasks, or click Skip if you want to get started configuring your project immediately.

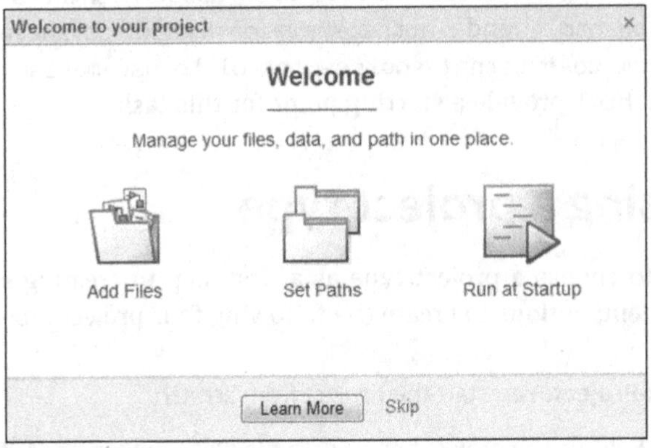

FIGURE 15-2: Review the welcome screen items before moving on if necessary.

3. **Click Skip.**

   You see the Project editor display, shown in Figure 15-3. This is where you start defining how your project is put together.

You need some files to use to work with your project. Rather than start completely from scratch, copy `Factorial1.m` from the `Chapter10` folder and paste it into the `Chapter15` folder. Open the file and remove the `fprintf()` and `disp()` lines so that you don't see the interim steps. Now create a new script file using the techniques found in Chapter 8, name it `TestFactorial1.m`, and add the following code to it:

```
for testValue = 1:5
    Output = Factorial1(testValue);
    fprintf('%d! equals %d\n', testValue, Output);
end
```

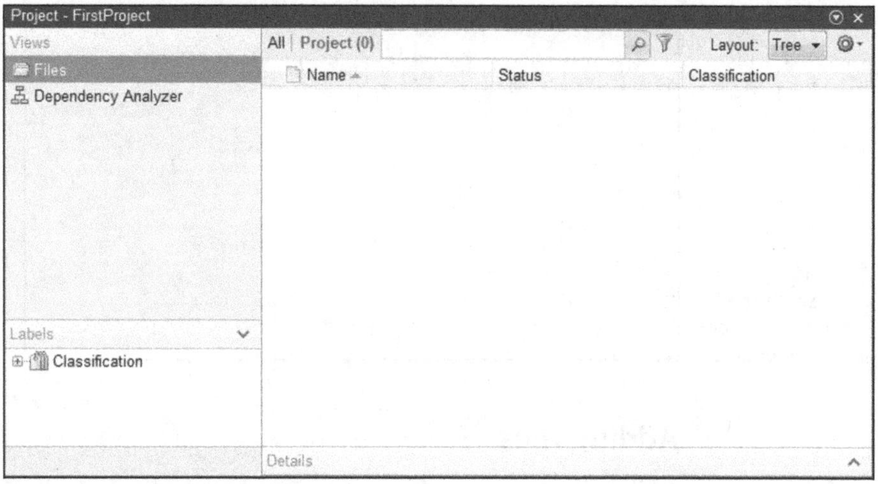

FIGURE 15-3: The Project editor helps you configure your project.

If you run `TestFactorial1.m` at this point, you see the following output:

```
1! equals 1
2! equals 2
3! equals 6
4! equals 24
5! equals 120
```

Using the Project editor

The Project editor (refer to Figure 15-3), enables you to configure your project. (The Project editor was formerly called the Project Live editor.) One of the odd parts of this process is that you must tell Project editor where to find your files before adding files to the project (even if these files are in the current directory). That's why the following sections have you setting the project path first, before you add files.

Setting paths

Your project must have at least one path to work correctly. To add a path, click Project Path in the Environment section of the Project tab. You see the Manage Project Path dialog box, shown in Figure 15-4.

Click Add with Subfolders because the `Chapter15` folder contains a `resources` subfolder, which MATLAB automatically creates as part of creating the project. You see an Open dialog box, where you can highlight the `Chapter 15` folder and click Select Folder. You see a new `[project root]` entry in the list of folder, which is all you need. Click Close to close the Manage Project Path dialog box.

Manage Project Path

Specify folders to add to the project path. These folders are added to the MATLAB search path when you open the project, and removed when you close the project.

Add Folder...
Add with Subfolders...
Remove
Close
Help

FIGURE 15-4: Set the path information for your project.

Adding files

Adding files to your project is easy. Just drag the file from the Current Folder pane and drop it in the right pane of your project. This project contains two files: `Factorial1.m` and `TestFactorial1.m`. Because `TestFactorial1.m` depends on `Factorial1.m`, add `Factorial1.m` first, and then drag `TestFactorial1.m` into the project.

REMEMBER

Notice the Classification column of the project. This column tells you the purpose of a particular file. The default classification is Design because MATLAB assumes that you use it most often. However, `TestFactorial1.m` isn't a design file; it's a test file. Select `TestFactorial1.m`. Drag `Classification\Test` from the Labels pane and drop it on `TestFactorial1`. The display changes, as shown in Figure 15-5. You can now sort your files by classification when needed.

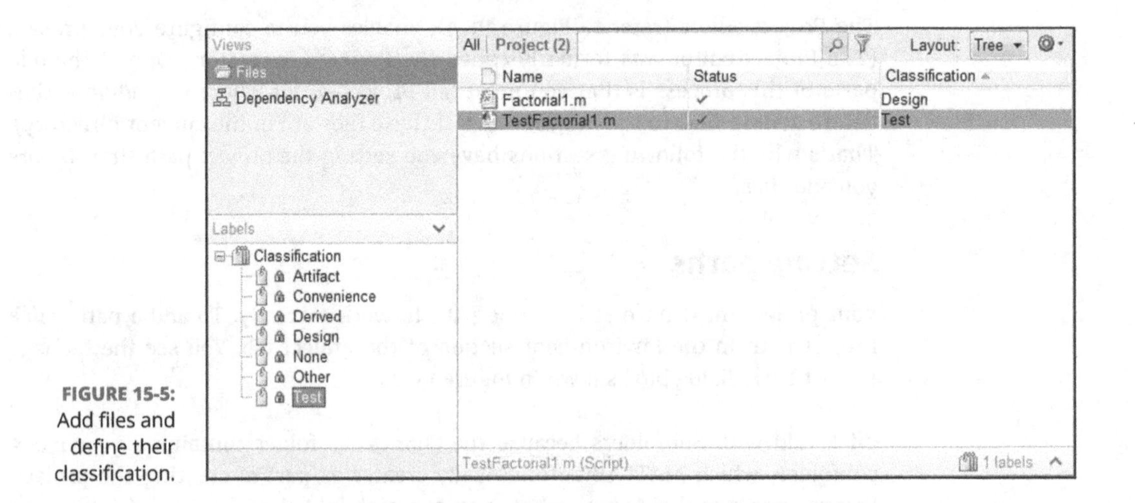

FIGURE 15-5: Add files and define their classification.

TIP

Note that you can drag and drop files from other folders. For example, try opening the `Chapter10` folder and dragging `Factorial2.m` to the right pane of the project. You see `Factorial2.m` added to the list and to the Current Folder pane. If you want to test `Factorial2.m`, you can add this code to `TestFactorial1.m`:

```
for testValue = 1:5
    Output = Factorial2(testValue);
    fprintf('%d! equals %d\n', testValue, Output);
end
```

Determining when to run your project

You use the startup and shutdown settings to ensure that your project works as anticipated. The startup files will typically perform any required configuration, so you don't have to depend on the project user doing it. The shutdown files will perform any required cleanup so that you don't end up with bits and pieces of old material hanging around. To configure your project to provide startup and shutdown support, click Startup Shutdown in the Environment section of the Project tab to display the Manage Project Startup and Shutdown dialog box, shown in Figure 15-6.

Manage Project Startup and Shutdown

Specify project files to automate startup tasks. Startup files automatically run (.m and .p files), and load (.mat files), when you open the project.

Startup files:

Add Remove

Shutdown files:

Add Remove

OK Cancel Apply

FIGURE 15-6: Select files to run to perform startup and shutdown tasks.

To add a script or Live Script file to either list, click Add. You wouldn't use other file types here because the file needs to execute by itself. When you decide that you no longer need to perform a startup or shutdown task, select the associated file in the list and click Remove.

Adding project details

At some point, you want to describe your project so that other people know what it does. To do this, click Details in the Environment section of the Project tab. You see the Project Details dialog box, shown in Figure 15-7.

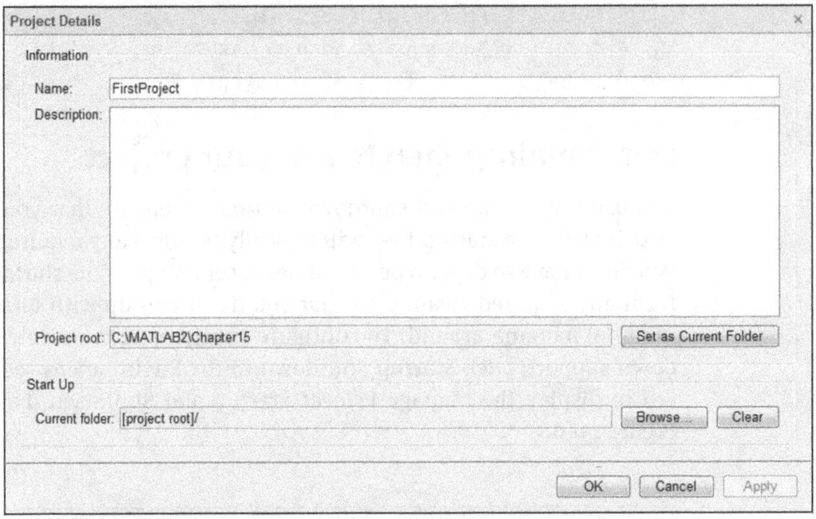

FIGURE 15-7: Describe your project so that others know what it does.

You do want to provide a new entry in the Name field, such as My First Project. A good description will help others know what your project does and why you created it. If your project requires special resources or has unique usage instructions, you might want to include them here as well.

Normally, you don't need to change the Project Root field unless you change the location of the resource files used with the project. Likewise, you normally leave the Current Folder entry alone. After you finish describing your project, click OK to save the settings and clear the Project Details dialog box.

Referencing other projects

Projects can (and should) be modular. You can create a project that does just one thing well, and use it as a standalone project. However, you might also want to use that project with other projects. Perhaps a master project calls on features of a subordinate project. In this case, you add a *project reference*, which is a pointer to another project that you use for some other purpose. To add a project reference, click References in the Environment section of the Project tab to display the Add Reference dialog box, shown in Figure 15-8.

PROGRAMMATICALLY CREATING PROJECTS

The easiest way to work with projects is to use the MATLAB GUI, as this chapter does. However, you might need to work with projects programmatically as part of a script or function. In this case, you can use the functions listed at `https://www.mathworks.com/help/matlab/projects.html`, which divide project management into these topics:

- Creating and exporting projects
- Setting projects up
- Managing project files
- Working with project objects

FIGURE 15-8: Add references to other projects as needed.

REMEMBER

This dialog box lets you create references to other projects, not other resources. If possible, use a relative reference to create flexible references to the other projects. Using absolute makes the reference brittle because now you're depending on a specific hard-drive location. You should also keep the checkpoint intact so that MATLAB automatically looks for future changes for you.

Understanding the Project Dependencies

A *project dependency* defines a relationship between two resources. For example, you can't run `TestFactorial1.m` by itself—you must have `Factorial1.m` as well. However, it's not a two-way dependency. You can run `Factorial1.m` by itself without any problem. (If you added `Factorial2.m` in the "Adding files" section, the same dependency results, but there is no dependency between `Factorial1.m` and `Factorial2.m`.) The point is that you don't want to send this project to anyone without all the required files, which can become a difficult proposition because of any of these actions:

- The project had files added to it by several people.
- Files are removed as the project changes, only to be added back later.
- The version of a file changes, as indicated by its filename.
- Some files are located in other folders.

Fortunately, MATLAB provides the means for performing automatic dependency checks. Click Dependency Analyzer in the Tools section of the Project tab to see the Dependency Analyzer window, shown in Figure 15-9. This particular display shows the effects of various changes throughout this chapter, such as adding `Factorial2.m` and modifying the project name (see the "Adding project details" section, earlier in this chapter, for more information).

FIGURE 15-9: The Dependency Analyzer shows which files are dependent on others.

If you find that your output looks incorrect, choose Analyze ⇨ Reanalyze All in the Analyze section of the Analyzer tab. The Dependency Analyzer will require some time to recheck all the dependencies, so be patient.

Clicking various elements provides you with details in the Properties pane. For example, clicking `TestFactorial1.m` will show you its location and the type of file it is, in addition to showing you the files it requires. You can also click the various arrows. For example, the arrow showing that `Factorial1.m` calls itself tells you that this script is recursive.

Running Required Checks

You may want to verify the status of your project before you send it to anyone. MATLAB provides two checks that you can perform to ensure that your project will work, as described in the following sections.

Checking project integrity

A project integrity check ensures that the project will run. It looks for issues like missing files or paths that are incorrectly set. To use this check, choose Run Checks⇨Check Project in the Tools section of the MATLAB Project tab. You see a Project Integrity Checks dialog box, like the one shown in Figure 15-10, as output.

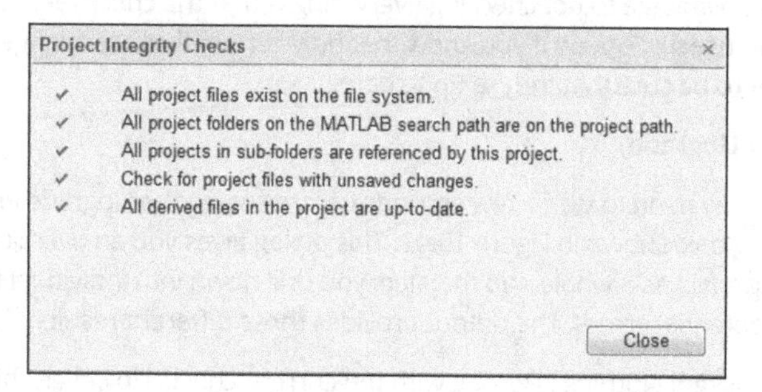

FIGURE 15-10: Locate any potential project integrity issues before sending it out.

In this case, the project passes with flying colors. However, if there were issues with the project, such as dependency files being outdated, MATLAB would provide you with details on what went wrong.

Looking for potential updates

If you want your project to run on other systems, it has to be compatible with those other systems. The best way to achieve this goal is to verify that all files use current calls and that the underlying files are up to date. The following steps help you verify that your project is current in every possible way to reduce the risk of compatibility issues with other systems.

1. **Choose Run Checks ⇨ Upgrade Project in the Tools section of the Project tab.**

   You see the Upgrade Project dialog box, shown in Figure 15-11. This dialog box tells you how many files MATLAB will check and what checks it will perform. The default is to check everything.

Upgrade Project
Check for compatibility issues in your project.
Check: Everything (3 files)
Change Options
Upgrade Cancel

FIGURE 15-11: Define the parameters of the checks you want to perform.

   Clicking the Change Options link displays the Upgrade Options dialog box, which allows you to control which files are checked and the kind of checks performed on them. However, you normally want to check the entire project; otherwise, you can't be sure that you won't encounter compatibility issues. The only advantage to not checking everything is that the check takes less time. You could use this option if you know that only certain files are changed and you want to be sure that they're up to date.

2. **Click Upgrade.**

   You may need to wait a few seconds before seeing the Upgrade Project Report dialog box, shown in Figure 15-12. This dialog gives you an overall report on the project as a whole and then lets you drill down into individual files to look for potential errors. The output provides three different results:

   - **Passed:** Nothing is wrong with the current check. However, this entry may include future issue warnings. For example, Figure 15-12 shows that `Factorial1.m` uses some calls that MATLAB will remove in the future. Highlight the entry and click the Learn More link to discover whether a fix for the issue exists.
   - **Passed with Fixes:** There was a problem with the current check, but MATLAB was able to provide an automatic fix for you. The problem with this approach is that you can't always be certain that MATLAB fixed the problem in the same manner as you would, so you need to verify the fix.

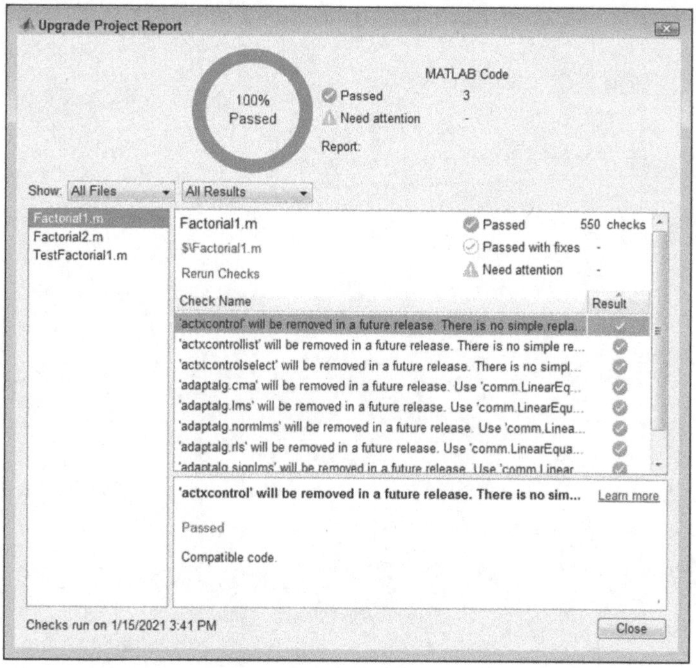

FIGURE 15-12: Generate a report telling you whether the project is up to date.

- **Need Attention:** You must manually fix the problem found by MATLAB. In this case, you click the file link shown in the display, such as `$\Factorial1.m` in Figure 15-12, to open the file in the editor and perform the correction. (Note that you may need to close the Upgrade Project Report dialog to use the editor.)

3. **Select a file to fix; then click its link in the display.**

   MATLAB opens the file for you in the editor or Live Editor.

4. **Perform any needed fixes or check any automatic updates; then click Rerun Checks in the Upgrade Project Report dialog, or click Run Checks ⇨ Upgrade Project in the Tools section of the MATLAB Project tab.**

   MATLAB performs the required checks on the selected file. If the changes you make are successful, the file will pass and you can move on to the next file in the list.

5. **Perform Steps 3 and 4 for each file in the project until your overall score is 100% Passed.**

6. **Click Close.**

   The Upgrade Project Report will verify that you want to close the dialog box before closing.

4

Employing Advanced MATLAB Techniques

IN THIS PART . . .

Importing and exporting data from various sources

Presenting data as reports

Outputting data in printed form

Detecting and handling errors

IN THIS CHAPTER

» Obtaining data from other sources

» Outputting data to other recipients

» Interacting with image data

Chapter 16
Importing and Exporting Data

An application isn't of much use if it can't interact with data — in the case of MATLAB, the formulas, scripts, Live Scripts, functions, Live Functions, and plots that you create interact with variables. In fact, even applications that you might think have nothing to do with data manage quite a lot of it. For example, you might be tempted to think that games don't work with data, but even the lowliest Solitaire game saves statistics, which means that it interacts with data. So you can easily see that most applications interact with at least their own data.

Larger, more complex applications, such as MATLAB, also need some method of interacting with data from other applications. For example, you may need to use Excel data from a colleague to perform a calculation. If MATLAB didn't provide a means to access that data, to *import* it into MATLAB, you couldn't use it to perform the calculation.

After you complete the calculation, you may need to send it back to your colleague, but the only application available at the other site is Excel. Now you must *export* the data from MATLAB into an Excel file that your colleague can use. An Excel data file will help your colleague a lot more than printouts you could send instead because the data is directly accessible.

Plain data — text and numbers — is one thing, but importing and exporting images is quite another. Images are complex because they present graphics — a visual medium — as a series of 0s and 1s (raster graphics) or in terms of points on a Cartesian plane (vector graphics). In addition, some image formats have quirks, such as data compression, that make them hard to work with. This chapter provides a deeper look into working with image files of various sorts.

TIP

Data import and export are among the few activities for which many people find using the GUI easier than typing commands. (The choice you make depends on the complexity of the data and just what you want to achieve by importing or exporting it.) Yes, you can type commands to perform the tasks, and you can add import and export commands to your applications, but importing or exporting complex data manually is often easier using the GUI. This chapter focuses on working with commands. However, you can see how to use the GUI in the "Importing" and "Exporting" sections of Chapter 4. These sections also discuss issues such as which file formats MATLAB supports.

REMEMBER

You don't have to type the source code for this chapter manually. In fact, using the downloadable source is a lot easier. You can find the source for this chapter in the `\MATLAB2\Chapter16` folder of the downloadable source. When using the downloadable source, you may not be able to work through some of the hands-on examples because the example files will already exist on your system. In this case, you can create an alternative folder, `Chapter16a`, to hold the results of the hands on exercises. See the Introduction for details on how to obtain the downloadable source.

Importing Data

For most people, importing data from various sources is almost a daily chore because our world is based on interconnectivity. Having as much data as possible to perform a task is critical if you want to obtain good results. That's why knowing just how to get the data into MATLAB is so important. It's not just a matter of getting the data, but getting it in such a manner that it can be truly useful. In addition, the import process can't damage the data in any way; otherwise it could become useless to you in the long run.

REMEMBER

A lot of people get the whole business of the importing and exporting of data confused. *Importing* data always involves taking outside information — something generated externally — and bringing it into a host application, such as MATLAB (as contrasted to *exporting*, which sends information from MATLAB to an external target). So, when someone sends you a file with Excel data, you must import it into MATLAB in order to use it.

The following sections describe the essentials for importing data into MATLAB from various sources.

Avoiding older import/export function calls

You find code online all the time that talks about using particular functions to perform import and export tasks, but these functions are outdated. The code still might work in your version of MATLAB, but it likely won't work very long. Consequently, you need to verify functions if you have any doubt about their reliability and longevity (such as code found on older sites). The following list shows common older function calls and their replacements:

Older Function	Replacement
`csvread()`	`readmatrix()`
`csvwrite()`	`writematrix()`
`dlmread()`	`readmatix()`
`dlmwrite()`	`writematrix()`
`xlsfinfo()`	`sheetnames()`
`xlsread()`	`readtable()`, `readmatrix()`, or `readcell()`
`xlswrite()`	`writetable()`, `writematrix()`, or `writecell()`

Performing import basics

A basic import uses all the default settings, which works fine for many kinds of data. MATLAB can determine the correct data format relatively often.

An essential part of importing data is to use the correct import function. Each import function has features that make it more suitable to a particular kind of data. Here are some of the text-specific import functions and how they differ:

- `readmatrix()`: Can import numbers, strings, and other data formats, unlike `csvread()` and `dlmread()`, which MathWorks no longer recommends and which don't appear in this chapter because they were data type-specific. You may encounter these functions in online examples and need to avoid the examples if you do. This function automatically detects the file import parameters and works with these file types: `.txt`, `.dat`, or `.csv` for delimited text files, and `.xls`, `.xlsb`, `.xlsm`, `.xlsx`, `.xltm`, `.xltx`, or `.ods` for spreadsheet files.

- `textscan()`: Can import both numbers and strings. You must provide a format specification to read the data correctly.
- `readtable()`: Can import both numbers and strings. The output from this function is always a table, even when the source doesn't contain tabular data. This function supports the same file formats as `readmatrix()`.

The output you receive depends on the function you use. For example, when working with `readtable()`, you actually get a table as output, not a matrix or a cell array. On the other hand, using `readmatrix()` results in a matrix as output. There are ways to obtain the kind of output you want, but you need to understand that you start with a specific kind of output data from these functions.

The examples found in the sections that follow each use a different method of reading the data from the disk. However, they all use the same data so that you can compare the results. Here's the data found in the `NumericData.csv` file supplied with the downloadable source code:

```
15,25,30
18,29,33
21,35,41
```

Using readmatrix()

Using `readmatrix()` enables you to read a Comma-Separated Value (CSV) or spreadsheet file without a problem. For this example, all you do is type **CSVOutput = readmatrix('NumericData.csv')** and press Enter. The output is a matrix that contains the following results:

```
CSVOutput =
    15    25    30
    18    29    33
    21    35    41
```

TIP

Each of the file formats that `readmatrix()` supports offers options that you can add as a second argument. These options appear as part of the objects listed here for the particular file type:

- **Fixed-width text files:** `FixedWidthImportOptions` (see `https://www.mathworks.com/help/matlab/ref/matlab.io.text.fixedwidthimportoptions.html` for details)
- **Spreadsheet files:** `SpreadsheetImportOptions` (see `https://www.mathworks.com/help/matlab/ref/matlab.io.spreadsheet.spreadsheetimportoptions.html` for details)

» **Text files:** `DelimitedTextImportOptions` (see `https://www.mathworks.com/help/matlab/ref/matlab.io.text.delimitedtextimportoptions.html` for details)

The options are remarkably complete, and if you don't supply options of your own, MATLAB supplies them for you. To see the default options, simply type the name of the option object, such as **delimitedTextImportOptions**, and press Enter. Figure 16-1 shows typical output.

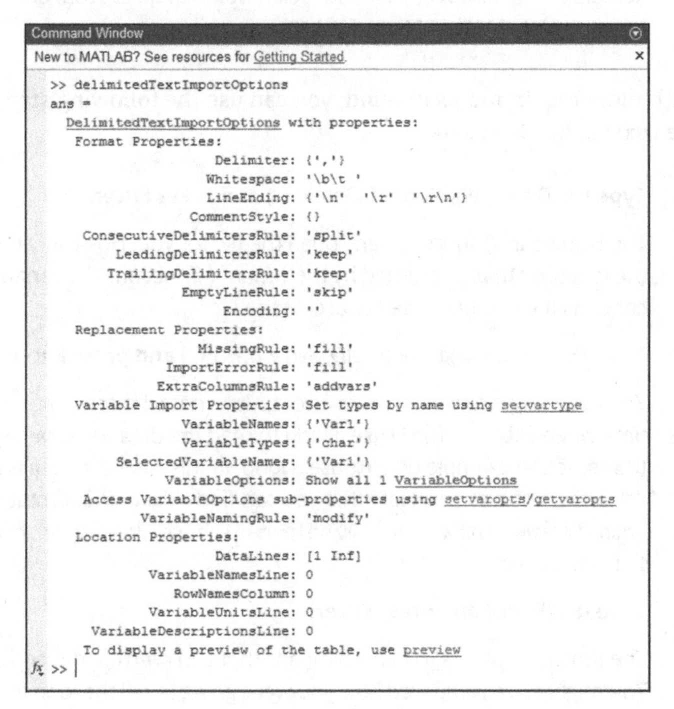

FIGURE 16-1: The list of options for a `readmatrix()` call are remarkably complete.

To create your own list of options, type **opts = delimitedTextImportOptions;** and press Enter; then modify individual property values. For example, if the file has the variable names in the first line, then you type **opts.VariableNamesLine = 1** and press Enter.

Using textscan()

The `textscan()` function can read both strings and numbers in the same data set. In addition, it can read some types of data that `readmatrix()` might encounter difficulty with, such as binary numbers. However, you must define a format specification to use this function. Also, you can't simply open the file and work with it without knowing the following:

- Whether the file is encoded in some way (perhaps using UTF-7 bit, see a discussion of UTF-8 versus UTF-7 at `https://www.techwalla.com/articles/utf-7-vs-utf-8`)
- Whether the file contains special control characters (some delimited files use a vertical tab for columns and form feeds for new lines as described at `https://community.denodo.com/answers/question/details?questionId=9060g000000L6vVAAS`)

With these requirements in mind, you can use the following steps to help you use the `textscan()` function.

1. **Type** FileID = fopen('NumericData.csv') **and press Enter.**

   The `textscan()` function can't open the file for you. However, it does accept the identifier that is returned by the `fopen()` function. The variable, `FileID`, contains the identifier used to access the file.

2. **Type** TSOutput = textscan(FileID, '%d,%d,%d/n') **and press Enter.**

   You get a single row of the data as output — not all three rows. So, this is a time when you'd normally use a loop to read the data. However, there is more to see, so the example doesn't use a loop. In this case, the data is read into a cell array, not a matrix. Note that you can find a list of the format specifiers at `https://www.mathworks.com/help/matlab/ref/textscan.html#btghhyz-1-formatSpec`.

3. **Type** feof(FileID) **and press Enter.**

   The function outputs a 0, which means that you aren't at the end of the file yet. You might have wondered how you were going to tell the loop to stop reading the file. A simple test using the `feof()` function takes care of that problem.

4. **Type** TSOutput = [TSOutput; textscan(FileID, '%f,%f,%f/n')] **and press Enter.**

   You now see the second row of data read in. However, look at the format specification. These numbers are read as floating-point values rather than integers. Using `textscan()` gives you nearly absolute control over the appearance of the data in your application.

5. **Type** isinteger(TSOutput{1,1}) **and press Enter.**

   The output value of 1 tells you that the element at row 1, column 1 is indeed an integer.

6. **Type** isinteger(TSOutput{2,1}) **and press Enter.**

   This step verifies that the element at row 2, column 1 isn't an integer because the output value is 0. It pays to ensure that the data you have in MATLAB is the type you actually expected.

7. **Type** TSOutput = [TSOutput; textscan(FileID, '%2s,%2s,%2s/n')] **and press Enter.**

REMEMBER

   This time, the data is read in as individual strings. However, notice that the format specification includes a field width value. If you had simply told `textscan()` to read strings, it would have read the entire row as a single string into one cell.

8. **Type** textscan(FileID, '%d,%d,%d/n') **and press Enter.**

   This read should take you past the end of the file. The output is going to contain blank cells because nothing is left to read.

9. **Type** feof(FileID) **and press Enter.**

   This time, the output value is 1, which means that you are indeed at the end of the file.

10. **Type** fclose(FileID) **and press Enter.**

    MATLAB closes the file.

WARNING

    Failure to close a file can cause memory leaks and all sorts of other problems. Not closing the file could quite possibly cause data loss, access problems, or a system crash. The point is that you really don't want to leave a file open after you're done using it.

Now that you have a better idea of how a `textscan()` should work, it's time to see an application that uses it. Listing 16-1 shows how you might implement the preceding procedure as a function. You can also find this function in the `UseTextscan.m` file supplied with the downloadable source code.

LISTING 16-1: Using textscan() in an Application

```
function [ ] = UseTextscan( )
%UseTextscan: A demonstration of the textscan() function
%   This example shows how to use textscan() to scan
%   the NumericData.csv file.

    FileID = fopen('NumericData.csv');
    TSOutput = textscan(FileID, '%d,%d,%d/n');
    while not(feof(FileID))
```

(continued)

LISTING 16-1: (continued)

```
        TempData = textscan(FileID, '%d,%d,%d/n');
        if feof(FileID)
            break;
        end
        TSOutput = [TSOutput; TempData];
    end
    disp(TSOutput);
    fclose(FileID);
end
```

You have already used most of this code as you worked through the exercise, but now you see it all put together. Notice that you must verify that you haven't actually reached the end of the file before adding the data in `TempData` to `TSOutput`. Otherwise, you end up with the blank row that `textscan()` obtains during the last read of the file.

Using readtable()

The `readtable()` function is a lot easier to use than `textscan()`, but it also has a few quirks, such as its assumption that the first row of data is actually column names. To use `readtable()` with the `NumericData.csv` file, type **RTOutput = readtable('NumericData.csv', 'ReadVariableNames', false)** and press Enter. You see the following output:

```
RTOutput =
  3x3 table
    Var1    Var2    Var3
    ____    ____    ____
     15      25      30
     18      29      33
     21      35      41
```

The output actually is a table rather than a matrix or cell array. The columns have names attached to them, as shown in Figure 16-2. As a consequence, you can access individual members using the variable name, such as `RTOutput{1, 'Var1'}`, which outputs a value of 15 in this case.

Notice that `readtable()` accepts property name and value pairs as input. In this case, `'ReadVariableNames'` is a property. Setting this property to `false` means that `readtable()` won't read the first row as a heading of variable names. You use `readtable()` where the output file does contain variable names, because having them makes accessing the data easier in many situations.

Variables - RTOutput

RTOutput

3x3 table

	1 Var1	2 Var2	3 Var3
1	15	25	30
2	18	29	33
3	21	35	41
4			

FIGURE 16-2: Tables provide names for each of the columns.

Importing mixed strings and numbers

Life isn't all about numbers. In some situations, you need to work with a mix of strings and numbers. Each function has its own particular capability. For example, `readmatrix()` makes it easy to specify the output type of the data with some independence over the data file (you can read an integer as a float or a string, as an example; see the earlier "Using readmatrix()" section for details). Likewise, `textscan()` provides absolute control over how the data is converted (as described in the "Using textscan()" section, earlier in this chapter).

The `readtable()` function is designed more for work with database output, for which the output file likely has header names. The database could reside in a Database Management System (DBMS) or as part of a spreadsheet. The source of the data doesn't matter — only the format does. For this example, you have an output file that contains both row and column headings, as shown here (you can also find this data in the `MixedData.csv` file supplied with the downloadable source code):

```
ID,Name,Age,Married
1234,Sam,42,TRUE
2345,Sally,35,TRUE
3456,Angie,22,FALSE
4567,Dan,55,FALSE
```

The first row isn't part of the data used for analysis because it contains the row headers. The `readtable()` function has features to handle extras like row headers. You can see these features described at `https://www.mathworks.com/help/matlab/ref/readtable.html`, which also demonstrates a few in the example.

To see `readtable()` in action with the `MixedData.csv` file, type **MixedData = readtable('MixedData.csv')** and press Enter. You see the following output:

```
MixedData =
  4×4 table
```

```
ID         Name        Age      Married
----     ---------     ---     ---------
1234     {'Sam'   }     42     {'TRUE' }
2345     {'Sally'}      35     {'TRUE' }
3456     {'Angie'}      22     {'FALSE'}
4567     {'Dan'   }     55     {'FALSE'}
```

Notice that the columns have the appropriate names and that each row has the expected identifier (as the `ID` column). However, the `Married` column is of the wrong type (it's currently a string). Follow these steps to fix this problem:

1. **Type** opts = detectImportOptions('MixedData.csv') **and press Enter.**

   You see the default options that MATLAB will use to read the data from the file. More important, `opts` now contains those options.

2. **Type** getvaropts(opts, 'Married') **and press Enter.**

   This function outputs the specific options for the `Married` column, which includes a Type field value of `'char'`. By default, `readtable()` converts logical values to `'char'` equivalents.

3. **Type** opts = setvartype(opts, 'Married', 'logical') **and press Enter.**

   The `Married` column will now appear as a logical value when you use `opts` with `readtable()`.

Now when you type **MixedData = readtable('MixedData.csv', opts, 'ReadRow Names', true)** and press Enter, `MixedData` contains `Married` as a logical value. In addition, by adding the `'ReadRowNames'` argument, the table in memory contains an index value as shown here:

```
              ID            Name        Age     Married
           --------      ---------      ---     -------
1234       {'1234'}      {'Sam'   }     42       true
2345       {'2345'}      {'Sally'}      35       true
3456       {'3456'}      {'Angie'}      22       false
4567       {'4567'}      {'Dan'   }     55       false
```

You can modify the formats of any other import columns using the same technique. For example, notice that `ID` is now a string column, but you could change it back to a numeric column should you wish to do so. The index column, which contains the row names, always appears in the leftmost output column. The table has a few interesting features. For example, type **MixedData('1234', 'Age')** and press Enter. You see the following output:

```
ans =
  table
                Age
                ---
    1234        42
```

The output is actually a table that contains just the value you want. Notice the use of parentheses for the index. Using identifiable names rather than numeric indexes is also quite nice.

However, you can access the information as actual data rather than as a table. Type **MixedData{'1234', 'Age'}** and press Enter. In this case, you obtain a simple output of 42. The use of curly braces means that you get a data value rather than a table as output. It is still possible to use numeric indexes if you want. Type **MixedData{1, 2}** and press Enter. You obtain the same output value of 42 as you did before.

If you use versions of MATLAB before R2020a, you might encounter issues when the delimiters or other formatting used in a file was other than expected. For example, the use of semicolons, rather than commas, between fields could cause problems. Newer versions of MATLAB, such as the one used for this book, automatically work well with all sorts of file formatting, so you may need to remove some of the old properties, such as `'Delimiter'`, from your code to keep it up to date.

Importing selected rows or columns

Sometimes you don't need an entire file, only certain rows and columns of it. All the functions described earlier in the chapter provide some means of selecting specific information, but the `readmatrix()` function supplies a straightforward example of how to perform this task. Use these steps to see just a range of data:

1. **Type** opts = detectImportOptions('NumericData.csv'); **and press Enter.**

   The `opts` variable now contains the options used to import data, which includes importing the entire file.

2. **Type** preview('NumericData.csv', opts) **and press Enter.**

   The `preview()` function lets you obtain a quick view (up to eight rows) of what the various options will do for you. It isn't meant to give you a full view of the import, and using it can save you time as you experiment. In this case, you see the following output:

   ```
   ans =
     3×3 table
   ```

```
Var1    Var2    Var3
----    ----    ----
 15      25      30
 18      29      33
 21      35      41
```

3. **Type** opts.SelectedVariableNames = [1:2]; **and press Enter.**

   This command selects Var1 and Var2 for the output, but excludes Var3.

4. **Type** opts.DataLines = [1:2]; **and press Enter.**

   This command selects the first two rows of data (1:2 means rows 1 through 2).

5. **Type** CSVOutput = readmatrix('NumericData.csv', opts) **and press Enter.**

   You see the following output:

```
CSVOutput =
    15    25
    18    29
```

TIP

If you want to keep the variable names, you use `readtable()` instead. Oddly enough, you can use the same `opts` variable, so you can play around with the data with greater ease than ever before.

Exporting Data

After you perform the calculations that you want to perform, you often need to put them in a form that others can use. However, not everyone has a copy of MATLAB on his or her computer, so you need to export the MATLAB data in some other form. Fortunately, getting the data out in a usable form is actually easier than importing it. The following sections describe how to export data, scripts, and functions.

Performing export basics

Importing data often focuses on getting the right results. For example, you might use `textscan()` or `readtable()` on a comma-separated value file, even though a perfectly usable `readmatrix()` function exists to perform the task. The goal is to get the data from the `.csv` file into MATLAB and preserve both the content and layout of the original information. However, now that you have data inside MATLAB and want to export it, the goal is to ensure that the resulting file is standardized so that the recipient has minimal problems using it. Therefore, you'd use `writetable()` only if the recipient really did require a custom format rather than a standard `.csv` file, or if the MATLAB data was such that you had to use something other than `writematrix()`. Because of the difference in emphasis, the following sections of the chapter focus on standardized export techniques.

Working with matrices and numeric data

Before you can do anything with exporting, you need data to export. Type **ExportMe = [1, 2, 3; 4, 5, 6; 7. 8, 9]** and press Enter. You see the following result:

```
ExportMe =
     1     2     3
     4     5     6
     7     8     9
```

The result is a matrix of three rows and three columns. Exporting matrices is simple because the majority of the functions accept a matrix as a default. To see how exporting matrices works, type **writematrix(ExportMe, 'ExportedData1.csv')** and press Enter. MATLAB creates the new file, and you see it appear in the Current Folder window. When you open the file, you see something like the output shown in Figure 16-3. (What you see precisely will vary, depending on the application you use to view `.csv` files.)

FIGURE 16-3: Matrices provide the easiest source for export.

TIP

Not all MATLAB data comes in a convenient matrix. When you use `writematrix()`, you must supply a matrix. To get a matrix, you may have to convert the data from the existing format to a matrix using a conversion function. For example, when the data appears as a cell array, you can use the `cell2mat()` function to convert it. However, some conversions aren't so straightforward. For example, when you have a table as input, you need to perform a two-step process:

1. Use the `table2cell()` function to turn the table into a cell array.
2. Use the `cell2mat()` function to turn the cell array into a matrix.

Fortunately, you can skip some conversions if you want. For example, instead of converting your cell array into a matrix, you can output it using `writecell()`.

Working with mixed data

Exporting simple numeric data is straightforward because you have a number of functions to choose from that create the correct formats directly. The problem comes when you have a cell array or other data form that doesn't precisely match the expected input for one of the output functions. To see how mixed data works, start by typing **MyCellArray = {'Andria', 42, true; 'Michael', 23, false; 'Zarah', 61, false}** and pressing Enter. You see the following result:

```
MyCellArray =
  3×3 cell array
    {'Andria' }    {[42]}    {[1]}
    {'Michael'}    {[23]}    {[0]}
    {'Zarah'  }    {[61]}    {[0]}
```

The easiest way to export this data to a CSV file is to type **writecell(MyCellArray, 'ExportedData2.csv')** and press Enter. Note that `writecell()` converts the third column from logical to text in the process.

REMEMBER

If you want to preserve the data format, you must use some other export format, such as Excel. In this case, you type **writecell(MyCellArray, 'ExportedData2.xls')** and press Enter. Figure 16-4 shows that the exported data now appears in the correct types. The point is that MATLAB preserves the data format only if you use the correct output file format.

ExportedData2.xls [Compatibi...

	A	B	C	D
1	Andria	42	TRUE	
2	Michael	23	FALSE	
3	Zarah	61	FALSE	
4				
5				
6				

Sheet1

FIGURE 16-4: Using the correct export format makes it possible to preserve the data format.

Using a nontext file format can have other advantages as well. For example, when working with the Excel format, you gain access to the ability to store different data on different sheets. You can also specify a range in which to store the data. The documentation at `https://www.mathworks.com/help/matlab/ref/writecell.html` provides additional insights into the properties you can use to modify how MATLAB exports your data.

Exporting scripts and functions

To export scripts and functions, you must actually publish them using the `publish()` function. MATLAB supports a number of output formats for this purpose. For example, if you want to publish the `UseTextscan()` function, which appears earlier in the chapter in HTML format, you type **publish('UseTextscan.m', 'html')** and press Enter. MATLAB provides the following output:

```
ans =
    'C:\MATLAB2\Chapter16\html\UseTextscan.html'
```

The actual location varies by system, but you also obtain the location of the published file. Notice that MATLAB places the published file in an `html` subdirectory. Figure 16-5 shows typical output. (What you see may differ based on your platform and the browser that you use.)

Publishing is a much larger topic than can fit in a single section of a chapter. Chapter 17 discusses publishing in considerably greater detail.

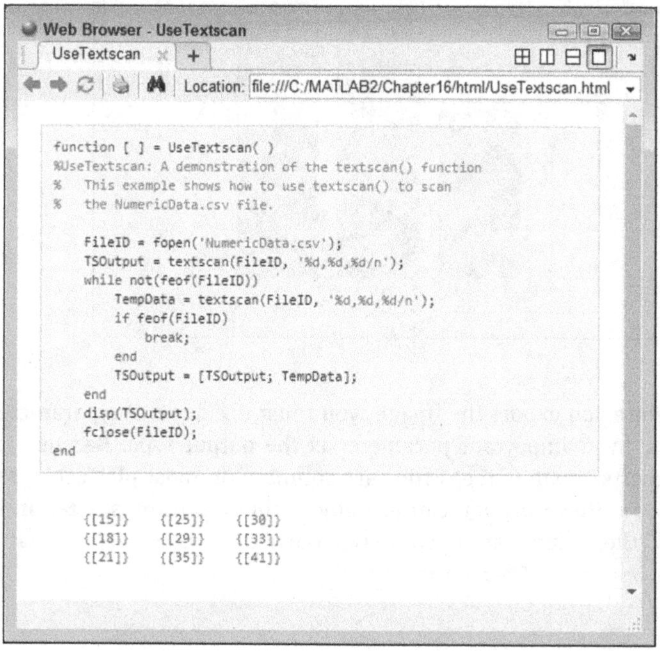

FIGURE 16-5: To export scripts and functions, you must publish them.

Working with Images

Images are more complex than text files because they use binary data that isn't easy for humans to understand, and the format of that data is intricate. Small, hard-to-diagnose errors can cause the entire image to fail. However, the process of exporting and importing images is relatively straightforward, as described in the following sections.

Exporting images

Before you can export an image, you need an image to export. The "Using the bar( ) function to obtain a flat 3D plot" section of Chapter 7 describes how to create the 3D bar graph shown in Figure 16-6. (You can also create the bar chart using the `CreateBarChart.m` script found in the downloadable source code.) This is the image used for the remainder of this chapter.

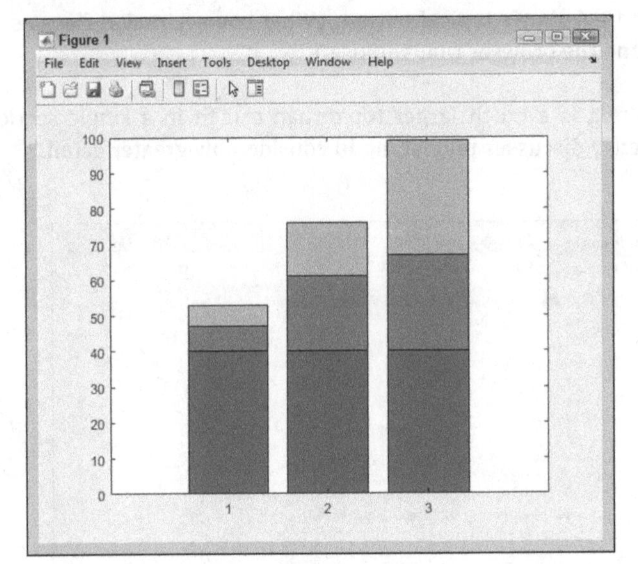

FIGURE 16-6: Create an image to export from MATLAB.

Before you export the image, you must decide on the parameters for the output. The most important parameter is the output type. Because Joint Photographic Experts Group (`.jpeg`) files are common on most platforms, the example uses a `.jpeg`. However, you can use any of the file formats listed in the Image section of the chart at `http://www.mathworks.com/help/matlab/import_export/supported-file-formats.html`.

After you decide on an export format, you can use the saveas() function to perform the task. In this case, you type **saveas(gcf(), 'Bar1.jpeg', 'jpg')** and press Enter. MATLAB exports the figure using whatever resolution is currently set for the figure. Remember that the gcf() function obtains the handle for the current figure. Figure 16-7 shows the plot in Windows Photo Viewer as Bar1.jpeg.

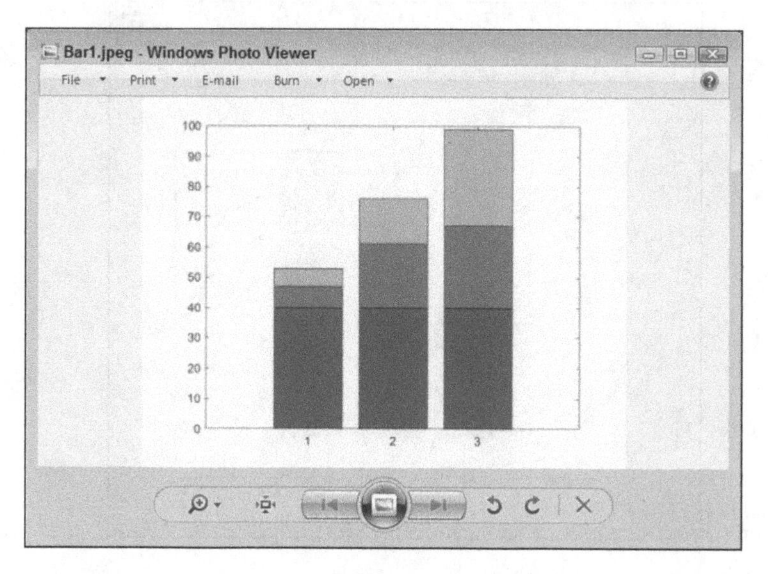

FIGURE 16-7: The output image shows the plot as Windows Photo Viewer sees it.

TIP

Use the saveas() function to save MATLAB objects, such as plots. However, when working with actual images, use the imwrite() function instead. The imwrite() function works essentially the same way that saveas() does, but it works directly with image files.

Importing images

MATLAB can also work with images that you import from other sources. The basic method of importing an image is to use imread(). For example, to import Bar1.jpeg, you type **ImportedImage = imread('Bar1.jpeg')** and press Enter. What you see as output is a matrix that has the same dimensions as the image. If the image has a resolution of 900 x 1200, the matrix will also be 900 x 1200. However, a third dimension is involved — the color depth. It takes red, green, and blue values to create a color image. So the resulting matrix is actually 900 x 1200 x 3.

REMEMBER

This is one of those situations in which the semicolon is absolutely essential at the end of the command. Otherwise, you may as well go get a cup of coffee as you wait for the numbers to scroll by. If you accidentally issue the command without the semicolon, you can always stop it by pressing Ctrl+C.

To display your image, you use the `image()` function. For example, to display the image you just imported, you type **image(ImportedImage)** and press Enter. Figure 16-8 shows the result of importing the image. You see the original plot, but in image form.

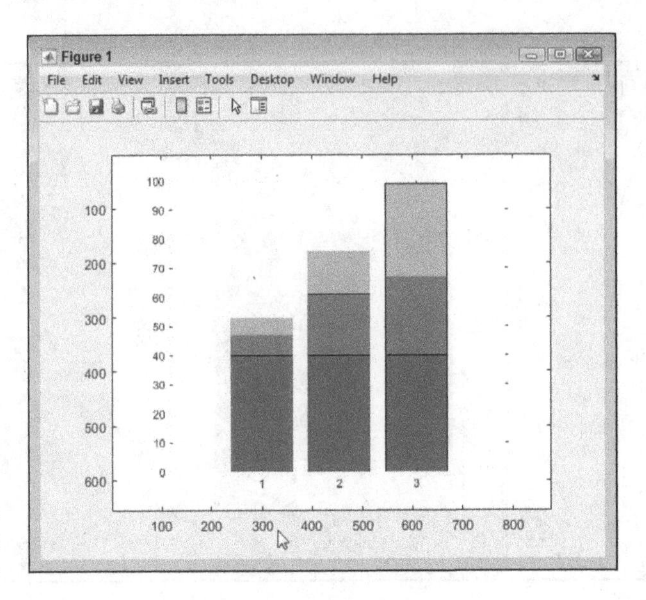

FIGURE 16-8: Seeing the plot in image form.

IN THIS CHAPTER

» Modifying the appearance of text

» Creating a MATLAB data publication

» Sending your data to the printer

Chapter 17 Printing and Publishing Your Work

After you have labored long to create the next engineering marvel or produce the next scientific breakthrough, you really do want others to know about it. Most people want to shout about their achievements from the mountaintops, but this chapter tells you only how to publish and print them. *Publishing* is electronic output of your data; *printing* is the physical output of your data. Whether you publish or print your data, it becomes available in a form that other people can use, even if they don't own a copy of MATLAB.

Part of the publishing and printing process is to make the information aesthetically appealing. It doesn't matter how impressive the data is if no one can read it. Formatting information is part of the documentation process because people need cues about the purpose of the text they're seeing. To that end, this chapter helps you create nicely formatted electronic publications and printed documents.

You don't have to type the source code for this chapter manually. In fact, using the downloadable source is a lot easier. You can find the source for this chapter in the `\MATLAB2\Chapter17` folder of the downloadable source. When using the downloadable source, you may not be able to work through some of the hands on examples because the example files will already exist on your system. In this case, you can create an alternative folder, `Chapter17a`, to hold the results of the hands on exercises.

Using Commands to Format Text

Presentation is a large part of how people receive and understand the material you provide. The same information, presented in two different ways, will elicit different responses from the audience. Little things, such as adding bold type to certain elements, can make a subtle but important difference in how the information is grasped by the audience.

REMEMBER

This book isn't about desktop publishing or creating professional-level presentations. In the following sections, you discover some ideas on how to dress up the appearance of your data so that an audience can see the data in a certain light. You can use various text effects to add emphasis where needed or to point out a particular element of interest. You can also use special symbols to add explanatory text so that the data is more easily understood.

Modifying font appearance

The fonts you use or misuse can say a lot about your presentation. Unfortunately, fonts are more often misused than used effectively to convey ideas. Here are the four ways in which you can modify the appearance of fonts in MATLAB to good effect:

- **Bold:** Adds emphasis so that the viewer sees the affected text as being more important than the text around it. However, bold is also used for headings to provide separation between elements. Always think of bold as grabbing the attention of the viewer in some way.
- **Monospace:** Creates an environment in which text elements line up, as when using a typewriter. Most people use monospace for code, or for numeric data presented free-form as a table, because monospace helps present a neat appearance. The viewer sees the data rather than being distracted by the data's lack of alignment.
- **Italic:** Defines elements that are special in some way but don't require emphasis. For example, if you define a term inline with the remainder of your data, the term should appear in italics to cue the viewer that it is explained in the material that follows. The point of italic type is to give your viewer cues about added material rather than to add emphasis.
- **Underline:** Provides pointers to additional resources or other external information (such as the URLs found in Web pages). Some people use an underline for emphasis as well, but this is actually a misuse of an underline because you already have bold to emphasize something. Combining bold and underline is even worse because the recipient can perceive it as shouting. Use underline as a means of pointing to other data that the viewer should know about but that doesn't appear in your presentation.

LISTING THE AVAILABLE FONTS

At some point, you need to know how to obtain a list of fonts available on the local system. To do so, you use the `listfonts()` function. To see how this function works, type **Fonts = listfonts();** and press Enter. The `Fonts` variable receives a sorted list of fonts on the current system, which you can then search for appropriate values. To find a specific font, such as Arial, type **Found = find(strcmp(Fonts, 'Arial'));** and press Enter. If Found isn't empty after the call (in other words, `length(Found)` returns a value greater than `0`), the system supports the font you want to use.

Some elements can have special fonts in MATLAB. To discover the identity of the special font, you call `listfonts()` with the handle of the element. For example, a user interface element may have a special font. If you obtain a handle to that user interface element and pass it to the `listfonts()` function, you receive not only a sorted list of the system fonts but also the identity of the special font used in the user interface element.

REMEMBER

You can probably find other opinions as to how to use various font styles, but this is the approach used by many (if not most) technical documents, so your viewer will already know the rules. Keeping things familiar and easy to understand will help the viewer focus on your data rather than on the fonts and styles you use. Always make the data king of the presentation and leave the beautiful text to the artists of the world.

Now that you have some idea of how to use the font styles, the following sections demonstrate how to add them to your presentation. The screenshots in each section build on the section before it so that you can see all the effects in action. You don't have to work through the sections in any particular order.

Bold

The use of emphasis, normally associated with bold type, can make data stand out. However, in MATLAB, the term *bold* actually refers to font weight. The strength of the font you use provides a level of emphasis. In fact, you can set a font to four different levels of emphasis:

- » Light
- » Normal (default)
- » Demi
- » Bold

REMEMBER

Not every font that you have installed on your system supports all four levels of emphasis, but at least some do. You may try to achieve a certain level of emphasis with a font and find your efforts thwarted. In many cases, it has nothing to do with your code and everything to do with the font you're using. To see what levels of support your font has, you can open the Fonts window in the Windows Control Panel to see the list of fonts that your system supports. Double-click the font you want to check, such as Arial (shown in Figure 17-1, your view may vary depending on your version of Windows), and you see that Arial doesn't support a light or demi font version. However, you could use this arrangement if you want to mimic having all four levels of emphasis:

- **Light:** Arial Narrow
- **Normal:** Arial Regular
- **Demi:** Arial Bold
- **Bold:** Arial Black

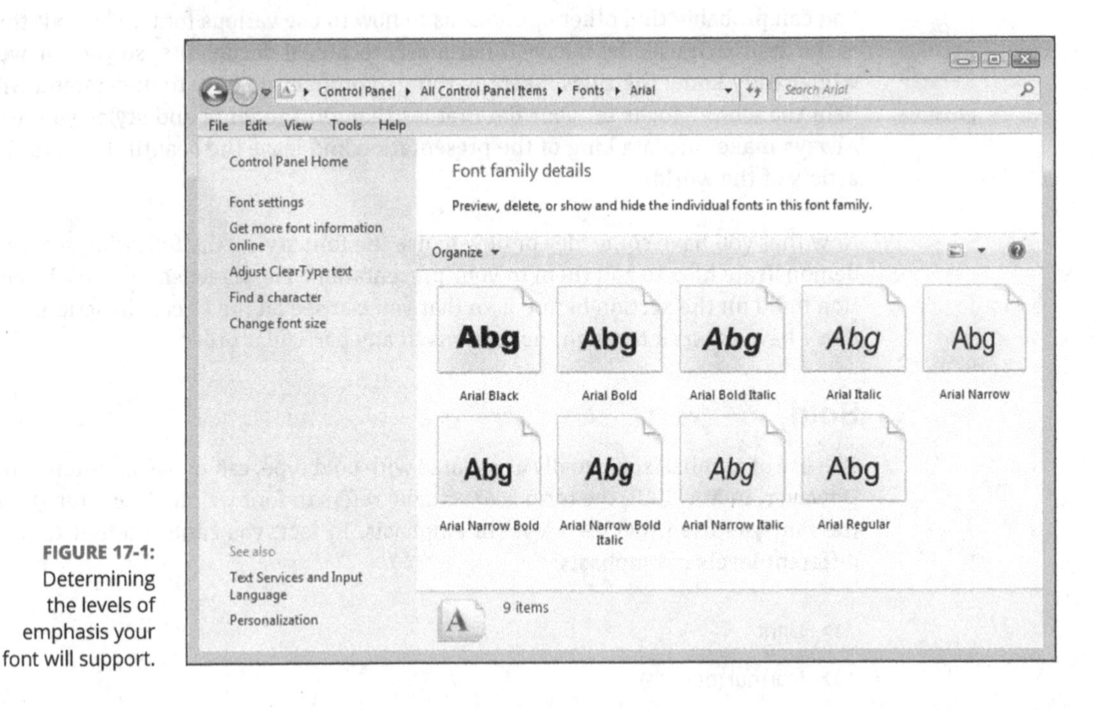

FIGURE 17-1: Determining the levels of emphasis your font will support.

With these font limitations in mind, the following steps help you see the varying levels of emphasis that you can achieve using MATLAB. (You can find all of the steps for this and the following sections in the `GeneratePlotFormatting.m` script provided with the downloadable source.)

1. **Type** Bar1 = bar([5, 15, 8, 2, 9]); **and press Enter.**

   MATLAB creates a new bar chart.

2. **Type** TBox1 = annotation('textbox', [.13, .825, .14, .075], 'String', 'Light', 'FontName', 'Arial Narrow', 'FontSize', 16, 'FontWeight', 'light', 'BackgroundColor', [1, 1, 0]); **and press Enter.**

   You see a new annotation of type `textbox` added to the plot. The various entries that you typed change the default font, the font size (so that you can more easily see the text), the font weight (as a matter of emphasis), and the background color (so that the textbox stands out from the bars in the background). To see the various weights side by side, the next few steps add three more textboxes, each with a different font weight.

3. **Type** TBox2 = annotation('textbox', [.29, .825, .14, .075], 'String', 'Normal', 'FontName', 'Arial', 'FontSize', 16, 'FontWeight', 'normal', 'BackgroundColor', [1, 1, 0]); **and press Enter.**

   You see the next annotation placed on the second bar. In most cases, you won't see much (or any) difference between the light and normal settings because few fonts support both light and normal. However, some do, so it's important to experiment.

4. **Type** TBox3 = annotation('textbox', [.45, .825, .14, .075], 'String', 'Demi', 'FontName', 'Arial', 'FontSize', 16, 'FontWeight', 'bold', 'BackgroundColor', [1, 1, 0]); **and press Enter.**

5. **Type** TBox4 = annotation('textbox', [.61, .825, .14, .075], 'String', 'Bold', 'FontName', 'Arial Black', 'FontSize', 16, 'FontWeight', 'normal', 'BackgroundColor', [1, 1, 0]); **and press Enter.**

   You see the final annotation placed on the plot. Looking at the four different annotations, you can see a progression of weights and emphasis. Even though you may not see much difference between light and normal, the weight differences among normal, demi, and bold are pronounced, as shown in Figure 17-2.

Monospace

A monospace font is one in which the characters take up precisely the same amount of space. It's reminiscent of the kind of text that typewriters put out. (If you don't know what a typewriter is, check out `https://site.xavier.edu/polt/typewriters/tw-history.html`.) Monospace is still useful because it makes getting elements to line up easy. That's why code is often presented in monospace: You can easily see the indentation that the code needs to look nice and work properly. The way that monospace is set for an element is through the font. Here are some commonly used monospace fonts:

- Anonymous Pro
- Courier
- Courier New
- Fixedsys
- Letter Gothic
- Lucida Sans Typewriter Regular
- Lucida Console
- Monaco
- Profont
- Ubuntu

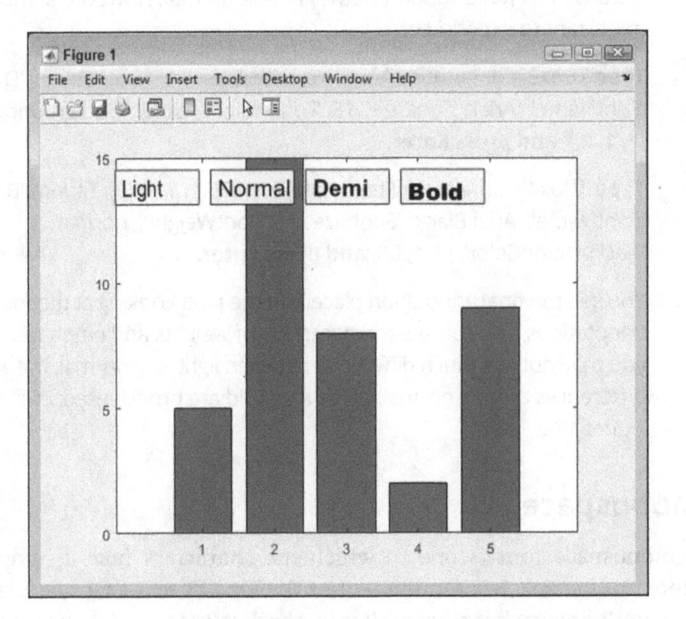

FIGURE 17-2: Different font weights provide different levels of text emphasis.

TIP

You can see some additional monospace fonts at `https://www.fontsquirrel.com/fonts/list/classification/monospaced`. The point is to use a font that provides the proper appearance for your application. To obtain a monospace font appearance, you simply change the `FontName` property to a monospace font. For example, type **TBox5 = annotation('textbox', [.13, .72, .15, .075], 'String',**

'Monospaced', 'FontName', 'CourierNew', 'BackgroundColor', [0, 1, 1]); and press Enter to produce a text box containing monospace text (using the plot generated in the preceding section of the chapter). Figure 17-3 shows the results of this command.

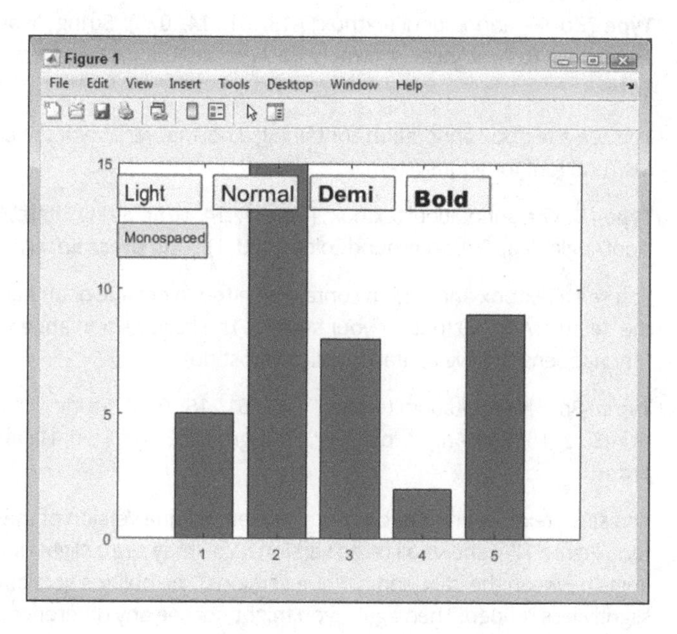

FIGURE 17-3: Monospace fonts make aligning text elements easy.

Italic

Fonts normally have a straight, up-and-down appearance. However, you can skew the font to give an angle to the upright characters and change how the font looks. The skewed version of a font is called *italic*. To create an italic font, the person creating the font must design a separate set of letters and place them in a file containing the italic version.

REMEMBER

Some fonts don't have an italic version. When you encounter such a situation, you can ask the computer to skew the font programmatically. The font file hasn't changed, and the font is still straight up and down, but it looks italicized onscreen. This version of italics is called *oblique*. The italic version of a font always gives a better visual appearance than the oblique version because the italic version is hand tuned — that is, individual pixels are modified so that the font appears smoother.

To configure a font for either italic or oblique, you use the `FontAngle` property. The following steps help you see the differences between the standard, italic, and oblique versions of a font. These steps assume that you have created the MATLAB plot found in the "Bold" section, earlier in this chapter.

1. **Type** TBox6 = annotation('textbox', [.13, .61, .14, .075], 'String', 'Normal', 'FontSize', 16, 'FontAngle', 'normal', 'BackgroundColor', [1, 0, 1]); **and press Enter.**

   You see a textbox annotation containing a normal version of the default MATLAB font for your system.

2. **Type** TBox7 = annotation('textbox', [.29, .61, .14, .075], 'String', 'Italic', 'FontSize', 16, 'FontAngle', 'italic', 'BackgroundColor', [1, 0, 1]); **and press Enter.**

   You see a textbox annotation containing either a normal or an italic version of the default MATLAB font for your system. The italic version appears only if the font happens to have an italic version (most do).

3. **Type** TBox8 = annotation('textbox', [.45, .61, .15, .075], 'String', 'Oblique', 'FontSize', 16, 'FontAngle', 'oblique', 'BackgroundColor', [1, 0, 1]); **and press Enter.**

   You see a textbox annotation containing an oblique version of the standard font. Figure 17-4 shows all three versions. You may see a slight difference in angle between the italic and oblique versions. The oblique version may seem slightly less refined. Then again, you might not see any difference at all between italic and oblique.

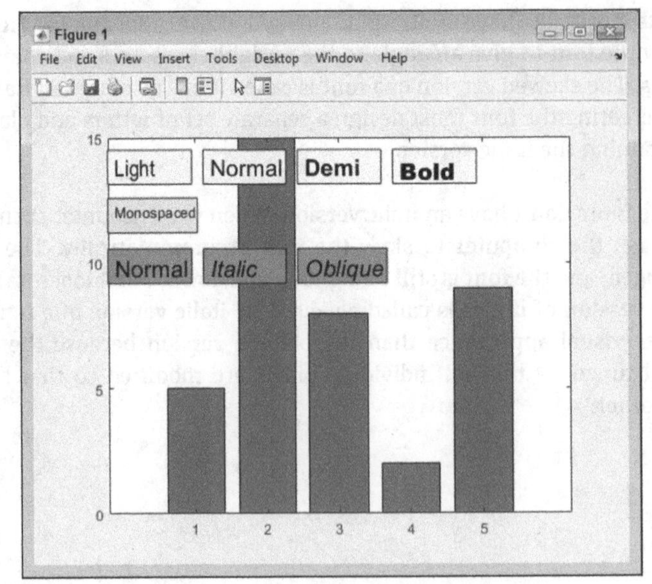

FIGURE 17-4: Italic and oblique versions of the same font could show subtle differences.

Underline

Interestingly enough, MATLAB doesn't provide a simple method to underline text in plots (you find a perfectly acceptable underline in the Live Editor). For example, you can't easily perform this particular task using the GUI. In fact, the default method of creating text using code doesn't provide a method for underlining text, either. However, you have a way to accomplish the goal using a special interpreter called LaTeX (`https://www.latex-project.org/`).

The LaTeX interpreter is built into MATLAB, but it isn't selected by default, so you must set it using the `Interpreter` property. In addition to underlining text, you use the `\underline()` LaTeX function. To see how this works, type **TBox9 = annotation('textbox', [.13, .5, .175, .075], 'String', '\underline{Underline}', 'FontSize', 16, 'BackgroundColor', [.5, 1, .5], 'Interpreter', 'latex');** and press Enter. You see the output shown in Figure 17-5.

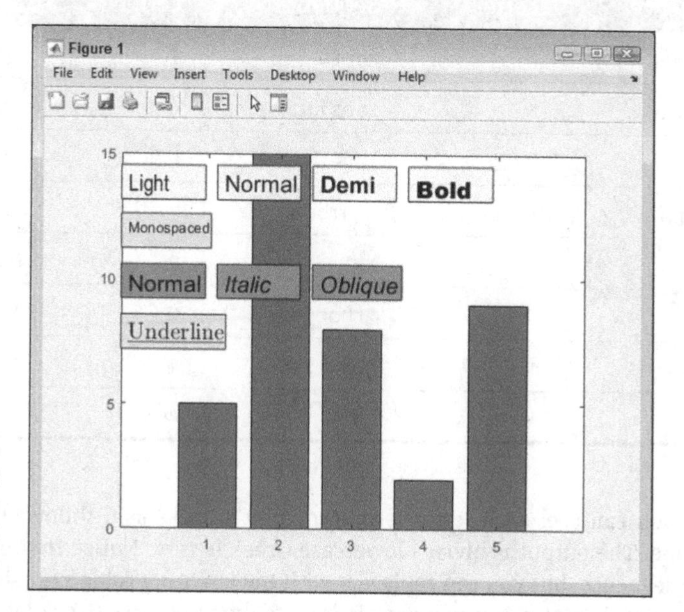

FIGURE 17-5: Underlining text is a little harder than performing other formatting tasks.

TIP

Notice that the output differs quite a bit when using the LaTeX interpreter. That's because LaTeX ignores many of the formatting properties supplied by MATLAB — you must set them using LaTeX functions. However, the problem is more serious than simply setting a font because MATLAB appears to lack the required font files for LaTeX. The bottom line is that you should avoid underlining text unless you truly need to do so. Use bold, italic, and font colors in place of the underline as often as possible.

Using special characters

Sometimes you need to use special characters and character formatting in MATLAB. The following sections describe how to add Greek letters to your output, as well as work with superscript and subscript as needed.

Greek letters

The 24 Greek letters are used extensively in math. To add these letters to MATLAB, you must use a special escape sequence, similar to the escape sequences, such as \n for the newline character, that you use in previous chapters. Table 17-1 shows the sequences to use for Greek letters.

TABLE 17-1 Adding Greek Letters to MATLAB

Letter	Sequence	Letter	Sequence	Letter	Sequence
α	`\alpha`	β	`\beta`	γ	`\gamma`
δ	`\delta`	ε	`\epsilon`	ζ	`\zeta`
η	`\eta`	θ	`\theta`	ι	`\iota`
κ	`\kappa`	λ	`\lambda`	μ	`\mu`
ν	`\nu`	ξ	`\xi`	ο	`Not Used`
π	`\pi`	ρ	`\rho`	σ	`\sigma`
τ	`\tau`	υ	`\upsilon`	φ	`\phi`
χ	`\chi`	ψ	`\psi`	ω	`\omega`

As you can see, each letter is preceded by a backslash, followed by the letter's name. The output is always lowercase Greek letters. Notice that omicron (`o`) has no sequence, but you can replace it with the letter `o`. (Table 17-1 shows the lowercase Greek letters, but you can also create the uppercase Greek letters by using an initial capital letter, such as `\Psi` for the uppercase version of `\psi`.) To see how the letters appear onscreen, type **TBox10 = annotation('textbox', [.13, .39, .51, .085], 'String', '\alpha\beta\gamma\delta\epsilon\zeta\eta\theta\iota\kappa\lambda\mu\nu\xi\pi\rho\sigma\tau\upsilon\phi\chi\psi\omega', 'FontSize', 16, 'BackgroundColor', [.5, .5, 1]);** and press Enter. Figure 17-6 shows how your sample should look at this point.

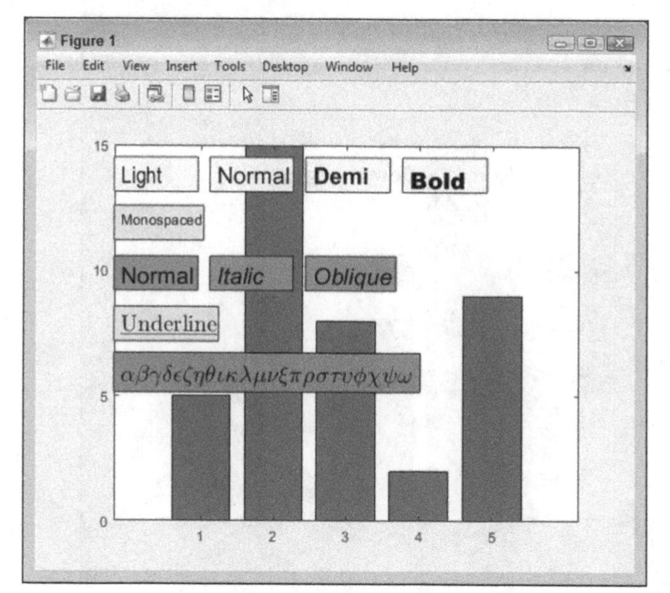

FIGURE 17-6: MATLAB provides access to the various Greek letters normally used in formulas.

TIP

Many of the Greek letters are also available in uppercase form. All you need to do is use initial caps for the letter name. For example, `\gamma` produces the lowercase letter, but `\Gamma` produces the uppercase version of the same letter. You can obtain additional information about text properties (including additional symbols that you can use) at `https://www.mathworks.com/help/matlab/creating_plots/greek-letters-and-special-characters-in-graph-text.html`.

Superscript and subscript

Using superscript and subscript as part of the output is essential when creating formulas or presenting certain kinds of other information. MATLAB uses the caret (^) to denote superscript and the underscore (_) to denote subscript. You enclose the characters that you want to superscript or subscript in curly brackets (`{}`). To see how superscript and subscript works, type **TBox11 = annotation('textbox', [.70, .39, .15, .075], 'String', 'Normal^{Super}_{Sub}', 'BackgroundColor', [.5, .5, 1]);** and press Enter. Figure 17-7 shows typical output from this command.

Notice that the superscript and subscript characters appear in the command without a space after the characters that are in normal type. The output shows these characters immediately after the normal type. In addition, the superscripted characters are over the top of the subscripted characters.

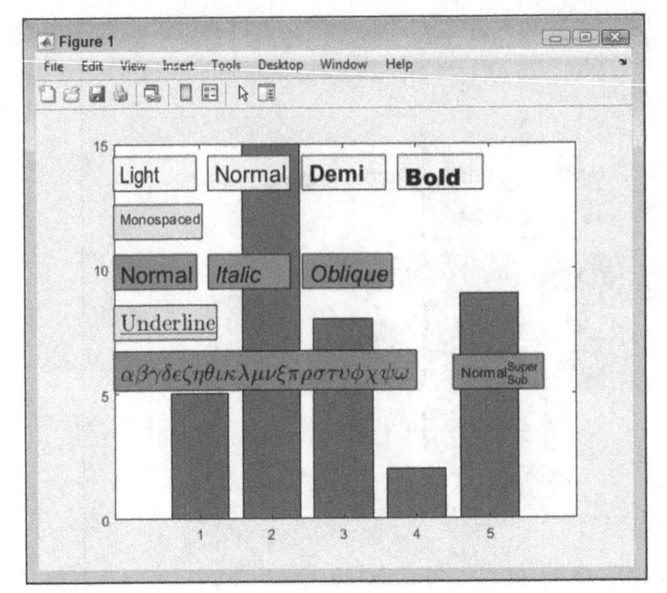

FIGURE 17-7: Working with superscript and subscript characters.

Adding math symbols

You would have a tough time presenting formulas to others without being able to use math symbols. MATLAB provides you with a wealth of symbols that you can use for output purposes. The following sections describe the most commonly used symbols and how you access them.

Fraction

Displaying what appears to be a fraction onscreen doesn't always mean writing an actual numeric fraction; it could be a formula that requires that sort of presentation, such as one showing division. Whatever your need, you can display fractions whenever needed. However, to do that, you must use the LaTeX interpreter mentioned in the "Underline" section, earlier in the chapter. This means that your formatting options are limited and that the output won't necessarily reflect the formatting choices you normally make when using MATLAB.

The fraction requires use of a LaTeX display style that you access using `$\displaystyle\frac{}`. The fraction itself appears in two curly brackets, such as `{1}{2}` for the symbol ½. The entry ends with another dollar sign ($). To see how fractions work with something a little more complex, type **TBox12 = annotation('textbox', [.13, .28, .15, .085], 'String', '$\displaystyle\frac{x-2y}{x+y}$', 'BackgroundColor', [1, .5, .5], 'Interpreter', 'latex');** and press Enter. Figure 17-8 shows typical output from this command.

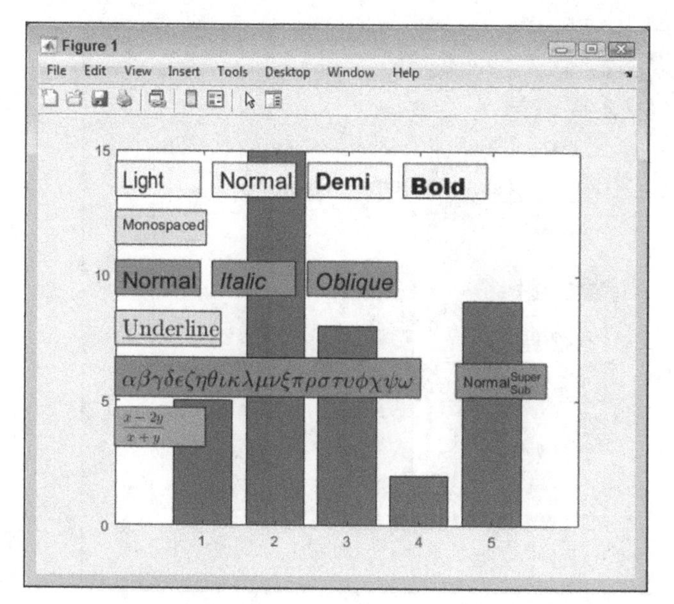

FIGURE 17-8: Using fractions to display formulas correctly.

Square root

MATLAB makes displaying a square root symbol easy. However, getting the square root symbol the right size, with the bar extended over the expression whose root is being taken, requires LaTeX. As with many LaTeX commands, you enclose the string that you want to format in a pair of dollar signs ($). The function used to perform the formatting is `\sqrt{}`, and the value that you want to place within the square root symbol appears within the curly brackets.

To see the square root symbol in action, type **TBox13 = annotation('textbox', [.29, .28, .15, .085], 'String', '$\sqrt{f}$', 'BackgroundColor', [1, .5, .5], 'Interpreter', 'latex');** and press Enter. The variable `f` will appear in the square root symbol, as shown in Figure 17-9.

Sum

Displaying a summation formula complete with sigma and the upper and lower limit involves using LaTeX with the `\sum_{}` function. You supply all three elements of the display in a single statement: the lower limit first; the upper limit second; and the expression third. Each element appears in separate curly brackets. The lower limit is preceded by the underscore used for subscripts, and the upper limit is preceded by the caret used for superscripts. The entire statement appears within dollar signs ($), as is normal for LaTeX. However, in this particular case, you must include a second set of dollar signs or the expression doesn't appear correctly onscreen. (The upper and lower limits don't appear in the correct places.)

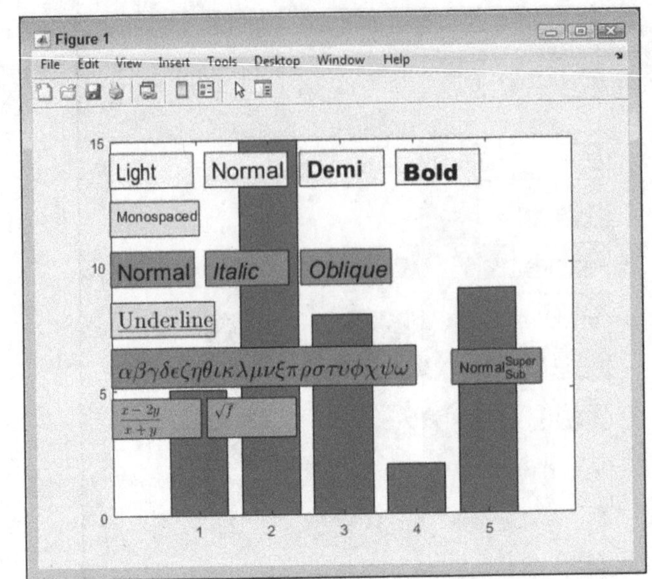

FIGURE 17-9: Displaying the square root symbol so that the bar extends over the f.

To see how summation works, type **TBox14 = annotation('textbox', [.45, .25, .15, .115], 'String', '$$\sum_{i=1}^{2n}{|k_i-k_j|}$$', 'BackgroundColor', [1, .5, .5], 'Interpreter', 'latex');** and press Enter. Notice the use of the double dollar signs in this case. In addition, be sure to include both the underscore and caret, as shown. Figure 17-10 shows the result of using this command.

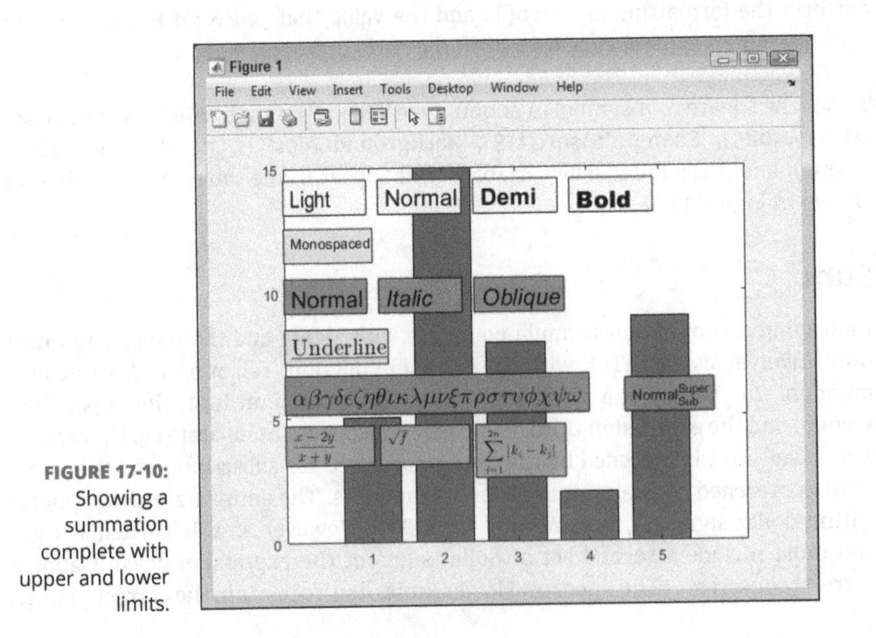

FIGURE 17-10: Showing a summation complete with upper and lower limits.

Integral

To display a definite integral, you use the LaTeX `\int_{}` function, along with the `\d{}` function for the slices. The `\int_{}` function accepts three inputs: two for the interval and the third for the function. In many respects, the format is the same as that used for summation. The beginning of the interval relies on the subscript underscore character, while the ending of the interval relies on the superscript caret character. You must enclose the entire command within double dollar signs ($$) or else the formatting of the superscript and subscript will fail.

To see how to create an integral, type **TBox15 = annotation('textbox', [.61, .25, .22, .115], 'String', '$$\int_{y1(x)}^{y2(x)}{f(x,y)}\d{dx}\d{dy}$$', 'BackgroundColor', [1, .5, .5], 'Interpreter', 'latex');** and press Enter. Notice that the two slices come after the `\int_{}` function and that each slice appears in its own `\d{}` function. Figure 17-11 shows the result of this command.

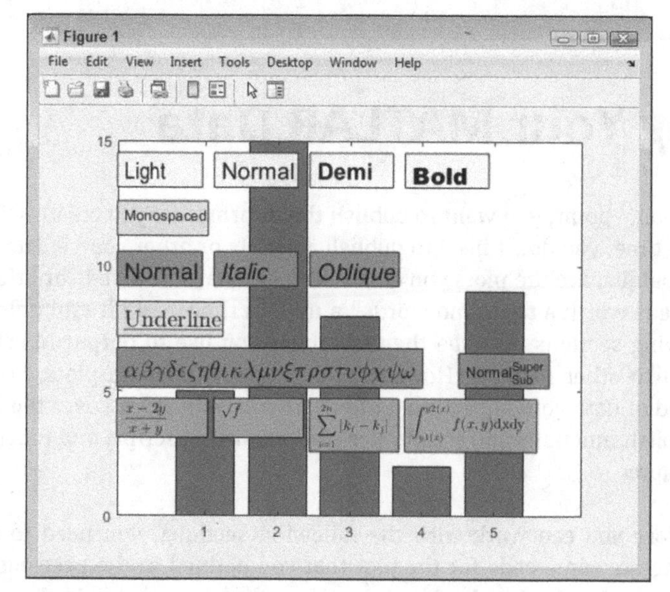

FIGURE 17-11: Defining a definite integral complete with interval.

Derivative

When creating a derivative, you use LaTeX to define a combination of a fraction with superscripts. So, in reality, you've already created a derivative in the past — at least in parts. To see how a derivative works, type **TBox16 = annotation('textbox', [.13, .17, .15, .085], 'String', '$\displaystyle\frac{d^2u}{dx^2}$', 'BackgroundColor', [1, .5, .5], 'Interpreter', 'latex');** and press Enter. Figure 17-12 shows the output from this command.

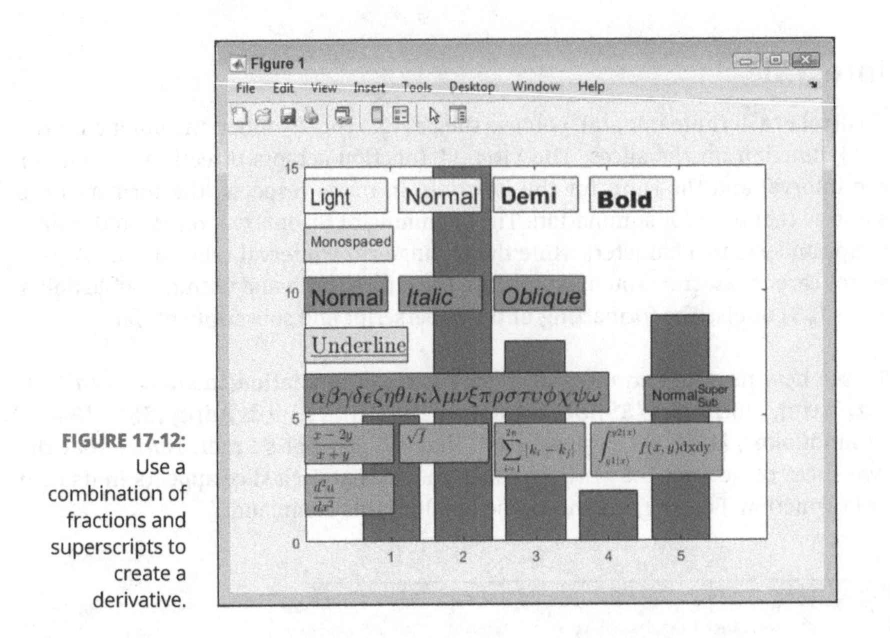

FIGURE 17-12: Use a combination of fractions and superscripts to create a derivative.

Publishing Your MATLAB Data

At some point, you want to publish the information you create. Of course, most of the time, you don't need to publish a matrix or other source data. What you want to publish are the plots you create from the data. You've heard the saying "a picture is worth a thousand words" a million times, and it still holds true. The following sections describe the techniques you use to output MATLAB data of any sort to other formats. However, the sections focus on plots, because that's the kind of data you output most often. (If you want to discover the basics of how to publish functions and scripts, see the "Exporting scripts and functions" section of Chapter 16.)

Before you can work with the following sections, you need to ask MATLAB to generate some code for the plot that you defined in the previous sections of the chapter. If you already closed the figure that you created in the previous sections or you didn't follow the steps, you can simply run the `GeneratePlotFormatting.m` script to generate the plot. Use the instructions in the upcoming "Saving your plot as a script" sidebar to save the resulting plot as a script named `Bar1.m`.

Performing advanced script and function publishing tasks

The `publish()` function can perform a number of different sorts of publishing tasks. The simplest way to publish a script or function is to call `publish()` with the name of the file. This approach produces an HTML file output. If you want to specify a particular kind of output, you provide the file format as the second input, such as `publish('Bar1.m', 'pdf')`. MATLAB supports the following output formats:

- `.doc`
- `.html` (default)
- `.latex`
- `.pdf`
- `.ppt`
- `.xml`

TIP

After you finish publishing a script or function in HTML format, you can view it by using the `web()` function or double-clicking its entry in the Current Folder pane. To test this feature, first publish the `Bar1.m` script by typing **publish('Bar1.m')** and pressing Enter. After you see the success message, type **web('html\Bar1.html')** and press Enter. You see the MATLAB browser output, shown in Figure 17-13. However, even if this output looks perfect, always test the published output using the applications that your viewers will use to ensure that everything displays correctly.

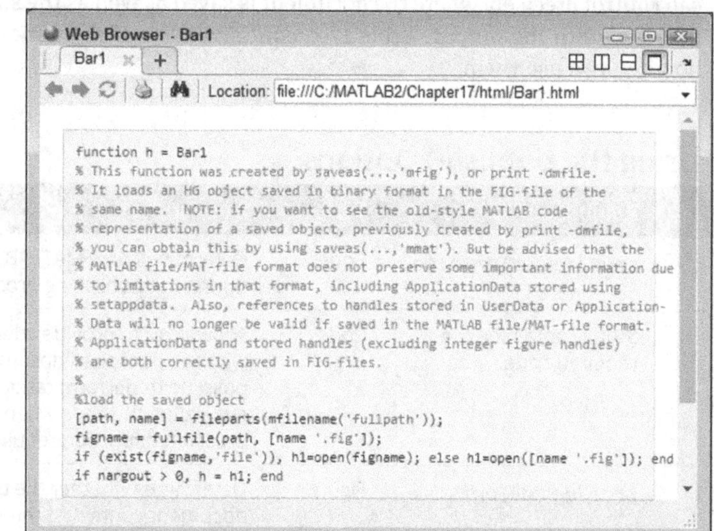

FIGURE 17-13: Display the results of the publication process using MATLAB's browser.

SAVING YOUR PLOT AS A SCRIPT

You may need to save your plot as a script for later use without resorting to hand coding. MATLAB provides a method that creates a script for programmatically loading the plot in `.fig` form from disk. To perform this task, type **saveas(Bar1, 'Bar1', 'm')** and press Enter. In this case, you get a script for loading the plot from disk. In addition, MATLAB saves a copy of the plot for you. However, it saves just the plot because you used the `Bar1` handle. If you want to save the entire figure instead, you must type **saveas(gcf(), 'Figure 1', 'm')** and press Enter (assuming that Figure 1 is indeed the current figure).

Another possibility is to use the GUI to create a function for your plot. To perform this task, choose File ⇨ Generate Code in the figure window. You see a message telling you that MATLAB is generating the code for you. When the message box disappears, the code is complete, and you see an editor window containing the resulting code. This technique creates a function to regenerate the entire figure. You can also click Edit Plot on the figure toolbar and then right-click individual elements. Choose Show Code from the context menu, and you see an editor window containing a function to generate just that element.

You can also use a version of `publish()` with name and value pair options. This version of `publish()` gives you the most control over the published output. You can control precisely where the document is saved as well as the size of any images provided with the output. Table 17-2 contains a list of the options and describes how you can use them.

TABLE 17-2 Using the publish() Options

Name	Values	Type	Description
`catchError`	`true` (default) or `false`	Code	Determines how MATLAB handles errors during the publishing process.
`codeToEvaluate`	String containing the required code	Code	Allows you to provide additional code with the published document so that it's possible to perform tasks such as evaluating the code when the associated function requires inputs.
`createThumbnail`	`true` (default) or `false`	Figure	Determines whether the output document contains a thumbnail version of the full image (when an image is part of the output).

Name	Values	Type	Description
evalCode	true (default) or false	Code	Forces MATLAB to evaluate the code as it publishes the script or function, which results in additional details in the output file in some cases.
figureSnapMethod	entireGUIWindow (default), print, getframe, or entire FigureWindow	Figure	Defines the technique used to obtain the figure contained within the published document.
format	doc, html (default), latex, pdf, ppt, and xml	Output	Determines the format of the published document.
imageFormat	png, epsc2, jpg, bmp, tiff, eps, eps2, ill, meta, or pdf	Figure	Indicates the format of the figure contained within the published document. The default setting depends on the output document format. Some document formats won't access all the format types. For example, PDF output is limited to the bmp or jpg options, but XML output can accept any of the file formats.
maxHeight	'' (default) or positive integer value	Figure	Determines the maximum height of the figure contained within the published document.
maxOutputLines	Inf (default) or non-negative integer value	Code	Specifies the number of lines of code that MATLAB includes in the published document. Setting this value to 0 means that no code is in the output.
maxWidth	'' (default) or positive integer value	Figure	Determines the maximum width of the figure contained within the published document.
outputDir	'' (default) or full path to output directory	Output	Specifies where to place the published document on disk.
showCode	true (default) or false	Code	Determines whether the published document contains any source code.
stylesheet	'' (default) or full path and XSL filename	Output	Specifies the location and name of an XSL file to use when generating XML file output.
useNewFigure	true (default) or false	Figure	Specifies that MATLAB is to create a new figure prior to publishing the document.

Saving your figures to disk

You must save your figures to disk if you want to use them the next session. However, saving a figure to disk can also help you publish the information in a form that lets others use the information as well. The format you choose determines how the saved information is used. Only the MATLAB figure (`.fig`) format provides an editable form that you can work with during the next session. The following sections describe the GUI and command method of saving figures to disk.

Using the GUI to save figures

To save an entire figure, choose File⇨Save As in the figure window. You see the Save As dialog box. Type a name for the file in the File Name field, select the format that you want to use to save the file in the Save As Type field, and click Save to complete the process.

Using commands to save figures

The command version of saving a figure depends on the `saveas()` command. To use this command, you supply a handle to the figure that you want to save as the first argument. The second argument is a filename. When you provide a type of file format to use as the third argument, the filename need not include a file extension. However, if you provide just the filename, you must provide an extension as well. MATLAB supports these file formats:

- `.ai`
- `.bmp`
- `.emf`
- `.eps`
- `.fig`
- `.jpg`
- `.m`
- `.pbm`
- `.pcx`
- `.pdf`
- `.pgm`
- `.png`
- `.ppm`
- `.tif`

REMEMBER

The handle that you provide need not be the figure itself. For example, if you type **saveas(TBox1, 'TBox1.jpg')** and press Enter, MATLAB still saves the entire figure. (Note that you have no way to specify that you want to save just a portion of the figure.)

Printing Your Work

Printing is one of the tasks that most people use the GUI to perform, even the most ardent keyboard user. The issue is one of convenience. Yes, you can use the `printopt()` and `print()` functions to perform the task using the keyboard, but only if you're willing to perform the task nearly blind. The GUI actually shows you what the output will look like (or, at least, a close approximation). Using the commands is significantly harder in this case and isn't discussed in the book. The following sections describe how to use the GUI to output your document.

Configuring the output page

Before you print your document, you need to tell MATLAB how to print it. Choose File⇨Print Preview in the figure window to display the Print Preview dialog box, shown in Figure 17-14.

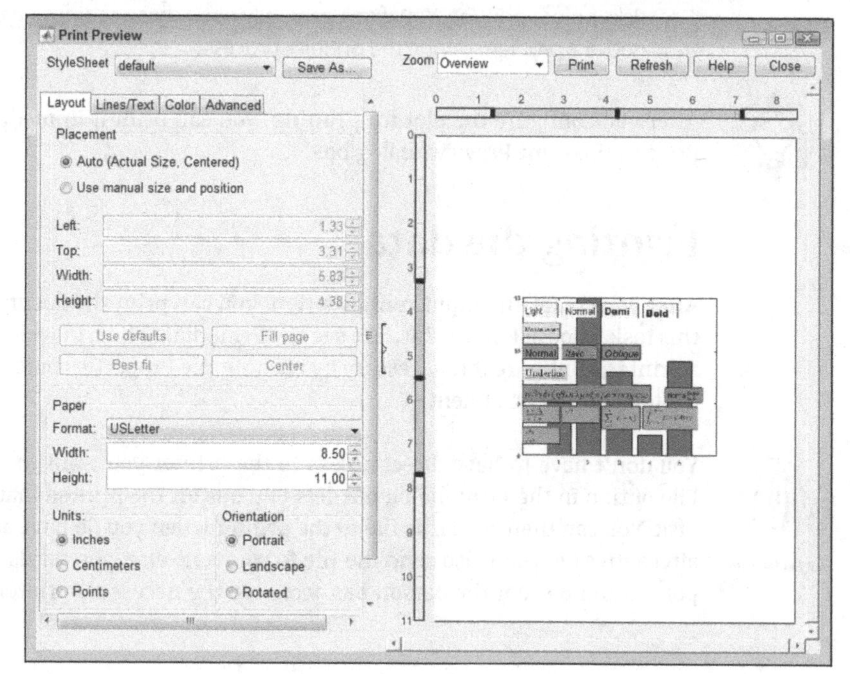

FIGURE 17-14: Configuring the page before you print it.

The right side of the dialog box shows an approximation of the changes you make on the left side. The settings are presented on four tabs with the following functions:

- **Layout:** Defines how the plot or other document will appear on the page. You specify the margins, size of the paper, and presentation of the information itself. The best way to work through the layout process is simply to try the settings and see how they affect the image shown in the right side of the dialog box.
- **Lines/Text:** Specifies the line widths and types of text used to print the documentation. In some cases, you find that using a different line weight or font for printing improves the output appearance of the document. Some of these settings are printer specific. For example, a built-in font is likely to provide a more pleasing appearance than a software font provided as part of the system.
- **Color:** Fixes the color changes between the display and the printer. What you see onscreen is unlikely to be what you get from the printer. The reasons for the difference are complex, but from an overview perspective, printers use subtractive color mixing, while displays use additive color mixing. The two kinds of color presentation don't always work in sync with each other. Read the article at `https://www.worqx.com/color/color_systems.htm` for additional information.
- **Advanced:** Provides access to features such as recomputing the limits and ticks before printing. You won't generally need these settings unless you notice a problem during the printing process.

TIP

After you configure the plot for printing, you can print it immediately by clicking Print in the Print Preview dialog box.

Printing the data

After you create an output configuration, you can print the document. To perform this task, choose File ⇨ Print. You see the Print dialog box, in which you can choose a printer, configure it if necessary by clicking the Properties button, and then click OK to print the document.

REMEMBER

You don't have to have direct access to the printer you want to use. The Print to File option in the Print dialog box lets you output the printed material to a file on disk. You can then send this file to the printer when you do have access to it. As an alternative, you can also send the file to someone else who needs the printed output (assuming that the person has access to the necessary printer).

IN THIS CHAPTER

» Understanding error messages

» Considering the uses of quick alerts

» Managing both common and custom error messages

» Embracing good coding practices

Chapter 18 Recovering from Mistakes

A lot of people associate mistakes with failure. Actually, everyone makes mistakes, and they're all learning opportunities. Some of the greatest minds the world has ever seen thrived quite nicely on mistakes. (Consider how many tries it took Edison to get the light bulb right; also, see my blog post at `http://blog.johnmuellerbooks.com/2013/04/26/defining-the-benefits-of-failure/` on the benefits of failure.) The point is that mistakes are normal, common, and even advantageous at times, as long as you get over them and recover. That's what this chapter is all about — helping you recover from mistakes.

MATLAB can't talk to you directly, so it communicates through error messages. Part of the problem is that MATLAB and humans speak two different languages, so sometimes the error messages aren't quite as helpful as they could be. This chapter helps you understand the error messages so that you can do something about them. You can even find sources for common fixes to these error messages, so the work of overcoming any issues may have already been done by someone. All you'll have to do is go to the right source.

When you write your own applications, sometimes you have to tell other people that they have made a mistake. Unfortunately, you can't record your voice saying, "Oops, I think you meant to type a number instead of some text here" and expect

MATLAB to play it for someone using your application. Therefore, you need to create your own custom error messages so that you can communicate with others using your applications.

Of course, the best way to handle error messages is to not make mistakes. Even though some mistakes are unavoidable (and others have had positive benefits, such as the making of the glue found on the back of sticky notes), you can work in ways that tend to reduce errors of all sorts. The final section of this chapter discusses good coding practices to use to avoid trouble. No, this section can't prevent you from ever seeing an error message, but by following its advice, you can see fewer of them.

REMEMBER

You don't have to type the source code for this chapter; using the downloadable source is a lot easier. You can find the source for this chapter in the `\MATLAB2\Chapter18` folder of the downloadable source. When using the downloadable source, you may not be able to work through some of the hands-on examples because the example files will already exist on your system. In this case, you can create an alternative folder, `Chapter18a`, to hold the results of the hands-on exercises. See the Introduction for details on how to obtain the downloadable source.

Working with Error Messages

Error messages are designed to tell you that something has gone wrong. It seems like an obvious thing to say, but some people actually view error messages as the computer's way of telling them it dislikes them. Computers don't actually hate anyone — they simply like things done a certain way because they really don't understand any other way.

Error messages come in two levels:

- **Error:** When an actual error occurs, processing stops because the MATLAB application can't continue doing useful work. You can tell an error message because it appears in dark-red type and the system makes a sound. However, the most important aspect of an error is that the application stops running, and you need to fix the problem before it can proceed.
- **Warning:** When a warning condition occurs, the computer tells you about it but continues processing the application as long as doing so is safe. The message type shows up in a lighter red in this case, and you don't hear any sort of system sound. In addition, the error message is preceded by the word *Warning*. In fact, you might not know that a warning has happened until you look at the Command Window. Fixing a warning is still important because warnings have a habit of becoming errors over time.

REMEMBER

The most important thing to remember about error and warning messages is that someone created them based on what that person thought would happen, rather than what *is* happening. You sometimes see an error or warning message that doesn't quite make sense in the situation, or is possibly completely wrong. Of course, first you try reacting to the error message, but if you can't find a cause for the error, it's time to start looking in other places. The error or warning message is designed to provide you with help after something has gone wrong — and sometimes the help is absolutely perfect but at other times it isn't, and you need to work a little harder to resolve the situation.

Responding to error messages

Until now, every time an error has happened, it has simply displayed in the Command Window without much comment. However, you can do more about errors than simply wait for them to announce themselves. MATLAB lets you intercept errors and handle them in various ways using a special `try...catch` structure. Listing 18-1 shows how to create such a structure in a function. You can also find this function in the `Broken.m` file supplied with the downloadable source code.

LISTING 18-1: Using the try...catch Structure

```
function [ ] = Broken( )
%BROKEN A broken piece of code.
%   This example is designed to generate an error.

    try
        Handle = fopen('DoesNotExist.txt');
        Data = fread(Handle);
        disp(Data);
    catch exc
        disp('Oops, an error has occurred!');
        disp(exc)
    end
end
```

First, look at the source of the error. The call to `fopen()` uses a file that doesn't exist. Using this call isn't a problem; in fact, some calls to `fopen()` are expected to fail. When this failure happens, `fopen()` returns a handle with a value of `-1`. The problem with this code occurs in the *next* call. Because `Handle` doesn't contain a valid handle, the `fread()` call fails. Reading data from a file that doesn't exist isn't possible.

REMEMBER

The `try...catch` block contains the code that you want to execute between `try` and `catch`. When an *exception* (something other than what was expected) does occur, the information is placed in `exc`, where you can use it in whatever way is necessary. In this case, the error-handling code (*error handler*) — the code between `catch` and `end` — displays a human-readable message and a more specific MATLAB error message. The first is meant for the user; the second is meant for the developer.

To try this code, type **Broken()** and press Enter in the Command Window. You see the following output:

```
Oops, an error has occurred!
  MException with properties:

    identifier: 'MATLAB:FileIO:InvalidFid'
       message: 'Invalid file identifier. Use fopen to
                 generate a valid file identifier.'
         cause: {}
         stack: [1×1 struct]
    Correction: []
```

TIP

The exception information starts with the second line. It tells you that the exception is a member of the `MException` class and has certain properties. Because these properties are standardized, you can often search the Internet for help in fixing a problem, in addition to the help that the MATLAB product provides. Here's the additional information you receive:

- `identifier`: A short, specific description of the error. An identifier provides a category of errors, and you can use it to find additional information about the error as a whole.
- `message`: A longer message that provides details about the problem in this particular instance. The `message` is generally easier for humans to understand than is the other information.
- `cause`: When establishing a cause for the problem is possible, this property contains a list of causal sources.
- `stack`: The path that the application has followed to get to this point. By tracing the application path, you can often find a source for an error in some other function — the caller of the current function (or one of its callers all the way up to the main application in the call hierarchy).

- `Correction`: Contains a suggested fix for the error when MATLAB can guess at one. This fix isn't guaranteed to work, but you do find that the fixes do work a high proportion of the time.

Understanding the MException class

The `MException` class is the basis for the exception information that you receive from MATLAB. It provides you with the properties and functions needed to work with exceptions; you then use the exceptions to overcome, or at least reduce, the effect of errors on the user. The previous section of the chapter, "Responding to error messages," acquaints you with the five properties that the `MException` class provides. Here's a list of the functions that you use most often:

- `addCause()`: Appends the cause that you provide to the list of causes already provided with the exception. You can use this feature to provide ancillary information about an exception.
- `getReport()`: Outputs a formatted report of the exception information in a form that matches the output that MATLAB provides.
- `last()`: Obtains the last exception that the application threw. Note that this is a static function, so you use it as `MException.last()` any time you need the previous error.
- `rethrow()`: Sends an exception to the next higher level of the application hierarchy when handling the error at the current level isn't possible. (After an exception is accepted by an application, it's no longer active, and you must *rethrow* it to get another part of the application to act on it.)
- `throw()`: Creates an exception that either displays an error message or is handled by another part of the application.
- `throwAsCaller()`: Creates an exception using the caller's identifier that either displays an error message or is handled by another part of the application. (When one part of an application accesses a function, the part that performs the access is named the caller, so this function makes the exception appear as it if were created by the caller rather than the current function.)

One of the more interesting functions provided by the `MException` class is `getReport()`. Listing 18-2 shows how to use this function to obtain formatted output for your application. You can also find this function in the `Broken2.m` file supplied with the downloadable source code.

LISTING 18-2: Creating an Error Report

```
function [ ] = Broken2( )
%BROKEN2 A broken piece of code.
%   This example is designed to generate an error
%   and display a report about it.

   try
      Handle = fopen('DoesNotExist.txt');
      Data = fread(Handle);
      disp(Data);
   catch exc
      disp('Oops, an error has occurred!');
      disp(exc.getReport());
   end
end
```

REMEMBER

Notice that you must still use the disp() function to actually output the formatted string to screen. The getReport() output is nicely formatted, if rather inflexible, because you don't have access to the individual MException properties. However, the output works fine for most uses. Here's the output of this example:

```
Oops, an error has occurred!
Error using fread
Invalid file identifier.  Use fopen to generate a valid file
  identifier.

Error in Broken2 (line 8)
       Data = fread(Handle);
```

The output includes three of the properties. However, the cause isn't used in this case because the cause can't be identified by MATLAB. You need to use the addCause() function to add a cause when desired. In addition, MATLAB doesn't have a handy correction for this problem.

Creating error and warning messages

As previously mentioned, MATLAB supports both error and warning messages. You have a number of ways to create exceptions based on application conditions. The easiest way is to use the error() and warning() functions. The first creates an error condition, while the second creates a lesser, warning condition.

The example shown in Listing 18-3 presents a basic method of issuing an error or warning resulting from user input. However, you can use the same technique whenever an error or warning condition arises and you can't handle it locally. You can also find this function in the `ErrorAndWarning.m` file supplied with the downloadable source code.

LISTING 18-3: Creating Errors and Warnings

```
function [ ] = ErrorAndWarning( )
%ERRORANDWARNING Create Error and Warning Messages
%   This example shows how to create error and warning messages.

    NotDone = true;

    while NotDone
        try

            Value = input('Type something: ', 's');

            switch Value
                case 'error'
                    error('Input Error');
                case 'warning'
                    warning('Input Warning');
                case 'done'
                    NotDone = false;
                otherwise
                    disp(['You typed: ', Value]);
            end
        catch Exception
            disp('An exception occurred!');
            disp(Exception.getReport());
        end
    end

end
```

The example begins by creating a loop. It then asks the user to type something. If that something happens to be **error** or **warning**, the appropriate error or warning message is issued. When the user types **done**, the application exits. Otherwise, the user sees a simple output message. The example looks simple, and it is, but it has a couple of interesting features. The following steps help you work with the example:

1. **Type** ErrorAndWarning() **and then presses the Enter key, all in the Command window.**

   The application asks you to type something.

2. **Type** Hello World! **and press Enter.**

   You see the following output:

   ```
   You typed: Hello World!
   ```

   The application asks the user to type something else.

3. **Type** warning **and press Enter.**

   You see the following output (which may differ slightly from that shown):

   ```
   Warning: Input Warning
   > In ErrorAndWarning at 16
   ```

   Notice that the message doesn't say anything about an exception. A warning is simply an indicator that something could be wrong, not that something is wrong. As a result, you see the message, but the application doesn't actually generate an exception. The application asks the user to type something else.

4. **Type** error **and press Enter.**

   You see the following output:

   ```
   An exception occurred!
   Error using ErrorAndWarning (line 14)
   Input Error
   ```

   This time, an exception is generated. If the exception handler weren't in place, the application would end at this point. However, because an exception handler is in place, the application can ask the user to type something else. Adding exception handlers makes recovering from exceptions possible, as happens in this case. Of course, your exception handler must actually fix the problem that caused the exception.

5. **Type** done **and press Enter.**

   The application ends.

The example application uses the simple form of the `error()` and `warning()` functions. Both the `error()` and `warning()` functions can accept an identifier as the first argument, followed by the message as the second. You can also add the cause, stack trace, and correction elements as arguments. The point is, all you really need in most cases is a simple message.

Setting warning message modes

Error messages always provide you with any details that the application provides. Warning messages are different — you can tell MATLAB to provide only certain details as needed. You can set the configuration globally or base it on the message identifier (so that warnings with some identifiers provide more information than others do). To set the global configuration, you provide two arguments. The first is the warning state:

- on: Sets the configuration element on.
- `off`: Turns the configuration element off.
- `query`: Determines whether the configuration element is currently on or off.

At a global level, you have access to two settings, which are provided as the second argument to the `warning()` function:

- `backtrace`: Determines whether the output includes stack trace information.
- `verbose`: Determines whether the output includes a short message or one with all the details as well.

To see how this works, type **warning('query', 'backtrace')** and press Enter. You should see an output message telling you the current status of the `backtrace` configuration element. (The default setting turns it on so that you can see the stack trace as part of the warning message when a stack trace is provided.)

The message identifier-specific form of the `warning()` function starts with a state. However, in this case, the second argument is a message identifier. For example, if you type **warning('query', 'MATLAB:FileIO:InvalidFid')** and press Enter, you see the current state of the `MATLAB:FileIO:InvalidFid` identifier.

TIP

To suppress a warning, you use the `warning()` function with `'off'` as the first argument. You supply the name of the warning as the second argument. For example, to suppress all warnings, you use `warning('off', 'all')`, but to suppress only warning messages associated with the attempted removal of a nonexistent path, you use `warning('off', 'MATLAB:rmpath:DirNotFound')`. You can obtain warning identifiers a number of ways, but the easiest method is to create the condition in which the warning occurs and then query it using `warning('query', 'last')`. To turn warnings back on, you simply use `'on'` as the first argument.

Setting warnings for particular message identifiers to `off` is usually a bad idea because the system can't inform you about problems. This is especially true for MATLAB-specific messages (rather than custom messages that you create, as described in the "Making Your Own Error Messages" section, later in the chapter). However, setting them to `off` during troubleshooting can help you avoid the headache of seeing the message all the time.

Understanding Quick Alerts

Errors can happen at any time. In fact, there seems to be an unwritten law that errors must happen at the most inconvenient time possible, and only when anyone who can possibly help is out of the building. When you face this problem, at least you have the assurance that every other person to ever write an application has faced the same problem. Waiting by your phone for a call to support your application that you don't want to hear (because it's always bad news) is one approach, but probably not the best because the person on the other end of the line is unlikely to have the information you so desperately need to resolve the issue.

When you're stuck with an error, sending yourself a note is probably a better option than waiting for a call from someone. Fortunately, MATLAB provides the `sendmail()` function for this purpose. It's possible for you to make one of the responses in your error-handling code be to send an email that you can pick up on your smartphone. That way, you get information directly from the application, which enables you to fix the problem right where you are, rather than have to go into work. The `sendmail()` function accepts these arguments:

- **Recipients:** Provides a list of one or more recipients for the message. Each entry is separated from the other with a semicolon.
- **Subject:** Contains the message topic. If the problem is short enough, you can actually present the error message in the subject line.
- **Message (optional):** Details the error information.
- **Attachments (optional):** Specifies the path and full filename of any attachment you want to send with the message.

Before you can send an email, you must configure MATLAB to recognize your Simple Mail Transfer Protocol (SMTP) server and provide a From address. To do this, you must use a special `setpref()` function. For example, if your server is `smtp.mycompany.com`, you type **setpref('Internet', 'SMTP_Server', 'smtp.mycompany.com')** in the Command Window and press Enter. After you set the SMTP address, you set the From address by providing your email address as input to the setpref() function, as in setpref('Internet', 'E_mail', '*myaddress@mycompany.com*'). You can also verify your settings using the getpref() function. Use rmpref() to remove preferences that you don't need, and use addpref() to add more preferences.

Listing 18-4 shows a technique for sending an email. The code used to create the error is similar to the Broken example used earlier in the chapter (see the "Responding to error messages" section). However, this time the example outputs an email message rather than a message onscreen. You can also find this function in the Broken3.m file supplied with the downloadable source code. (For this example to work, you must use setpref() to set the address of a real email server, as well as change the email address found in the sendmail() call in the code.)

LISTING 18-4: Sending an Email Alert

```
function [ ] = Broken3( )
%BROKEN3 A broken piece of code.
%   This example is designed to generate an error
%   and send an e-mail about it.

    try
        Handle = fopen('DoesNotExist.txt');
        Data = fread(Handle);
        disp(Data);
    catch exc
        disp('Sending for help!');
        sendmail('myaddress@mycompany.com',...
            'Broken3',...
            ['Identifier: ', exc.identifier,10,...
             'Message: ', exc.message]);
    end
end
```

Notice how the example uses the sendmail() function. The address and subject appear much as you might think from creating any other email message. The message is a concatenation of strings. The first line is the error identifier, and the second is the error message. Notice the number 10 between the two lines. This value actually creates a new line so that the information appears on separate lines in the output message. Figure 18-1 shows a typical example of the message (displayed using Outlook in this case).

FIGURE 18-1: A typical email message containing MATLAB error information.

WARNING

Working with email can be tricky because different email servers handle message requests in different ways, and modern email systems have multiple levels of security in many cases. Consequently, you may need to employ some specialized code to make the Broken3 example work with your particular server. For example, if you use Google Mail, you may need to take the helpful information at `https://support.google.com/accounts/answer/185833?hl=en` into account when putting this example together.

Relying on Common Fixes for MATLAB's Error Messages

MATLAB does try to inform you about errors whenever it finds them. Of course, your first line of defense in fixing those errors is to rely on MATLAB itself. Previous chapters of the book outline a number of these automatic fixes. For example, when you make a coding error, MATLAB usually asks whether you meant to use some alternative form of the command, and it's right quite often about the choice it provides.

The editor also highlights potential errors in orange. When you hover the mouse over these errors, you see a small dialog box telling you about the problem and offering to fix it for you. The article at `https://www.mathworks.com/help/matlab/matlab_prog/check-code-for-errors-and-warnings.html` discusses a number of other kinds of automatic fixes that you should consider using.

REMEMBER

Sometimes an automatic fix doesn't make sense, but the combination of the error message and the automatic fix provides you with enough information to fix the problem on your own. The most important things to do are to read both the error message and the fix carefully. Humans and computers speak different languages, so there is a lot of room for misunderstanding. After you read the information carefully, look for typos or missing information. For example, MATLAB does understand A*(B+C) but doesn't understand A(B+C). Even though a human would know that the A should be multiplied by the result of B+C, MATLAB can't make that determination. Small bits of missing text have a big impact on your application, as do seemingly small typos, such as typing Vara instead of VarA.

Don't give up immediately, but at some point you need to start consulting other resources rather than getting bogged down with an error that you can't quite fix. The MATLAB documentation can also be a good source of help, but knowing where to look (when you can barely voice the question) is a challenge. That's where MATLAB Answers (https://www.mathworks.com/matlabcentral/answers/) comes into play. You can use this resource to obtain answers from MATLAB professionals, in addition to the usual peer support. If you can't find someone to help you on MATLAB Answers, you can usually get script and function help on Code Project (https://www.codeproject.com/script/Answers/List.aspx?tab=active&tags=922) and other third-party answer sites.

Fortunately, there are other documentation alternatives when the MATLAB documentation can't help. For example, the MATLAB Programming/Error Messages article at https://en.wikibooks.org/wiki/MATLAB_Programming/Error_Messages describes a number of common errors and how to fix them. Another good place to look for helpful fixes to common problems is MATLAB Tips (https://www3.nd.edu/~nancy/Math20550/Matlab/tips/matlab_tips.html). In short, you have many good places to look online for third-party assistance if your first line of help fails.

Although you can find a lot of MATLAB information online, be aware that not all of it is current. Old information may work fine with a previous version of MATLAB, but it may not work at all well with the version installed on your system. When looking for help online, make sure that the help you obtain is for the version of MATLAB that you actually have installed on your machine, or test the solution with extreme care to ensure that it does work.

Making Your Own Error Messages

At some point, the standard error messages that MATLAB provides will fall short, and you'll want to create custom error messages of your own. For example, MATLAB has no idea how your custom function is put together, so the standard messages can't accommodate a situation in which a user needs to provide a specific kind of input. The only way you can tell someone that the input is incorrect is to provide a custom error message.

Fortunately, MATLAB provides the means for creating custom error messages. The following sections describe how to create the custom error messages first, and then how to ensure that your custom error messages are as useful as possible. The most important task of an error message is to inform others about the problem at hand in a manner that allows them to fix it. So creating custom error messages that really do communicate well is essential.

Developing the custom error message

The example in this section is a little more complex than earlier examples in this book. When you develop something like custom error messages, you want to create the code itself, followed by a means to test that code. Developers use a fancy term, *testing harness*, to describe the code used to test other code. The odd name isn't necessary, though. One file contains the code that you use to check for a condition, and another file contains the test code. The following sections describe the two files used for this example.

Creating the exception code

Testing user inputs is usually a good idea because you never know what a user will provide. In Listing 18-5, the code performs a typical set of checks to ensure that the input is completely correct before using it to perform a task — in this case, displaying the value onscreen. The technique used in this example is a good way to ensure that no one can send you data that isn't appropriate for a particular application need. You can also find this function in the `CustomException.m` file supplied with the downloadable source code.

LISTING 18-5: Checking for Exceptional Conditions

```
function [ ] = CustomException( Value )
%CUSTOMEXCEPTION Demonstrates custom exceptions.
%   This example shows how to put a custom exception
%   together.

    if nargin < 1
        NoInput = MException('MyCompany:NoInput',...
            'Not enough input arguments!');
        NoInput.throw();
    end
```

```
    if not(isnumeric(Value))
        NotNumeric = MException('MyCompany:NotNumeric',...
            'Input argument is type %s number needed!',...
            class(Value));
        NotNumeric.throw();
    end

    if (Value < 1) || (Value > 10)
        NotInRange = MException('MyCompany:NotInRange',...
            'Input argument not between %d and %d!',...
            1, 10);
        NotInRange.throw();
    end
    fprintf('Successfully entered the value: %d.\r',...
        Value);

end
```

The code begins by checking the number of arguments. This example contains no default value, so not supplying a value is an error, and the code tells the caller about it. The `NoInput` variable contains an `MException` object that has an identifier for `MyCompany:NoInput`. This is a custom identifier, the sort you should use when creating your own exceptions. An identifier is a string, such as your company name, separated by a colon from the exception type, which is `NoInput` in this case.

The message following the identifier provides additional detail. It spells out that the input doesn't provide enough arguments in this case. If you wanted, you could provide additional information, such as the actual input requirements for the application.

After `NoInput` is created, the code uses the `throw()` method to cause an exception. If the caller hasn't placed the function call in a `try...catch` block, the exception causes the application to fail. The exception does cause the current code to end right after the call to `throw()`.

TIP

The second exception works much the same as the first. In this case, the code checks to ensure that the input argument (now that it knows there is one) is a numeric value. If `Value` isn't numeric, another exception is thrown. However, notice that this exception detects the kind of input actually provided and returns it as part of the message. The messages you create can use the same placeholders, such as `%d` and `%s`, as the `sprintf()` and `fprintf()` functions used in earlier chapters.

REMEMBER

Note the order of the exceptions. The code tests to ensure that there is an argument before it tests the argument type. The order in which you test for conditions that will stop the application from running properly is essential. Each step of testing should build on the step before it.

The third exception tests the range of the input number (now that the code knows that it is indeed a number). When the range is outside the required range, the code throws an exception.

When everything works as it should, the code ends by displaying `Value`. In this case, the application uses `fprintf()` to make displaying the information easier than it would be when using `disp()`, because `disp()` can't handle numeric input.

Creating the testing code

Testing your code before using it in a full-fledged application is essential. This is especially true for error-checking code, such as that found in `CustomException()`, because you rely on such code to tell you when other errors occur. Any code that generates exceptions based on errant input must be held to a higher standard of testing, which is why you need to create the testing harness shown in Listing 18-6. You can also find this function in the `TestCustomException.m` file supplied with the downloadable source code.

LISTING 18-6: Testing the Exception Code

```
function [ ] = TestCustomException( )
%TESTCUSTOMEXCEPTION Tests the CustomException() function.
%   Performs detailed testing of the CustomException() function
%   by checking for input type and ranges.

    % Check for no input.
    try
        disp('Testing no input.');
        CustomException();
    catch Exc
        disp(Exc.getReport());
    end
    % Check for logical input.
    try
        disp('Testing logical input.');
        CustomException(true);
    catch Exc
        disp(Exc.getReport());
    end
```

```
    % Check for string input.
    try
        disp('Testing string input.');
        CustomException('Hello');
    catch Exc
        disp(Exc.getReport());
    end
    % Check for number out of range.
    try
        disp('Testing input too low.');
        CustomException(-1);
    catch Exc
        disp(Exc.getReport());
    end
    try
        disp('Testing input too high.');
        CustomException(12);
    catch Exc
        disp(Exc.getReport());
    end
    % Check for good input.
    try
        disp('Testing input just right.');
        CustomException(5);
    catch Exc
        disp(Exc.getReport());
    end
end
```

This code purposely creates exceptions and then outputs the messages generated. By running this code, you can ensure that `CustomException()` works precisely as you thought it would. Notice that the test order follows the same logical progression as the code in the `CustomException.m` file. Each test step builds on the one before it. Here's the output you see when you run `TestCustomException()`:

```
Testing no input.
Error using CustomException (line 9)
Not enough input arguments!
Error in TestCustomException (line 9)
        CustomException();

Testing logical input.
Error using CustomException (line 16)
```

```
Input argument is type logical number needed!
Error in TestCustomException (line 17)
       CustomException(true);

Testing string input.
Error using CustomException (line 16)
Input argument is type char number needed!
Error in TestCustomException (line 25)
       CustomException('Hello');

Testing input too low.
Error using CustomException (line 23)
Input argument not between 1 and 10!
Error in TestCustomException (line 33)
       CustomException(-1);

Testing input too high.
Error using CustomException (line 23)
Input argument not between 1 and 10!
Error in TestCustomException (line 39)
       CustomException(12);

Testing input just right.
Successfully entered the value: 5.
```

The output shows that each phase of testing ends precisely as it should. Only the final output provides the desired result. Notice how the incorrect input types generate a custom output message that defines how the input is incorrect.

Creating useful error messages

Creating useful error messages can be hard. When you find yourself scratching your head, trying to figure out just what's wrong with your input, you're experiencing a communication problem. The error message doesn't provide enough information in the right form to communicate the problem to you. However, creating good error messages really is an art, and it takes a bit of practice. Here are some tips to make your error-message writing easier:

- Keep your messages as short as possible, because long messages tend to become difficult to understand.
- Focus on the problem at hand, rather than what you think the problem might be. For example, if the error message says that the file is missing, focus on the missing file, rather than on something like a broken network connection. It's

more likely that the user mistyped the filename than it is that the network is down. If the filename turns out to be correct, it could have been erased on disk. You do need to eventually check out the network connection, but focus on the problem at hand first and then move out from there so that your error trapping is both procedural and logical.

- Provide specific information whenever possible by returning the errant information as part of the error message.
- Ask others to try your testing harness, read the messages, and provide feedback.
- Make the error message a more detailed version of the message identifier and ensure that the message identifier is unique.
- Verify that every message is unique so that users don't see the same message for different conditions. If you can't create unique wording, perhaps you need to create a broader version of the message that works for both situations.
- Ensure that each message is formatted in a similar way so that users can focus on the content rather than the format.
- Avoid humorous or irritating language in your messages — make sure that you focus on simple language that everyone will understand and that won't tend to cause upset rather than be helpful.

Using Good Coding Practices

A lot of places online tell you about good coding practice. In fact, if you ask five developers about their five best coding practices, you get five different answers, partly because everyone is different. The following list represents the best coding practices from a number of sources (including personal experience) that have stood the test of time.

- **Get a second opinion:** Many developers are afraid to admit that they make mistakes, so they keep looking at the same script or function until they're bleary-eyed, and they usually end up making more mistakes as a result. Having someone else look at the problem could save you time and effort, and will most definitely help you discover new coding practices more quickly than if you go it alone.
- **Write applications for humans, not machines:** As people spend more time writing code, they start to enjoy what they do and start engaging in writing code that looks really cool but is truly horrible to maintain. In addition, the

code is buggy and not very friendly to the people using it. Humans use applications. No one uses cool code — people use applications that are nearly invisible and that help them get work done quickly, without a lot of fuss.

- **Test often and make small changes:** A few people actually try to write an entire application without ever testing it, even once, and then they're surprised when it doesn't work. The best application developers work carefully and test often. Making small changes means that you can find errors faster and fix them faster still. When you write a whole bunch of code without testing it, you really don't have any way to know where to start looking for problems.
- **Don't reinvent the wheel:** Take the opportunity to use someone else's fully tested and bug-free code whenever you can (as long as you don't have to steal the code to do so). In fact, actively look for opportunities to reuse code. Using code that already works in another application saves you time in writing your application.
- **Modularize your application:** Writing and debugging a piece of coding takes time and effort. Maintaining that code takes even longer. If you have to make the same changes in a whole bunch of places every time you discover a problem with your code, you waste time and energy that you could use to do something else. Write the code just one time, place it in a function, and then access that piece of code everywhere you need it.
- **Plan for mistakes:** Make sure your code contains plenty of error trapping. It's easier to catch a mistake and allow your application to fail gracefully than it is to have the application crash and lose data that you must recover at some later time. When you do add error-trapping code, make sure to write it in such a manner as to actually trap the sorts of errors that you expect, and then add some general-purpose error trapping for the mistakes you didn't expect.
- **Create documentation for your application:** Every application requires documentation. Even if you plan to use the application to meet just your own needs, you need documentation, because all developers eventually forget how their code works. Professionals know from experience that good documentation is essential. When you do create the documentation, make sure that you discuss why you designed the software in a certain manner, what you were trying to achieve by creating it, problems you encountered making the software work, and fixes you employed in the past. In some cases, you want to also document how something works, but keep the documentation of code mechanics (how it works) to a minimum.

REMEMBER

- **Ensure that you include documentation within your application as comments:** Comments within applications help at several different levels, the most important of which is jogging your memory when you try to figure out how the application works. However, it's also important to remember that typing `help('ApplicationName')` and pressing Enter will display the comments as help information to people using your application.

- **Code for performance after you make the application work:** Performance consists of three elements: reliability, security, and speed. A reliable application works consistently and checks for errors before committing changes to data. A secure application keeps data safe and ensures that user mistakes are caught. A fast application performs tasks quickly. However, before you can do any of these things, the application has to work in the first place. (Remember that you can use the `profile()` command to measure application performance and determine whether changes you implement actually work as intended.)
- **Make the application invisible:** If a user has to spend a lot of time acknowledging the presence of your application, your application will eventually end up collecting dust. For example, the most annoying application in the world is the one that displays those "Are you sure?" messages. If the user wasn't sure, there would be no reason to perform the act. Instead, make a backup of the change automatically so that the user can reverse the change later. Users don't even want to see your application — it should be invisible, for the most part. When a user can focus on the task at hand, your application becomes a favorite tool and garners support for things like upgrades later.
- **Force the computer to do as much work as possible:** Anytime you can make something easier for the user, you reduce the chance that the user will make a mistake that you hadn't considered as part of the application design. Easy doesn't mean fancy. For example, many applications that try to guess what the user is going to type next usually get it wrong and only end up annoying everyone. Let the user type whatever is needed, but then you can check the input for typos and ensure that the input won't cause the application to fail. In fact, rather than try to guess what the user will type next, create an interface that doesn't actually require any typing. For example, instead of asking the user to enter a city and state in a form, have the user type a zip code and obtain the city and state from the zip code information.

5
Specific MATLAB Applications

IN THIS PART . . .

Installing and using the Symbolic Math Toolbox

Solving problems using algebra and statistics

Performing analysis using linear algebra, calculus, and differential equations

IN THIS CHAPTER

» Using the Symbolic Math Toolbox

» Solving algebraic problems

» Solving statistical problems

Chapter 19

Solving Equations and Finding Roots

MATLAB is amazing when it comes to helping you solve equations and find roots. Of course, getting the right answer happens only when you know how to ask the right question. Communicating with MATLAB in a manner it understands is an important part of solving the questions you have. This chapter demonstrates how to solve specific kinds of equations and how to find roots. The important thing to consider as you read is how the information creates patterns that you can use to solve your specific algebraic or statistical problem.

In most cases, there are multiple ways to obtain an answer to any question. This chapter demonstrates one method for each kind of equation or root. However, you can find additional solutions online in locations such as MATLAB Answers (`https://www.mathworks.com/matlabcentral/answers/`). The point is that MATLAB can provide an answer as long as you have a viable means to ask the question.

REMEMBER

You don't have to type the source code for this chapter; in fact, using the downloadable source is a lot easier. You can find the source for this chapter in the `\MATLAB2\Chapter19` folder of the downloadable source. When using the downloadable source, you may not be able to work through some of the hands-on examples because the example files will already exist on your system. In this case, you can create an alternative folder, `Chapter19a`, to hold the results of the hands-on exercises. See the Introduction for details on how to obtain the downloadable source.

Working with the Symbolic Math Toolbox

The Symbolic Math Toolbox immensely reduces the work required to solve equations. In fact, it might almost seem like magic to some people. Of course, no magic is involved — the clever programmers at MathWorks just make it look that way.

However, before you can begin using the Symbolic Math Toolbox to perform amazing feats, you need to have it installed. If you have the student version, the Toolbox is installed by default, and you can skip the first two sections that follow (going right to the "Working with the GUI" section). Otherwise, start with the first section that follows to get your copy of the Symbolic Math Toolbox and install it on your system.

Obtaining your copy of the Symbolic Math Toolbox

You need to obtain either a trial version or a purchased version of the Symbolic Math Toolbox before you can do anything else. (When getting a trial version, you must discuss the download with someone from MATLAB before you can actually download the product or at least go through a rather confusing series of forms.) Check out the product information at `https://www.mathworks.com/products/symbolic.html#try-buy` and click one of the links in the Try or Buy section of the page. The following steps (with possibly some small changes) assume that you're using the trial version for installation purposes:

1. **Click Download Now.**

   You see the page shown in Figure 19-1 (or one similar to it), which provides options for identifying yourself and determining whether you actually need a copy of the Symbolic Math Toolbox. For example, college students may already have the required toolbox installed.

2. **Provide the required information, select I Agree, and click Submit.**

   The website will ask you to choose a trial release. Click the button that matches the version of your MATLAB installation. You're asked to choose a platform: Windows, macOS, or Linux.

3. **Click a platform option.**

   You see a list of trial products.

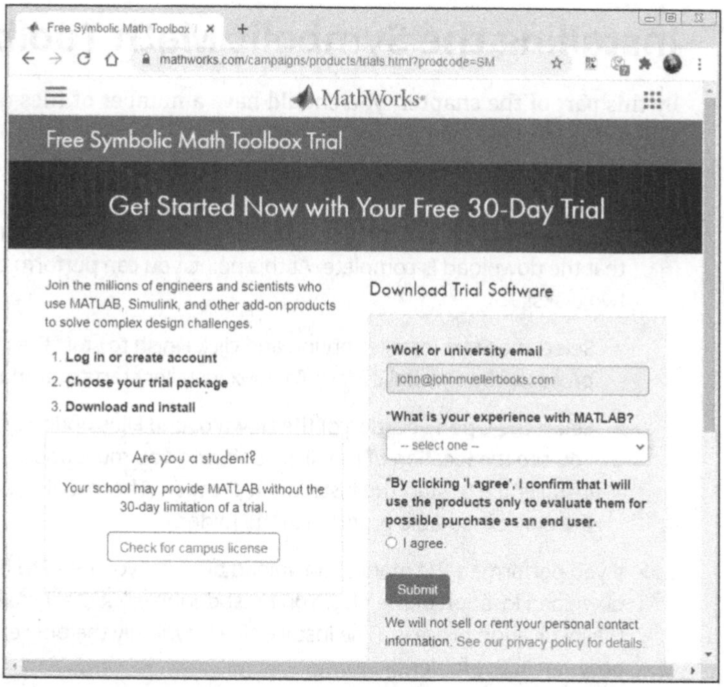

FIGURE 19-1: Verify the need for the toolbox, then supply the required information.

4. **Select the Symbolic Math Toolbox option; then click the Trial Products button that appears near the bottom of the page.**

   The website shows the products you require when using the product you've selected. For example, all the toolboxes require that you have MATLAB installed. You select an option only if you don't have the product installed on your system already. Otherwise, leave the list cleared. You shouldn't have to check any of the items if you've been working through the book because you already have MATLAB, the only required product for the Symbolic Math Toolbox, installed.

5. **Click Continue.**

   You see an informational page for the product. In the upper-left corner, you also see three buttons: Windows, macOS, and Linux.

6. **Click the button that matches your platform.**

   The product and associated installer will download to your system.

Installing the Symbolic Math Toolbox

By this part of the chapter, you should have a number of files on your hard drive. These files provide everything needed to install the Symbolic Math Toolbox. You have two ways to interact with the files:

- If you were able to use the download agent, you see a dialog box telling you that the download is complete. At this point, you can perform one of these two tasks:
  - Select the Start Installer option and click Finish to start the installation process. The Symbolic Math Toolbox installer starts automatically.
  - Select the Open Location of the Downloaded Files option and click Finish. You see the location of the files open, and you must double-click the installer file to start the installation process. (The installer file is typically the only executable program in the folder.)
- If you performed the manual download process, you need to find the download location of the files. You must double-click the installer file to start the installation process. (The installer file is typically the only executable program in the folder.)

Windows platform users may see a User Account Control (UAC) dialog box when starting the installer. Click Yes to give the installer permission to install the Symbolic Math Toolbox. Otherwise, the installation will fail.

No matter how you start the installer, eventually you see a MathWorks Product Installer wizard. This wizard asks questions about how to install the toolbox. The following steps help you complete the installation process. Make sure you have no copies of MATLAB open before you begin this process.

1. **Enter your email address and click Next.**

   The wizard asks for your password.

2. **Enter your password and click Sign In.**

   The wizard shows the licensing terms for using the toolbox.

3. **Select Yes to accept the licensing agreement; then click Next.**

   You see the license selection page, shown in Figure 19-2. It's important to select the correct license, or to enter an activation key, depending on what you have. Select the Self Serve R2020b Trial option only if you don't have a license or activation key to use.

If you see an error message at this point, it could be because the application can't access the MathWorks website. Ensure that your firewall will let MATLAB connect to outside networks, restart your copy of MATLAB, and try again.

FIGURE 19-2: Choose the licensing option that best matches your setup.

4. **Choose a Licensing option and then click Next.**

   The wizard asks for confirmation of your identity.

5. **Enter the required information and then click Next.**

   The wizard asks for a destination folder for the toolbox. This folder should match the folder of your MATLAB installation.

6. **If necessary (normally it's not, because the default is correct), supply a destination for the installation and click Next.**

   At this point, it may appear that the MathWorks Product Installer has frozen. However, the wizard is simply checking all the installation conditions. After performing the required checks, the wizard shows a list of detected products.

7. **Select the Select All option and click Next.**

   You see a list of installation options that include adding a shortcut to the desktop and allowing MATLAB to send usage information to MathWorks.

8. **Select the options you want to use and click Next.**

   You see a list of the installation selections you have made.

9. **Click Begin Install.**

   The wizard performs the required installation tasks. You see a completion page when the process completes.

10. **Click Finish.**

    The installation is complete.

Working with the GUI

When performing tasks with the Symbolic Math Toolbox, you create a Live Script, or possibly a Live Function, as described in Chapters 11 and 12 of the book. The focus when working with the toolbox basics is the `sym()` function. For example, when you type **sym(1/3)** and run the code, you see the symbolic version of the code as a symbolic number. You can see how this use of the `sym()` function works in the `Symbolic_Basics.mlx` file provided with the downloadable source.

You can also create symbolic variables, such as `y = sym('y', [1 5])`, and symbolic expressions, such as:

```
phi = (1 + sqrt(sym(5)))/2;
f = phi^2 - phi - 1
```

In every case, the Live Script display will show the result of the symbolic expression, as shown in Figure 19-3. Notice that each entry is a neatly formatted version that you could easily use for a report.

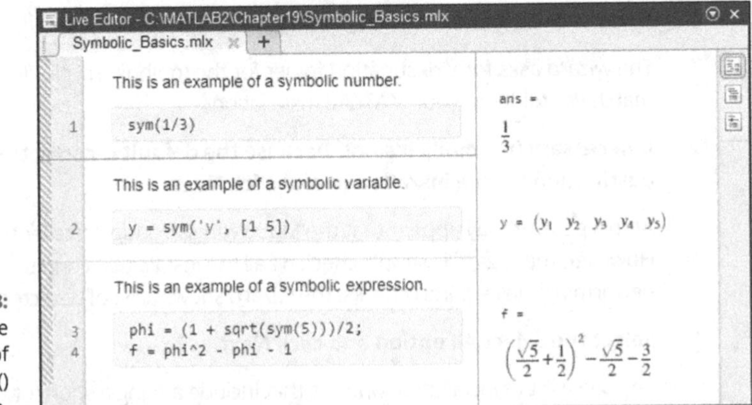

FIGURE 19-3: Viewing the Live Script output of the sym() function.

WHERE HAS MuPAD GONE?

Older versions of the Symbolic Math Toolbox provided an app named MuPAD. However, MathWorks removed this app from MATLAB starting with R2020a. The recommendation now is to use Live Script instead. Fortunately, you can update your MuPAD files to their Live Script form by using the instructions found at `https://www.mathworks.com/discovery/mupad.html`.

Typing a simple command in the Command Window

The essentials of the MATLAB Command Window haven't changed since you installed the Symbolic Math Toolbox, and you may decide to try things there instead of creating a Live Script. This section shows how to use features such as `solve()` (which you can also find in the `UsingSolve.mlx` file). You can get all the details about the Symbolic Math Toolbox functionality at `https://www.mathworks.com/help/symbolic/index.html`. The details about the new functions that MATLAB can access from the Symbolic Math Toolbox appear at `https://www.mathworks.com/help/symbolic/referencelist.html?type=function`. The following steps show how to perform the same task using MATLAB at the Command Window:

1. **Type** syms x y **and press Enter.**

   MATLAB creates two symbolic variables, `x` and `y`. You use symbolic variables when working with `solve()`. When you look in the Workspace window, you see that the variables are actually defined as being symbolic.

2. **Type** solve(2 * x + 3 * y - 22 == 0) **and press Enter.**

   You see the following output:

   ```
   ans =
   11 - (3*y)/2
   ```

   This code solves the equation for x in terms of y.

3. **Type** solve(2 * x + 3 * y - 22 == 0, y) **and press Enter.**

   This time, `solve()` solves for `y` rather than `x`. The output is now:

   ```
   ans =
   22/3 - (2*x)/3
   ```

4. **Type** solve(11 - (3*y)/2) **and press Enter.**

   You now have the value of x, which is 22/3.

5. **Type** solve(2 * 22/3 + 3 * y - 22 == 0) **and press Enter.**

   The value of y is 22/9. The next step is to test the solution by plugging in both values.

6. **Type** 2 * 22/3 + 3 * 22/9 - 22 == 0 **and press Enter.**

   You see an output of

```
ans =
  logical
   1
```

   The output is now logically equal to 0. Figure 19-4 shows this example in Live Script form.

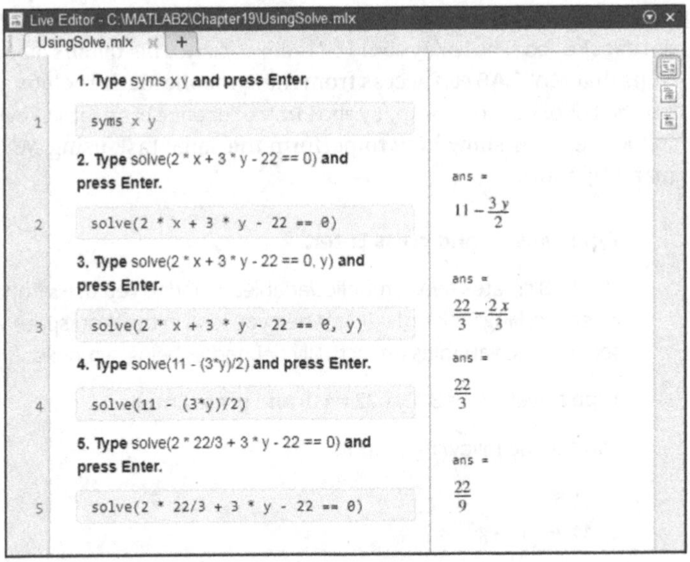

FIGURE 19-4: Seeing the Command Window steps as a Live Script.

Performing Algebraic Tasks

MATLAB lets you perform a wide range of algebraic tasks even without the Symbolic Math Toolbox installed, but adding the toolbox makes performing the tasks easier. The following sections discuss using the Symbolic Math Toolbox to perform a variety of algebraic tasks. You also discover a few alternatives for performing these tasks.

Differentiating between numeric and symbolic algebra

The essential difference between symbolic and numeric algebra is that the first is used by computer science to explore principles of algebra using symbols in place of values, while the second is used by science to obtain approximations of equations for real-world use. In the "Typing a simple command in the Command Window" section, earlier in this chapter, you type equations to perform symbolic algebra. In that case, you use `solve()`, which outputs a precise number (which is why you see a value of `x` that is `22/3`). When you want to perform numeric algebra, you use the `vpasolve()` function instead. (*VPA* stands for Variable Precision Arithmetic.) The following steps demonstrate how to perform this task (you can also see the Live Script form in `UsingVpaSolve.mlx`):

1. **Type** syms x y **and press Enter.**

   Even when performing numeric algebra, you must define the variables before you use them.

2. **Type** vpasolve(2 * x + 3 * y - 22 == 0, x) **and press Enter.**

   You see the following output:

   ```
   ans =
   11.0 - 1.5*y
   ```

   The output is simpler this time, but notice that it also relies on floating-point numbers. To ensure precision, symbolic algebra relies on integers or integer fractions. A floating-point number is an approximation in the computer — an integer is precise.

   REMEMBER

   When working with `vpasolve()`, you must specify which variable to solve for. There is no assumption, and if you don't provide a variable when working with multiple variables, the output is less than useful. In this case, `vpasolve()` solves for `x`.

3. **Type** vpasolve(11.0 - 1.5*y) **and press Enter.**

   You see the following output:

   ```
   ans =
   7.3333333333333333333333333333333
   ```

   The output is a floating-point number. You aren't dealing with a fraction any longer, but the number is also an approximation. You need to note that `vpasolve()` defaults to providing 32-digit output — a double is 16 digits.

4. **Type** vpasolve(2 * 7.3333333333333333333333333333333 + 3 * y - 22 == 0) **and press Enter.**

   You see the following output:

   ```
   ans =
   2.4444444444444444444444444444444
   ```

   Again, the output is a floating-point number. The result is imprecise. However, seeing whether the computer can actually show you how much it's off might be interesting.

5. **Type** 2 * 7.3333333333333333333333333333333 + 3 * 2.444444444444444444 4444444444444 - 22 **and press Enter.**

   MATLAB likely outputs a value of 0. The point is that the two output values truly aren't precise values, but the computer lacks the precision to detect just how much of an error exists. Figure 19-5 shows the Live Script output of this example.

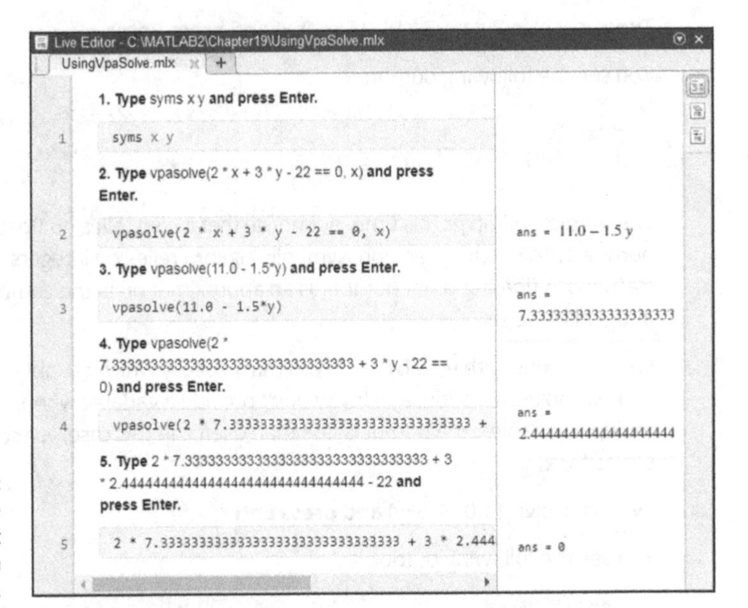

FIGURE 19-5: Seeing the algebraic computation as a Live Script.

Solving quadratic equations

There are times when using the Symbolic Math Toolbox makes things easier, but using it isn't absolutely necessary. This is the case when working with quadratic equations. You actually have a number of ways to solve a quadratic equation, but two straightforward methods exist: `solve()` and `roots()`.

CONVERTING BETWEEN SYMBOLIC AND NUMERIC DATA

Symbolic and numeric objects aren't compatible. You can't directly use one with the other. To make the two coexist, you must perform a conversion. Fortunately, MATLAB makes converting between symbolic and numeric data easy. The following functions perform the conversions for you:

- `double()`: Converts a symbolic matrix to a numeric form.
- `char()`: Converts symbolic objects to plain strings.
- `int8()`: Converts a symbolic matrix into 8-bit signed integers.
- `int16()`: Converts a symbolic matrix into 16-bit signed integers.
- `int32()`: Converts a symbolic matrix into 32-bit signed integers.
- `int64()`: Converts a symbolic matrix into 64-bit signed integers.
- `poly2sym()`: Converts a polynomial coefficient vector to a symbolic polynomial.
- `single()`: Converts a symbolic matrix into single-precision floating-point values.
- `sym()`: Defines new symbolic objects.
- `sym2poly()`: Converts a symbolic polynomial to a polynomial coefficient vector.
- `symfun()`: Defines new symbolic functions.
- `uint8()`: Converts a symbolic matrix into 8-bit unsigned integers.
- `uint16()`: Converts a symbolic matrix into 16-bit unsigned integers.
- `uint32()`: Converts a symbolic matrix into 32-bit unsigned integers.
- `uint64()`: Converts a symbolic matrix into 64-bit unsigned integers.
- `vpa()`: Performs a conversion between symbolic and numeric output. For example, `vpa(22/3)` results in an output of `7.3333333333333333333333333333333`.

REMEMBER

The `solve()` method is actually easier to understand, so type **solve(x^2 + 3*x - 4)** and press Enter. This approach finds the *roots* of the equation, the values of `x` that make the equation equal zero. You see the following output:

```
ans =
-4
 1
```

In this case, you work with a typical quadratic equation. The equation is entered directly as part of the `solve()` input. You don't necessarily need to use a double equals sign (==); using `solve(x^2 + 3*x - 4)` results in the same output as `solve(x^2 + 3*x - 4 == 0)`. However, you must remember to add the multiplication operator. Otherwise, the equation looks precisely as you might write it manually.

The `roots()` approach isn't quite as easy to understand by just viewing it. Type **roots([1 3 -4])** and press Enter. As before, you get the following output:

```
ans =
    -4
     1
```

The outputs are the same. However, when working with `roots()`, you pass a vector containing just the constants (coefficients) for the equation in a vector. Nothing is wrong with this approach, but six months from now, you may look at the `roots()` call and not really understand what it does.

Working with cubic and other nonlinear equations

The Symbolic Math Toolbox makes it easy to solve cubic and other nonlinear equations. The example in this section explores the cubic equation, which takes the form of `ax^3+bx^2+cx+d=0`. Each of the coefficients takes the following values (you can see the Live Script version of this example in `SolvingCubic.mlx`):

```
>> a=2
>> b=-4
>> c=-22
>> d=24
```

Now that you have the parameters for the cubic equation, it's time to solve it. The following steps show you how to solve the problem in the Command Window:

1. **Type** syms x **and press Enter.**

   MATLAB creates the required symbolic object.

2. **Type each of the following coefficients in turn, pressing Enter after each coefficient:**

```
a=2;
b=-4;
c=-22;
d=24;
```

3. **Type** Solutions = solve(a*x^3 + b*x^2 + c*x + d == 0) **and press Enter.**

   You see the following output:

```
Solutions =
-3
 1
 4
```

TIP

Of course, you can get far fancier than the example shown here, but the example gives you a good starting point. The main thing to consider is the coefficients you use. (If you ever want to check your answers, the Cubic Equation Calculator at `http://www.1728.org/cubic.htm` can help.)

Understanding interpolation

MATLAB supports a number of types of *interpolation*, a statistical method that uses related known values to estimate an unknown value. (See `https://whatis.techtarget.com/definition/extrapolation-and-interpolation` for a description of interpolation.) You can see an overview of support for interpolation at `https://www.mathworks.com/help/matlab/interpolation.html`. For this section, you work with 1D interpolation using the `interp1()` function. The following steps show how to perform the task (you can see the Live Script version of this example in `SolvingInterpolation.mlx`):

1. **Type** x = [0, 2, 4]; **and press Enter.**
2. **Type** y = [0, 2, 8]; **and press Enter.**

   These first two steps create a series of points to use for the interpolation.

3. **Type** x2 = [0:.1:4]; **and press Enter.**

   At this point, you need to calculate the various forms of interpolation: linear, nearest, spline, and pchip. Steps 4 through 7 take you through this process. (Older versions of MATLAB also had a cubic option that's been replaced by pchip.)

4. **Type** y2linear = interp1(x, y, x2); **and press Enter.**
5. **Type** y2nearest = interp1(x, y, x2, 'nearest'); **and press Enter.**
6. **Type** y2spline = interp1(x, y, x2, 'spline'); **and press Enter.**

7. **Type** y2pchip = interp1(x, y, x2, 'pchip'); **and press Enter.**

   At this point, you need to plot each of the interpolations so that you can see them onscreen. Steps 8 through 11 take you through this process.

8. **Type** plot(x,y,'sk-') **and press Enter.**

   You see a plot of the points, which isn't really helpful, but it's the starting point of the answer.

9. **Type** hold on **and press Enter.**

   The plot will contain several more elements, and you need to put the plot into a hold state so that you can add them without clearing the previous results.

10. **Type** plot(x2, y2linear, 'g--') **and press Enter.**

    You see the interpolation added to the figure (along with the others as you plot them).

11. **Type** plot(x2, y2nearest, 'b--') **and press Enter.**

12. **Type** plot(x2, y2spline, 'r--') **and press Enter.**

13. **Type** plot(x2, y2pchip, 'm--') **and press Enter.**

14. **Type** legend('Data','Linear', 'Nearest', 'Spline', 'PCHIP', 'Location', 'West') **and press Enter.**

    You see the result of the various calculations, as shown in Figure 19-6.

15. **Type** hold off **and press Enter.**

    MATLAB removes the hold on the figure.

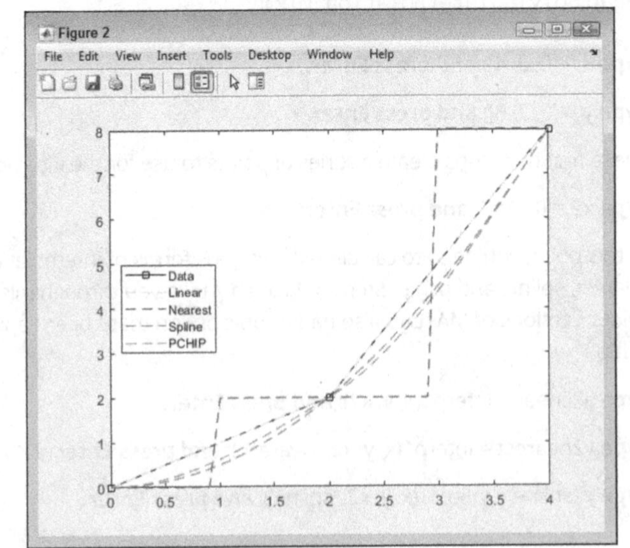

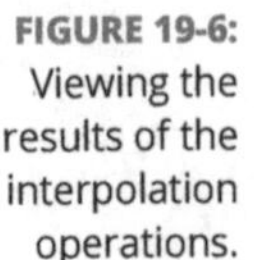

FIGURE 19-6: Viewing the results of the interpolation operations.

Working with Statistics

Statistics is an interesting area of math that deals with the collection, organization, analysis, interpretation, and presentation of data. You use it to determine the probability of the next customer buying your new widget instead of the obviously inferior widget offered by your competition. The fact is that modern business couldn't exist without the use of statistics.

REMEMBER

MATLAB provides basic statistical support. However, if your living depends on working with statistics and you find the default MATLAB support a little lacking, you can check out the Statistics and Machine Learning Toolbox at `https://www.mathworks.com/products/statistics/` to gain additional functionality. Likewise, if you perform a lot of curve fitting, you may find that the Curve Fitting Toolbox found at `https://www.mathworks.com/products/curvefitting/` comes in handy. (Admittedly, you *can* perform elementary ad hoc curve fitting in the figure window, but it's usually not sufficient to get the results you want.) The following sections don't use either the Statistics Toolbox or the Curve Fitting Toolbox for examples — they demonstrate how you can perform these tasks using MATLAB alone.

Understanding descriptive statistics

When working with *descriptive statistics,* the math quantitatively describes the characteristics of a data collection, such as the largest and smallest values, the mean value of the items, and the average. This form of statistics is commonly used to summarize the data, thus making it easier to understand. MATLAB provides a number of commands that you can use to perform basic statistics tasks. The following steps help you work through some of these tasks (you can see the Live Script version of this example in `DescriptiveStatistics.mlx`):

1. **Type** rng('shuffle', 'twister'); **and press Enter.**

   You use the `rng()` function to initialize the pseudo-random number generator to produce a sequence of pseudo-random numbers. Older versions of MATLAB use other initialization techniques, but you should rely on the `rng()` function for all new applications.

   The first value, `shuffle`, tells MATLAB to use the current time as a seed value. A *seed* value determines the starting point for a numeric sequence so that the pattern doesn't appear to repeat. If you want to exactly repeat the numeric sequence for testing purposes, you should provide a number in place of `shuffle`.

The second value, twister, is the number generator to use. MATLAB provides a number of these generators so that you can further randomize the numeric sequences you create. The upcoming "Creating pseudo-random numbers" sidebar discusses this issue in more detail.

2. **Type** w = 100 * rand(1, 100); **and press Enter.**

   This command produces 100 pseudo-random numbers that are uniformly distributed between the values 0 and 1. The numbers are then multiplied by 100 to bring them up to the integer values used in Steps 4 and 5.

3. **Type** x = 100 * randn(1, 100); **and press Enter.**

   This command produces 100 pseudo-random numbers that are normally distributed. The numbers can be positive or negative, and multiplying by 100 doesn't necessarily ensure that the numbers are between –100 and 100 (as you see later in the procedure).

4. **Type** y = randi(100, 1, 100); **and press Enter.**

   This command produces 100 pseudo-random integers (the first argument) that are uniformly distributed between the values of 1 (the second argument) and 100 (the third argument).

5. **Type** z = randperm(200, 100); **and press Enter.**

   This command produces 100 unique pseudo-random integers between the values of 1 and 200. There is never a repeated number in the sequence, but the 100 values are selected from the range of 1 to 200.

6. **Type** AllVals = [w; x; y; z]'; **and press Enter.**

   This command creates a 100 x 4 matrix for plotting purposes. Combining the four values lets you create a plot with all four distributions without a lot of extra steps.

7. **Type** hist(AllVals, 50); **and press Enter.**

   You see a histogram created that contains all four distributions.

8. **Type** legend('rand', 'randn', 'randi', 'randperm'); **and press Enter.**

   Adding a legend helps you identify each distribution, as shown in Figure 19-7. Notice how the various distributions differ. Only the randn() distribution provides both positive and negative output.

9. **Type** set(gca, 'XLim', [0, 200]); **and press Enter.**

   Figure 19-8 shows a close-up of the rand(), randi(), and randperm() distributions, which are a little hard to see in Figure 19-7. Notice the relatively even lines for randperm(). The rand() and randi() output has significant spikes.

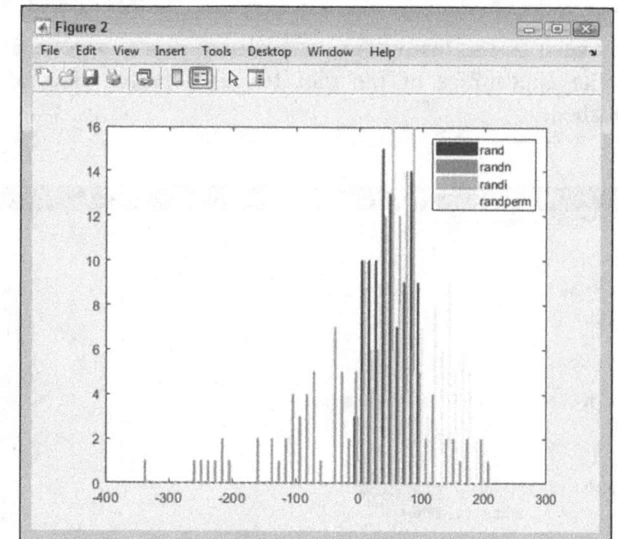

FIGURE 19-7: The histogram shows the distribution of the various numeric values.

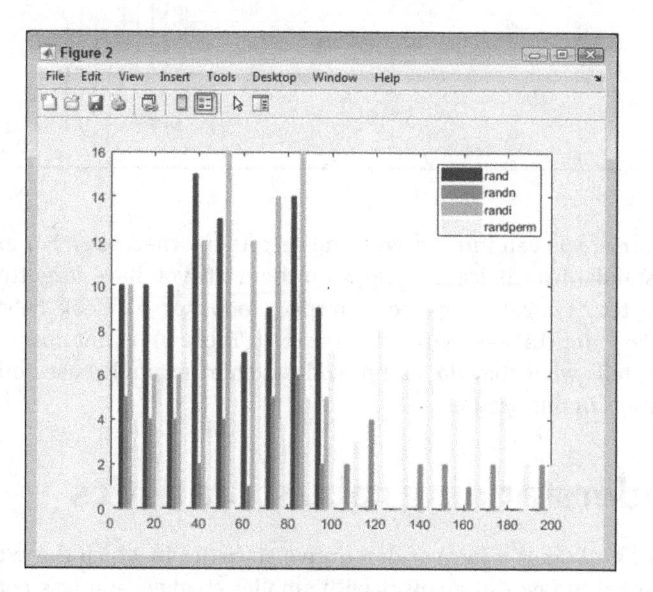

FIGURE 19-8: Zoom in to see the differences in distributions better.

TIP

When working with a Live Script, you must place a section break between Steps 8 and 9 to see both versions of the histogram. Otherwise, you see just the version produced by Step 9. Figure 19-9 shows part of the Live Script for this example.

This procedure has demonstrated a few aspects of working with statistics, the most important of which is that choosing the correct function to generate your random numbers is important. When viewing the results of your choices, you can use plots such as the histogram. In addition, don't forget that you can always modify the appearance of the plot to get a better view of what you have accomplished.

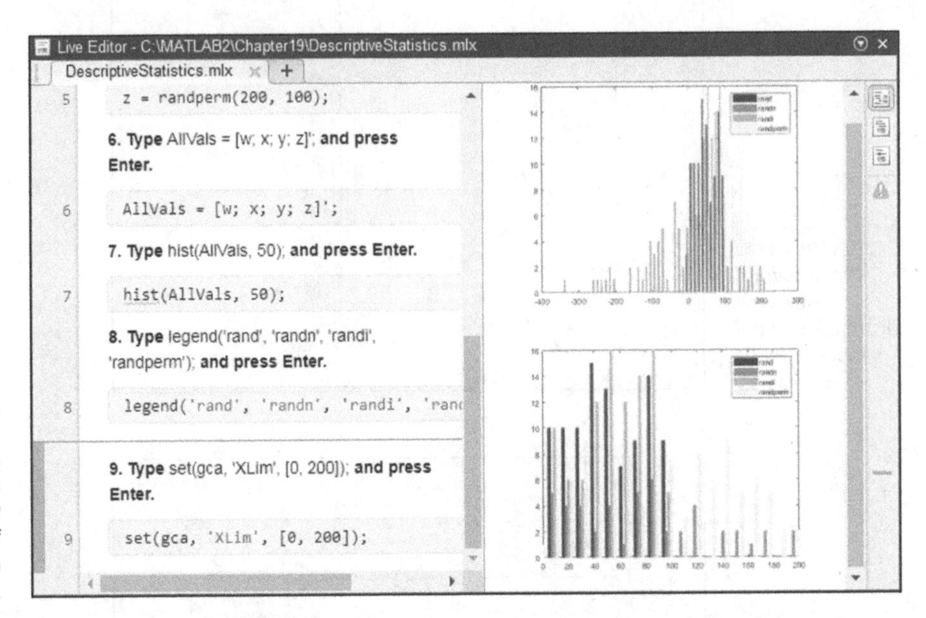

FIGURE 19-9: Use section breaks to allow the display of multiple graph versions.

REMEMBER

Of course, you can interact with the vectors in other ways. For example, you can use standard statistical functions on them. (If you have forgotten what some of these terms mean, check out `https://anothermathgeek.hubpages.com/hub/How-to-calculate-simple-statistics`.) Table 19-1 contains a list of the functions, tells what they do, and provides a short example based on the example in the steps in this section.

Understanding robust statistics

Robust statistics is a form of descriptive statistics in which the extreme values are discarded in favor of analysis with smaller changes and less potential for error. You use robust statistics when you have a potential for error in the extreme values. For example, you might use it when trying to figure out the average height of drivers today compared with those of 1940. However, you wouldn't use it when building a bridge, because the extreme values are important in this second case. When working with MATLAB without any of the specialized toolkits, the best way to create robust statistics is to simply eliminate the largest and smallest values from a vector.

TABLE 19-1 MATLAB Basic Statistical Functions

Function	Usage	Example
`corrcoef()`	Determines the correlation coefficients between members of a matrix.	`corrcoef(AllVals)`
`cov()`	Determines the covariance matrix for either a vector or a matrix.	`cov(AllVals)`
`max()`	Specifies the largest element in a vector. When working with a matrix, you see the largest element in each row.	`max(w)`
`mean()`	Calculates the average or mean value of a vector. When working with a matrix, you see the mean for each row.	`mean(w)`
`median()`	Calculates the median value of a vector. When working with a matrix, you see the median for each row.	`median(w)`
`min()`	Returns the smallest element in a vector. When working with a matrix, you see the smallest element in each row.	`min(w)`
`mode()`	Determines the most frequent value in a vector. When working with a matrix, you see the most frequent value for each row.	`mode(w)`
`std()`	Calculates the standard deviation for a vector. When working with a matrix, you see the standard deviation for each row.	`std(w)`
`var()`	Determines the variance of a vector. When working with a matrix, you see the variance for each row.	`var(w)`

CREATING PSEUDO-RANDOM NUMBERS

Creating truly random numbers on a computer is impossible without resorting to some really exotic techniques. All random sequences on a computer are generated by algorithms, making them pseudo-random. The numeric sequence has a pattern that eventually repeats itself. Depending on the algorithm used, the sequence can be quite long and, to a human, nearly indistinguishable from a random sequence. However, other computers aren't fooled, and any computer can eventually detect the repetition of the sequence.

MATLAB provides three methods of making the pseudo-random numbers created by it appear more random. The first is the seed value. Choosing a different starting point in the numeric sequence (which is quite large) means that people are less likely to actually notice any repetition. The second method provides four different distributions: `rand()` produces a uniform distribution of random numbers; `randn()` produces a normalized distribution of random numbers; `randi()` produces a uniform distribution of integers

(continued)

(continued)

(where only whole numbers are needed); and `randperm()` produces a random permutation (where each number appears only once). The third method provides a number of randomizing generators (or engines) — essentially different algorithms — to produce the numeric sequence:

- `'combRecursive'`: Relies on the combined Multiple Recursive Generator (MRG) (see `https://www.value-at-risk.net/multiple-recursive-generators/` for details).
- `'multFibonacci'`: Uses the Multiplicative Lagged Fibonacci Generator (MLFG) variant of the Lagged Fibonacci Generator (LFG) (see `https://northstar-www.dartmouth.edu/doc/sprngsv1.0/DOCS/www/paper/node13.html` for details).
- `'twister'`: Relies on the Mersenne Twister algorithm (see `http://www.math.sci.hiroshima-u.ac.jp/~m-mat/MT/emt.html` for details).
- `'simdTwister'`: Defines a faster, Single Instruction/Multiple Data (SIMD), version of the 'twister' algorithm (see `https://walkingrandomly.com/?p=6502` for details.) You can discover more about SIMD at `https://us.fixstars.com/products/opencl/book/OpenCLProgrammingBook/parallel-computing-hardware/`.
- `'philox'`: Relies on the 64-bit, counter-based, Philox algorithm (see `https://numpy.org/doc/stable/reference/random/bit_generators/philox.html` for details).
- `'threefry'`: Relies on either a 32-bit or 64-bit counter-based generator (see `https://bashtage.github.io/randomgen/bit_generators/threefry.html` for details).

You also have access to these legacy generators, but it's best to use one of the newer versions. Use the `rng()` function, rather than the `rand()` function used in the past, to randomize the generators whenever possible (see `https://www.mathworks.com/help/matlab/math/updating-your-random-number-generator-syntax.html` for details).

- `'v5uniform'`: Specifies the legacy MATLAB 5.0 uniform generator, which produces the same result as the `'state'` option with `rand()`.
- `'v5normal'`: Specifies the legacy MATLAB 5.0 normal generator, which produces the same result as the `'state'` option with `randn()`.
- `'v4'`: Specifies the legacy MATLAB 4.0 generator, which produces the same result as using the `'seed'` option.

REMEMBER

The easiest way to find and remove the largest and smallest values is to use the statistical functions found in Table 19-1. For example, to remove the largest value from a vector, a, you use a(a == max(a)) = [];. The max(a) part of the command finds the maximum value in vector a. The index (a == max(a)) tells MATLAB to find the index where the maximum value resides. You then set this element to an empty value, which deletes it.

Removing the smallest value from a vector is almost the same as removing the largest value. However, in this case you use a(a == min(a)) = [];. Notice that the min(a) function has taken the place of max(a).

To verify that the changes are successful, you use the std(), or standard deviation, function. As you remove large and small values, you start to see smaller std() output values.

Employing the Symbolic Math Toolbox for plotting

MATLAB provides access to a wide variety of plotting function, and you see them spread in various places throughout the book, such as in Chapters 6 and 7. The Symbolic Math Toolbox provides another means of plotting, fplot(), that involves using functions. This section shows how to use fplot() to visualize the output of symbolic functions you create (you can see this example in SMTPlot.mlx file).

One of the easiest ways to see fplot() in action is to use one of the built-in MATLAB functions, such as sin(). The simplest version of this plot calls fplot(), as shown here:

```
syms x
fplot(sin(x))
```

Of course, you might want to see a particular range of outputs, so you can add them as a second argument: fplot(sin(x), [-pi pi]). If you want, you can plot multiple functions simultaneously. For example, this call, fplot([sin(x), cos(x)], [-pi pi]), produces the output shown in Figure 19-10.

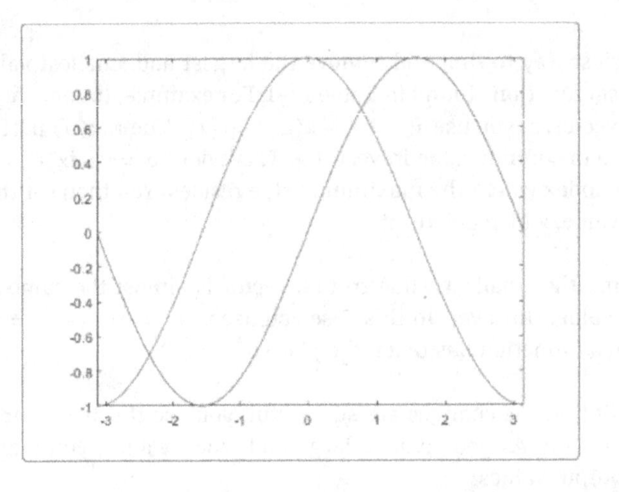

FIGURE 19-10: You can plot multiple functions in a single call to `fplot()`.

You can use all the same line styles, markers, and colors with `fplot()` that you do when working with other MATLAB plotting functions. Setting these visual elements up using a single call can be time consuming, so the best option is to make one call for each function, as shown here:

```
fplot(sin(x), [-pi pi], '--g')
hold on
fplot(cos(x), [-pi pi], ':ob')
hold off
```

It wouldn't be helpful if you couldn't create custom functions to work with `fplot()`. Here's an example of using a symbolic function:

```
syms f(x);
f(x) = tan(x);
fplot(f, [0 5])
```

The limitation on custom functions is that they can have only one input variable. However, you can plot as many custom functions as needed. Here's another example with two plot custom functions:

```
syms t1
x = (sin(t1^2));
y = (cos(t1^2));

fplot(x, [0 5], '--+g')
hold on
fplot(y, [0 5], ':*b')
hold off
```

Another way to plot multiple custom lines is to place the functions in a vector, like this: `fplot([x y], [0 5])`. However, using this form means that you would have to modify features like line style and markers outside the plotting call.

You can also create parametric curves using `fplot()`. A *parametric curve* is one in which the curve defined by an x and a y coordinate equation traces changes in a parameter, so that the curve has direction. Think about a curve that would trace Earth's orbit around the Sun. The article at `http://mathonline.wikidot.com/parametric-curves` provides more information about parametric curves. The following code creates a circle:

```
syms t2
x = (sin(t2^2));
y = (cos(t2^2));
fplot(x,y, [0 5])
```

IN THIS CHAPTER

» Working with linear algebra

» Using calculus

» Resolving differential equations

Chapter 20
Performing Analysis

Chapter 19 introduces you to the Symbolic Math Toolbox and shows you how to use it to perform a number of tasks. This chapter expands on some of the information presented in Chapter 19. In fact, before you proceed, make sure that you have the Symbolic Math Toolbox installed. The student version of MATLAB comes with the Symbolic Math Toolbox installed by default — otherwise, you must install it manually, using the instructions found in the first section of Chapter 19.

This chapter doesn't take a detailed look at any one particular area of solving equations, but it does give you a good start on working with linear algebra, calculus, and differential equations. Being able to solve these equations quickly and correctly can make a huge difference in the success of your project. Some of the concepts in this chapter are a little advanced, so you see links for additional information about many of them. In addition, Chapter 5 provides you with a lot of details about working with matrices in MATLAB, so a review of that chapter is helpful.

TIP

You may think that these kinds of math are used only in high-tech environments, such as building a sub that can safely traverse the Marianas Trench. (You can read about James Cameron's successful exploration of the trench using a custom submarine at `https://news.nationalgeographic.com/news/2012/03/120325-james-cameron-mariana-trench-challenger-deepest-returns-science-sub/`.) However, these equations are used in everyday life. For example, check out this story about how linear algebra can be employed to make a restaurant more profitable: `https://smallbusiness.chron.com/restaurants-use-linear-programming-menu-planning-37132.html`. Lest you think that this story is uncommon, check out this second story on calculating food costs at `https://smallbusiness.chron.com/calculate-food-cost-restaurant-39551.html`. The point is that you don't really know when or where you'll encounter these equations, so it's a good idea to be prepared to use them. In many cases, the person employing the high-level math doesn't even realize they're doing it, as in *13 Examples Of Algebra In Everyday Life* at `https://studiousguy.com/examples-of-algebra-in-everyday-life/`.

Using Linear Algebra

You use linear algebra to perform a number of tasks with matrixes in MATLAB. For example, you can determine whether a matrix is singular (a square matrix that doesn't have a matrix inverse) or unimodular (a square integer matrix having determinant +1 or −1) by using the `det()` function. You can also reduce a matrix to determine whether it's solvable — that it has an inverse (where matrix `A * inv(A)` results in an identity matrix; see Chapter 5 for additional matrix manipulation details). In fact, you can perform a relatively wide range of tasks using linear algebra with MATLAB and the Symbolic Math Toolbox; the following sections tell you how. (See Chapter 19 for details on the Symbolic Math Toolbox add-on.)

Working with determinants

Determinants are used in the analysis and solution of systems of linear equations. A nonzero value means that the matrix is nonsingular and that the system has a unique solution. A value of 1 usually indicates that the matrix is *unimodular* — that it's a real square matrix, in other words. The function used to obtain the determinant value is `det()`. You supply a matrix, and the output value tells you about the ability to create a solution for that matrix.

REMEMBER

Equally important is to know about the `cond()` function, which tests for singular matrices. Again, you supply a matrix as an input value, and the output provides a condition number that specifies the sensitivity of the matrix to error. An output value near 1 indicates a *well-conditioned* matrix. (The YouTube video at `https://www.youtube.com/watch?v=JODxbi9B3tg` provides an incredibly simplified illustration of the difference between a well-conditioned and an ill-conditioned matrix.)

To see how these two functions work together, type **A = [1, 2, 3; 4, 5, 6; 7, 8, 9];** and press Enter to create a test matrix. This is a singular matrix. Type **cond(A)** and press Enter. The result of 1.1439e+17 (an approximation of infinity) tells you that this is a highly sensitive matrix — a singular matrix. Type **det(A)** and press Enter. The incredibly small output value of 6.6613e-16 (an approximation of zero) tells you that this is a singular matrix. (If you want to see a perfect singular matrix, try `[0, 0, 0; 0, 0, 1; 0, 0, 0];`. The `cond()` value is `Inf`, or infinity, and the `det()` value is `0`.)

For comparison purposes, try a unimodal matrix. Type **B = [2, 3, 2; 4, 2, 3; 9, 6, 7];** and press Enter to create the matrix. Type **cond(B)** and press Enter to see the condition number of `313.1721`, which isn't a perfect unimodal matrix, but it's quite close. Type **det(B)** and press Enter to see the result of `1.0000`, which is good (doesn't approximate `0`) for a unimodal matrix.

Performing reduction

Reduction lets you see the structure of what a matrix represents, as well as to write solutions to the system equations that the matrix represents. MATLAB provides the `rref()` function to produce the Reduced Row Echelon Form (RREF). (You can find out more about RREF at `https://www.educba.com/matlab-rref/`.) You can find an interesting tool to see the steps required to produce RREF using any matrix as input at `https://www.emathhelp.net/calculators/linear-algebra/reduced-row-echelon-form-rref-caclulator/`. The point is that you can perform reduction using MATLAB, and doing so requires only a couple of steps.

The first step is to create the matrix. In this case, the example uses a magic square. Type **A = magic(5)** and press Enter. The `magic()` function will produce a magic square of any size for you. (You can read about magic squares at `https://mathworld.wolfram.com/MagicSquare.html`). The output you see is

```
A =
    17    24     1     8    15
    23     5     7    14    16
     4     6    13    20    22
    10    12    19    21     3
    11    18    25     2     9
```

The second step is to perform the reduction. Type **rref(A)** and press Enter. Any nonsingular matrix will reduce to identity, as follows:

```
ans =
     1     0     0     0     0
     0     1     0     0     0
     0     0     1     0     0
     0     0     0     1     0
     0     0     0     0     1
```

You can use rref() to solve linear equations. In this case, if A*x=y and y=[1;0;0;0;0], then B=rref([A,y]) solves the equation. The following steps demonstrate how this works:

1. **Type** y=[1;0;0;0;0]; **and press Enter.**
2. **Type** A=magic(5); **(in case you didn't type it before) and press Enter.**
3. **Type** B=rref([A,y]) **and press Enter.**

   You see the following output:

```
B =
1.0000        0        0        0        0   -0.0049
     0   1.0000        0        0        0    0.0431
     0        0   1.0000        0        0   -0.0303
     0        0        0   1.0000        0    0.0047
     0        0        0        0   1.0000    0.0028
```

4. **Type** x=B(:,6) **and press Enter.**

   You see the following output:

```
x =
   -0.0049
    0.0431
   -0.0303
    0.0047
    0.0028
```

   At this point, you want to test the equation.

5. **Type** A*x **and press Enter.**

   You see the following output:

```
ans =
    0.9999
   -0.0001
   -0.0001
   -0.0001
   -0.0001
```

   Notice that the output values match the original value of y to within a small amount. In other words, the steps have proven the original equation, A*x=y, true.

Using eigenvalues

An *eigenvalue* (v) is an eigenvector of a square matrix. You see this form of math used in transformations as described at https://www.mathsisfun.com/algebra/eigenvalue.html. The variable A is a nonzero matrix. When v is multiplied by A, it yields a constant multiple of v that is commonly denoted by λ. Eigenvalues are defined by the following equation:

```
Av = &#x03BB;v
```

Eigenvalues are used in all sorts of ways, such as for graphics manipulation (sheer mapping) and analytic geometry (to display an arrow in three-dimensional space). You can read more about eigenvalues at https://mathworld.wolfram.com/Eigenvalue.html and https://en.wikipedia.org/wiki/Eigenvalues_and_eigenvectors.

To see how this works, you first need to create a matrix. Type **A = gallery('riemann', 4)** and press Enter. The gallery() function produces test matrices of specific sizes filled with specific information so that you can repeat test results as needed. The output of gallery() depends on the matrix size and the function used to create the matrix. (You can read more about gallery() at https://www.mathworks.com/help/matlab/ref/gallery.html.) The output from this particular call is

```
     1    -1     1    -1
    -1     2    -1    -1
    -1    -1     3    -1
    -1    -1    -1     4
```

Obtaining the eigenvalue comes next. The output will contain one value for each row of the matrix. Type **lambda = eig(A)** and press Enter to see the eigenvalue of the test matrix, A, as shown here:

```
lambda =
   -0.1249
    2.0000
    3.3633
    4.7616
```

Understanding factorization

Factorization is the decomposition of an object, such as a number or polynomial. The idea behind factorization is to reduce the complexity of the object so that it's easier to understand and solve. In addition, it helps you determine how the object is put together, such as its use for prime factorization (see https://www.calculatorsoup.com/calculators/math/prime-factors.php). You can read more about factorization at https://mathworld.wolfram.com/Factorization.html.

You perform factorization in MATLAB using the `factor()` function. You can use the `factor()` function in a number of ways: working with numbers and working with polynomials.

When working with a number, you simply provide the number as input to obtain a list of prime number values. For example, type **factor(2)** and press Enter. The output is 2 because 2 is a prime number. Type **factor(12)** and press Enter. The output is

```
ans =
     2     2     3
```

because 2 * 2 * 3 equals 12. You can also handle the number symbolically. Type **factor(sym(12))** and press Enter to see an output of

```
ans =
[2, 2, 3]
```

REMEMBER

When working with negative numbers, you use the symbolic form or you receive an error message saying the following:

```
N must be a real nonnegative integer.
```

Instead of entering the number directly, you type **factor(sym(-12))** and press Enter to see an output of

```
ans =
[-1, 2, 2, 3]
```

You can use `factor()` on fractional values. Type **factor(sym(14/3))** and press Enter to see an output of `[2, 7, 1/3]` because `2 * 7 * 1/3` would equal `14/3`.

Polynomials require that you declare symbolic objects first by using `syms`. Type **syms x y** and press Enter to create the required objects. Type **factor(x^2 + 2*x*y + y^2)** and press Enter. The output is `[x + y, x + y]`.

It's possible to achieve particular results when using a factor mode with `factor()`. The following list shows what happens when you add the `'FactorMode'` argument and a variable to the `factor()` call.

Type	To Get
factor(3*x^2 - 2*x + 22)	3*x^2 - 2*x + 22
factor(3*x^2 - 2*x + 22, x, 'FactorMode', 'real')	[3.0, x^2 - 0.66666666666666666666666666666667*x + 7.3333333333333333333333333333333]
factor(3*x^2 - 2*x + 22, x, 'FactorMode', 'complex')	[3.0, x - 0.33333333333333333333333333333333 + 2.6874192494328498841222044101013i, x - 0.33333333333333333333333333333333 - 2.6874192494328498841222044101013i]
factor(3*x^2 - 2*x + 22, x, 'FactorMode', 'full')	[3, x + (65^(1/2)*1i)/3 - 1/3, x - (65^(1/2)*1i)/3 - 1/3]

You can combine the output of the `'full'` mode with `vpa()` to approximate a result. For example, using the `'full'` mode results of `3*x^2 - 2*x + 22` with `vpa()`, you obtain

```
[3.0, ...
 x - 0.33333333333333333333333333333333 + ...
     2.6874192494328498841222044101013i, ...
 x - 0.33333333333333333333333333333333 - ...
     2.6874192494328498841222044101013i]
```

as a result.

Employing Calculus

Calculus can solve myriad problems that algebra can't. It's really the study of how things change. This branch of math is essentially split into two pieces: differential calculus, which considers rates of change and slopes of curves; and integral calculus, which considers the accumulation of quantities and the areas between and under curves. The following sections show how you can use MATLAB with the Symbolic Math Toolbox to solve a number of relatively simple calculus problems.

Working with differential calculus

MATLAB offers good differential calculus support. The example in this section starts with something simple: univariate differentiation. (Remember that *univariate* differentiation has a single variable.) MATLAB supports a number of forms of differential calculus — each of which requires its own set of functions. In this case, you use the `diff()` function to perform the required tasks. The following steps help you perform a simple calculation:

1. **Type** syms x **and press Enter.**

   MATLAB creates a symbolic object to use in the calculation.

2. **Type** f(x) = sin(x^3) **and press Enter.**

   Doing so creates the symbolic function used to perform the calculation. Here's the output you see:

```
f(x) =
sin(x^3)
```

3. **Type** Result = diff(f) **and press Enter.**

   The output shows the result of the differentiation:

```
Result(x) =
3*x^2*cos(x^3)
```

REMEMBER

`Result(x)` is actually a symbolic function. You can use it to create a picture of the output.

4. **Type** plot(Result(1:50)) **and press Enter.**

   Figure 20-1 shows the plot created from the differentiation of the original symbolic function.

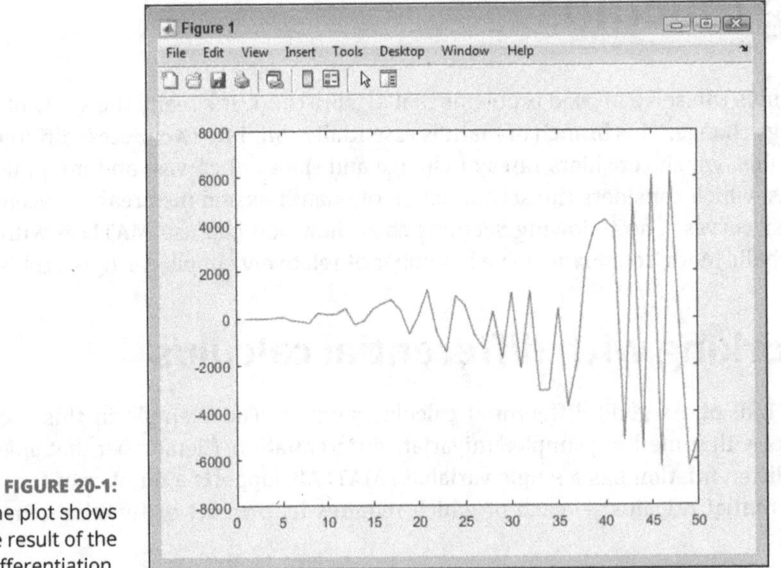

FIGURE 20-1: The plot shows the result of the differentiation.

Using integral calculus

You'll also find great integral calculus support in MATLAB. As does the example in the preceding section, the example in this section focuses on a univariate calculation. In this case, the example relies on the `int()` function to perform the required work. The following steps help you perform a simple calculation:

1. **Type** syms x **and press Enter.**

   MATLAB creates a symbolic object to use in the calculation.

2. **Type** f(x) = (x^3 + 3*x^2) / x^3 **and press Enter.**

   The symbolic function that you create produces the following output:

   ```
   f(x) =
   (x^3 + 3*x^2)/x^3
   ```

3. **Type** Result = int(f, x) **and press Enter.**

   Notice that you must provide a symbolic variable as the second input. The output shows the following symbolic function as the result of the integration:

   ```
   Result(x) =
   x + 3*log(x)
   ```

4. **Type** plot(Result(1:50)) **and press Enter.**

   Figure 20-2 shows the plot created from the integration of the original symbolic function.

Working with multivariate calculus

The "Working with differential calculus" section, earlier in the chapter, shows how to work with a single variable. Of course, many (if not most) problems don't involve just one variable. With this idea in mind, the following steps demonstrate a problem with more than one variable — a *multivariate* example:

1. **Type** syms x y **and press Enter.**

   MATLAB creates the two symbolic objects used for this calculation.

2. **Type** f(x, y) = x^2 * sin(y) **and press Enter.**

   This symbolic function accepts two inputs, x and y, and uses them to perform a calculation. Here's the output from this step:

   ```
   f(x, y) =
   x^2*sin(y)
   ```

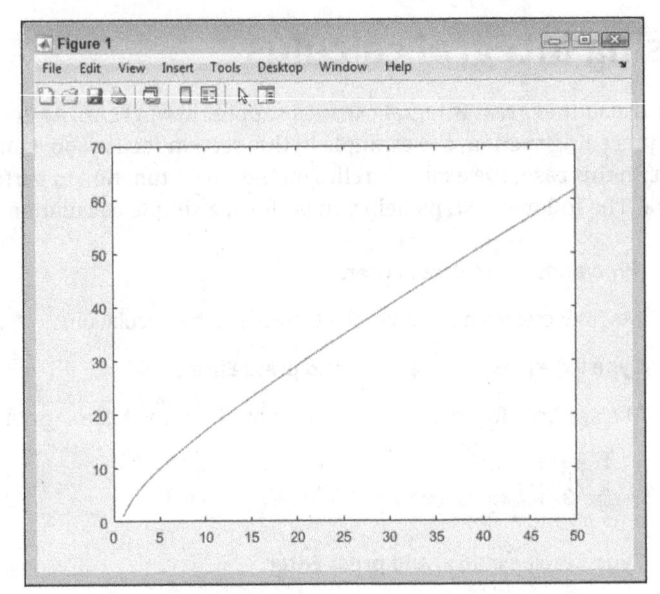

FIGURE 20-2: The output is usually a curve.

3. **Type** Result = diff(f) **and press Enter.**

   The output shows the result of the differentiation:

```
Result(x, y) =
2*x*sin(y)
```

   In this case, `Result(x, y)` accepts two inputs, `x` and `y`. As before, you can create a picture from the output of `Result()`.

TIP

The example shows the derivative with respect to `x`, which is the default. To obtain the derivative with respect to `y (df/dy)`, you type `diff(f, y)` instead.

4. **Type** plot(Result(1:50, 1:50)) **and press Enter.**

   Figure 20-3 shows the output of the plot created in this case.

REMEMBER

Notice that in this case, you must provide both `x` and `y` inputs, which isn't surprising. However, the two vectors must have the same number of elements or MATLAB will raise an exception.

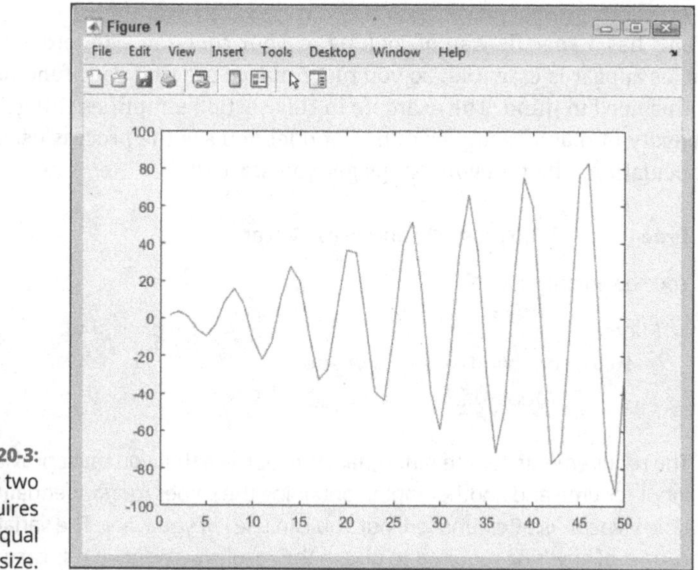

FIGURE 20-3: Using two variables requires vectors of equal size.

Solving Differential Equations

When working with differential equations, MATLAB provides two different approaches: numerical and symbolic. The following sections demonstrate both approaches to solving differential equations. Note that these sections provide just an overview of the techniques; MATLAB provides a rich set of functions to work with differential equations.

Using the numerical approach

When working with differential equations, you must create a function that defines the differential equation. This function is passed to MATLAB as part of the process of obtaining the result. There are a number of functions you can use to perform this task; each has a different method of creating the output. You can see a list of these functions at `https://www.mathworks.com/help/matlab/ordinary-differential-equations.html` and additional information at `https://www.mathworks.com/help/matlab/numerical-integration-and-differential-equations.html`. The example in this section uses `ode23()`, but the technique works for the other functions as well (see `https://www.mathworks.com/help/matlab/math/choose-an-ode-solver.html` for a list of standard functions).

REMEMBER

MATLAB has a specific way of looking at your function. The order in which the variables appear is essential, so you must make sure that your function is created with this need in mind. The example in this section simplifies things to avoid the complexity of many examples online and let you see the process used to perform the calculation. The following steps get you started:

1. **Type** Func = @(T, Y) cos(T*Y) **and press Enter.**

   You see an output of

   ```
   Func =
     function_handle with value:
       @(T,Y)cos(T*Y)
   ```

   The requirements for the differential function are that you must provide an input for time and another input containing the values for your equation. The time value, `T`, is often unused, but you can use it if you want. The variables can consist of anything required to obtain the result you want. In this case, you input a simple numeric value, `Y`, but inputs can be vectors, matrices, or other objects as well.

2. **Type** [TPrime, YPrime] = ode23(Func, [-10, 10], .2); **and press Enter.**

   When using `ode23()`, you must provide a function — `Func` in this case — as input. As an alternative, you provide the name of the file containing the function. The second argument is a vector that contains the starting and ending times of the calculation. The third argument is the starting input value for the calculation.

   The `TPrime` output is always a vector that contains the time periods used for the calculation. The `YPrime` output is a vector or matrix that contains the output value or values for each time period. In this case, `YPrime` is a vector because there is only one output value.

3. **Type** plot(TPrime, YPrime) **and press Enter.**

   You see the plotted result for this example, as shown in Figure 20-4.

Using the symbolic approach

When working with the symbolic approach, you rely on the functionality of the Symbolic Math Toolbox to speed the solution and make it easier to solve. Even though the solution in the previous section looks easy, it can become quite complicated when you start working with larger problems. The symbolic approach is straightforward because you use `dsolve()` (see `https://www.mathworks.com/help/symbolic/dsolve.html`). The following steps show a simple example of using dsolve() to create a differential solution and then plot it (this technique will generate a warning message that you can ignore; it's the easiest way to perform this task):

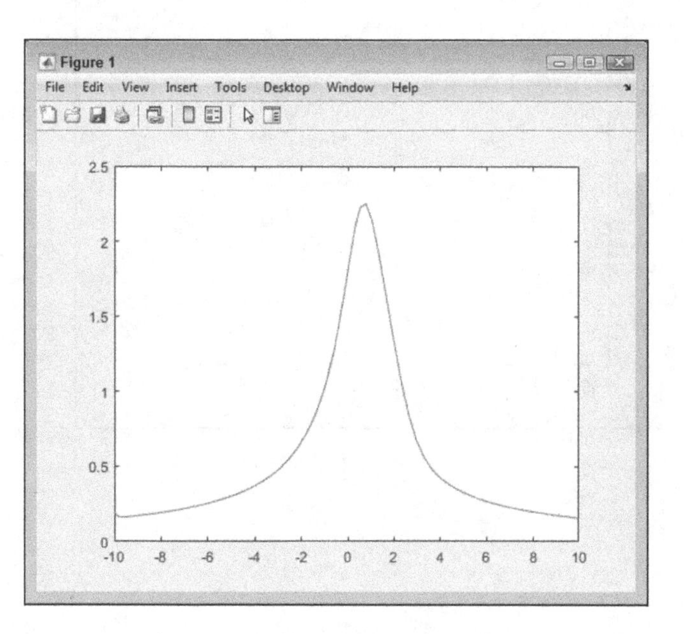

FIGURE 20-4: Obtaining a result using the numeric approach.

1. **Type** Solution = dsolve('Dy=(t^2*y)/y', 'y(2)=1', 't') **and press Enter.**

   The arguments to dsolve() consist of the equation you want to solve, the starting point for y (a condition), and the name of the independent variable. You see the following output from this entry:

   ```
   Solution =
   t^3/3 - 5/3
   ```

2. **Type** Values = subs(Solution, 't', -10:.1:10); **and press Enter.**

   Solution simply contains the solution to the equation given the conditions you provide. The subs() function substitutes values for t one at a time. In this case, the values range from –10 to 10 in 0.1 increments. When this command completes, Values contains a list of results for the values you provided that you can use as plot points.

3. **Type** plot(Values) **and press Enter.**

   You see the output shown in Figure 20-5.

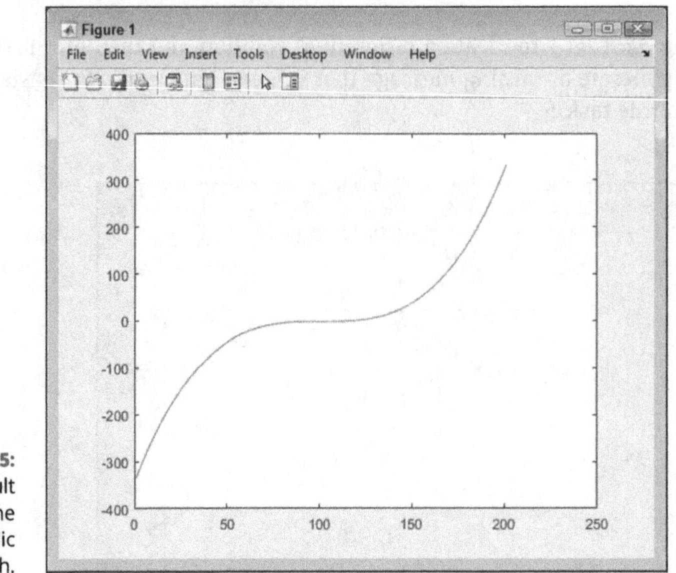

FIGURE 20-5: Obtaining a result using the symbolic approach.

6 The Part of Tens

IN THIS PART . . .

Performing analysis and conducting research

Delving into audio and graphic analysis

Working with sensors and machines

Finding your next occupation

IN THIS CHAPTER

» Performing analysis and solving problems

» Conducting research

» Working with audio and graphics

» Interacting with sensors and machines

Chapter 21

The Top Ten Uses of MATLAB

MATLAB is used in a lot of different ways by lots of people in occupations you might not necessarily think about when it comes to a math product. In fact, math is used in ways that many people don't consider. For example, video games simply wouldn't exist without a lot of relatively complex math. However, even the chair you're sitting in required some use of math to ensure that it would perform as expected. A mission to Mars or to the bottom of the ocean would be impossible without all sorts of different math applications. Even the mixtures of food we eat require an explanation based on math principles. In short, you might be surprised at just how many different ways MATLAB is being used, and this chapter tells you only about ten of the most popular uses. Explaining every potential MATLAB use would likely require an entire book.

Working with Linear Algebra

It may be hard to believe, but linear algebra really is part of the workplace (and not just for sending someone to the moon). For example, to calculate Return on Investment (ROI), you must know algebra. The same holds true for the following:

- Predicting the amount of turnover a company will have
- Determining how many items to keep in inventory
- Making life and business decisions, such as whether it's cheaper to rent a car or to buy one outright
- Creating a financial plan, such as determining whether it makes more sense to pay down a credit card or build up savings

No one would obtain MATLAB to perform these tasks just one time. However, if your job is helping people perform these sorts of tasks, you really do need something like MATLAB so that you can get the answers you need fast.

TIP

Uses for linear algebra often appear in places that you might never consider. For example, if you're a restaurant owner, you might use linear algebra on a regular basis to make your business more efficient. Check out the article at `https://smallbusiness.chron.com/restaurants-use-linear-programming-menu-planning-37132.html` for details. Imagine how surprised you might be if you walked into the back room of a restaurant sometime to find the manager poking away at a keyboard with a copy of MATLAB running!

Performing Numerical Analysis

Numerical analysis (see `https://www.scienceabc.com/eyeopeners/why-do-we-need-numerical-analysis-in-everyday-life.html` for a history and overview) relies on approximation rather than the precision you see in symbolic math. It seems that the world is filled with approximations, and so is the galaxy. Performing certain building construction tasks is impossible without applying numerical analysis, and astronomy seems to require heavy use of it as well. You probably won't see a carpenter applying numerical analysis on the job site with MATLAB, but you will see architects who might need to do so.

REMEMBER

Numerical analysis truly does apply to the natural world. For example, much of modern biology and medicine rest on principles described using numerical analysis. Your family physician probably doesn't require a copy of MATLAB, but the researcher who provides your physician with the information needed to diagnose any problems with your health does. When it comes to numerical analysis, you're better off thinking about the creative end of things rather than the application. The person inventing a new procedure needs MATLAB, but the person applying it on the job site doesn't.

Designing a Neural Network Simulation

A *neural network* makes it possible to simulate the functionality of the human mind and apply it to solving certain classes of problems. You often see neural networks used for sales forecasting, customer research, data validation, and risk management. Delving into the intricacies of neural networks is outside the scope of this book, but you can learn more about them in *Deep Learning For Dummies*, by John Paul Mueller and Luca Massaron. You can also read a short overview of the topic at `https://www.mathworks.com/discovery/neural-network.html`.

The MATLAB Deep Learning Toolbox (`https://www.mathworks.com/products/deep-learning.html`) helps you perform a wide variety of tasks with neural networks. Adding the Parallel Computing Toolbox (`https://www.mathworks.com/products/parallel-computing.html`) increases processing speed so that you can obtain the answers you need quickly. In fact, getting a timely response is a critical part of designing a neural network.

TIP

Trying to picture how a neural network does its work can be quite difficult because the concepts are abstract. One way to overcome this problem is to use MATLAB to create a neural network simulation. The article at `https://www.electronicsforu.com/electronics-projects/software-projects-ideas/artificial-neural-network-simulation-labview-matlab` provides one approach to creating such a simulation.

Getting Involved in Science

Science is a pretty broad term, but it does have specific applications. MATLAB is likely to be used to explore new theories. It's important to differentiate between science and engineering in this case. *Engineering* is the application of known principles and theories to a problem in a predictable and usually standardized manner. *Science*, on the other hand, is the act of creating, testing, and proving principles and theories to eventually use for solving problems. In other words, when applied to science, MATLAB helps you perform "what if" analysis that helps you confirm (or deny) the viability of a theory (see `https://docs.genesys.com/Documentation/DEC/9.0.0/UsrHlp/WhatIf` for examples of how what if analysis is used in specific tasks).

Of course, science is used in many different ways. For example, you might be involved in the health industry and use science to find a cure for cancer or a deadly virus. A computer scientist might look for a new way to use computer technology to aid those with accessibility needs. In fact, there are all sorts of ways in which MATLAB could figure into helping someone do something special for humanity.

Logging Sensor Data

Sensors provide the means to monitor events and data of all sorts at a level that no human can match. For example, cameras diligently record activities around a business for security reasons at a remarkable level of reliability. Industrial cameras can record emissions that humans can't even see. Other sensors record temperature, vibration, sound, and other forms of input that might be impossible to document otherwise.

Unfortunately, the best a sensor can do is to provide input. To use the data, a human or application must receive the input and interpret it, which is where the concept of logging comes into play. Sometimes the data

- Isn't needed in real time
- Requires preservation as a legal or other requirement
- Appears as raw data that requires analysis

TIP

You can see an example of how to log and interpret raw sensor data using a combination of MATLAB and Excel at `https://www.electronicsforu.com/electronics-projects/software-projects-ideas/logging-sensor-data-ms-excel-matlab-gui`.

Exploring Research

Researchers have the world's best job in many respects. As a researcher, you get to ask a question, no matter how absurd, and determine whether the question is both answerable and relevant. After the question is answered, a researcher needs to determine whether the answer is both useful and reliable. In short, some view research as a kind of play (and they are correct — it really is play for the creative and intelligent mind).

Of course, research isn't just fun and games. If it were, people would have flocked to research as they do to video games now. After a question is asked and an answer is given, the researcher must convince colleagues that the answer is correct and then viable to put into practice. MATLAB lets you check the answer and verify that it does, in fact, work as the researcher suggests. After an answer is proven, the researcher can use MATLAB further to define precisely how the answer is used.

REMEMBER

Although many of MATLAB's tasks require light use of graphics, research has a significant need for graphics because the researcher must often explain answers to people who don't have the researcher's skills. In most cases, the explanation will never work with a text-only presentation; the researcher must also include plenty of graphics that start with abstract concepts, and then turn them into concrete ideas that the audience can understand.

Creating Light Animations Using Arduino

Light animations are often used for advertising and presentation purposes. There is strong evidence showing that the use of light influences both decision making and perception (see `https://www.researchgate.net/publication/330258298_Ambient_Lights_Influence_Perception_and_Decision-Making` as an example). The YouTube video at `https://www.youtube.com/watch?v=QvrtneelQ8A` demonstrates a software-based animation — the animations you create using hardware are far more dramatic.

You can use MATLAB to design various patterns and control the output using Live Script (see Chapter 11). The benefit of using Live Script for this purpose is that you can create a GUI so that the presenter can perform other tasks on the PC while the audience views the various light patterns. You can see an example of this sort of MATLAB use at `https://www.electronicsforu.com/electronics-projects/hardware-diy/light-animations-arduino-matlab`.

Employing Image Processing

Images are made up of pixels. Each *pixel* defines a particular color in a specific location in the image. In short, a pixel is a dot of just one color. The color is actually a numeric value that defines how much red, blue, and green to use in order to create the pixel color. Because the pixel is represented as a number, you can use various math techniques to manage the pixel. In fact, images (see the "Working with Images" section of Chapter 16) as well as plots (see Chapters 6 and 7) are often managed as matrices. A matrix is simply a structure consisting of numeric information.

Image processing is the act of managing the pixels in an image using math techniques to modify the matrix values. Techniques such as adding two matrices together are common when performing image processing. In fact, probably any technique you can think of that applies to matrices is also employed in image processing in some way.

The one thing you should know by this point in the book is that MATLAB excels at matrix manipulation. Anyone involved in image processing needs the sort of help that MATLAB provides to create and test new image-processing techniques. The point is that you can test the math and then see the result right onscreen without changing applications. You can use MATLAB to both create the required math and then test it (at least in a simulated environment).

Controlling Industrial Equipment

The ability of MATLAB to use Live Scripts and Live Functions (see Chapters 11 and 12) to graphically model scenarios enables you to use it to create apps (see Chapter 14) that less skilled individuals can use to control equipment from a remote location, such as a control booth. By placing the individual in a control booth, it's easier to provide a full view of the equipment in a safe environment so that the individual can react when something unexpected happens. The article at `https://www.electronicsforu.com/electronics-projects/equipment-controller-using-matlab-based-gui` provides an overview of how this process might work with four devices.

Depending on the industry and the level of machine control required, many industrial situations see use of machine learning applications to perform essential machine control (you can download an e-book describing how machine learning works in MATLAB from `https://www.mathworks.com/campaigns/offers/mastering-machine-learning-with-matlab.html`). However, the idea that a combination of machine learning and advanced control hardware can do the entire job is ill founded. Unexpected situations still require human intervention because machine learning can't deal with the unexpected (see `https://www.lanner-america.com/blog/5-industries-artificial-intelligence-machine-learning-transforming/` and `https://usmsystems.com/machine-learning-in-manufacturing/` for an overview of the state of the art in five industries). In fact, machine learning is unlikely to ever be able to take over completely. However, the need for human interaction doesn't make today's robots any less interesting to learn about (see the article at https://www.educba.com/what-is-robotics/ for an overview).

The Internet of Things (IoT) presents another industrial-type application with consumer applications. For example, it doesn't matter to a MATLAB app-controlled device whether it configures lighting in a factory setting or in a home. To the app, it's just lights (or a thermostat or other device). You can read more about IoT and MATLAB at https://www.educba.com/iot-applications/. The MathWorks page at https://www.mathworks.com/solutions/internet-of-things.html provides additional implementation details.

Performing Audio Compression Using Wavelets

Working with sound can be difficult because not everyone hears sound in the same way. That's why it's important to have some means of measuring and defining sound in a specific manner. Fortunately, the MATLAB Signal Processing Toolbox (https://www.mathworks.com/products/signal.html) makes it easy to create Live Script apps that fully analyze sound in the manner you require. The overview at https://www.mathworks.com/help/signal/getting-started-with-signal-processing-toolbox.html provides you with an impressive list of ways to interact with signals of all sorts.

Part of the problem in working with sound, especially on the Internet, is that high-quality sound can consume a great deal of memory, so streaming it becomes nearly impossible. A way around this problem is to compress the sound wave at the source and then decompress the sound wave at the destination. MATLAB can help in this regard as well. You can see an example of this approach at https://www.electronicsforu.com/electronics-projects/audio-compression-haar-wavelet-matlab.

IN THIS CHAPTER

» Performing various kinds of research

» Developing industrial and consumer solutions

» Keeping things safe

Chapter 22

Ten Ways to Make a Living Using MATLAB

MATLAB is an excellent tool that performs a great many tasks. However, you might perceive it as just a tool, not a means by which to earn a living. Be ready to be amazed! Adding MATLAB proficiency to your résumé may be the thing that gets your foot in the door for that bigger, better job you've been wanting. In other words, MATLAB *is* the job, rather than a tool you use to perform the job, in at least some cases. Employers have learned to ask for MATLAB by name.

REMEMBER

Monster.com (`https://jobsearch.monster.com/search/?q=matlab`) clearly shows that proficiency in using MATLAB is the main educational requirement for getting some jobs. At the time of this writing, Monster.com had 5,106 postings specifically stating a need for MATLAB proficiency. Another good place to look is Indeed (`https://www.indeed.com/q-MATLAB-jobs.html`), which had 10,322 job listings that require some level of MATLAB experience. In fact, you can find online job recruiters, such as ZipRecruiter (`https://www.ziprecruiter.com/`), offering MATLAB-related work. You could also work at MathWorks (`https://www.mathworks.com/company/jobs/opportunities.html`). The following sections describe some of the more interesting ways (and just the tip of the iceberg, at that) to get a job with your newfound MATLAB skills.

Working with Green Technology

At an earlier point, energy came from one source — a power company. The grid could be simple because the energy went directly from one producer to multiple homes and businesses. Yes, there were interconnected grids, but the connections were relatively simple.

REMEMBER

Today the energy picture is far different because many homes and businesses now produce energy in addition to using it. A solar panel on a home might produce enough energy to meet the home's needs and the needs of a neighboring home, which means that the grid has to be able to deal with a surplus from a home at times. In addition, the power company sources may now include wind farms or other green-technology energy sources in diverse locations. The number of green-energy sources is constantly growing, and you can see some of the more beneficial sources at `https://www.epa.gov/greenpower/what-green-power`.

The old grid system doesn't have the intelligence to manage such diverse power sources and sinks (the article at `https://instrumentationtools.com/difference-between-the-sinking-and-sourcing/` provides several examples of sourcing and sinking), so new smart grids are used to replace the old system and make the new system more robust and flexible. Developing and implementing smart grids is a lot more difficult than working with the grids of old, and MATLAB can help you perform the math required to make these new grids happen. You can read about this exciting job at `https://www.modelit.nl/index.php/matlab-webserver-user-story`. (See `https://www.mathworks.com/matlabcentral/answers/561482-smart-grid-project-energy-management-system` to learn about a graduate student's research project.)

Creating Speech Recognition Software

The need for computers to recognize what humans are saying is increasing as humans rely more heavily on computers for help. Although products like Siri (`https://www.apple.com/ios/siri/`) are helpful, the uses for speech recognition go well beyond the basic need of turning human speech into something a computer can understand.

Consider the fact that robots are becoming part of daily life. In fact, robots may eventually provide the means for people to stay in their homes at a time when they'd normally move to a nursing home (see `https://waypointrobotics.com/blog/elder-care-robots/`, `https://www.robotics.org/blog-article.cfm/The-Future-of-Elder-Care-is-Service-Robots/262`, and `https://time.com/`

longform/senior-care-robot/ as examples). These robots must recognize everything a person is saying and react to it. That's why ongoing research is important (see https://emerj.com/ai-sector-overviews/ai-for-speech-recognition/ and https://ieeexplore.ieee.org/document/8632885 for details). There are also incredible hurdles to overcome, such as perceived racism (see https://analyticsindiamag.com/new-research-suggests-speech-recognition-technology-may-be-racist/ for details).

Expect to find a lot of jobs that work in the areas of speech in the future. Most of these jobs will require the ability to use math software, such as MATLAB, to speed up the process. Efficient use of math to help solve speech recognition problems will become more essential as the market for various types of computer-assisted technology increases.

Performing Antenna Analysis and Design

Antennas appear everywhere, and often you don't even realize that an antenna is present. An antenna doesn't necessarily appear on a tall mast and present an imposing figure. In fact, your contactless credit card (and other Near Field Communication [NFC] devices) has an antenna in it that would be tough to see even if you removed the plastic (see https://wallethub.com/answers/cc/inside-a-credit-card-2140719924/). The MATLAB Antenna Toolbox (https://www.mathworks.com/products/antenna.html) helps you perform a wide assortment tasks to make antennas work. The documentation at https://www.mathworks.com/help/antenna/gs/antenna-modeling-and-analysis.html and https://www.mathworks.com/help/antenna/ref/antennadesigner-app.html supplies details about the toolbox. No matter how large or small, you need good math skills to design antennas that work well.

TIP

Any time a technology becomes wireless, you can be sure that an antenna is involved. You might not see the antenna, but it's there somewhere. So, it's no surprise that a recruiter like Glassdoor (https://www.glassdoor.com/Job/antenna-design-engineer-jobs-SRCH_KO0,23.htm) would have 1,469 antenna-design engineer listings as of this writing. Lots of jobs are available now, and the need will only increase as more technologies become wireless.

Getting Disease under Control

Even though eliminating every disease may not be possible, MATLAB has an important role to play in the never-ending quest for the cure and management of disease. For example, the Centers for Disease Control and Prevention (CDC) uses MATLAB for poliovirus sequencing and tracking. (See the story at `https://www.mathworks.com/company/user_stories/centers-for-disease-control-and-prevention-automates-poliovirus-sequencing-and-tracking.html`.) Of course, the poliovirus has been around for a long time. MATLAB also gets used in Covid-19 research (see `https://www.mathworks.com/solutions/covid-19-research-and-development.html`) in many interesting areas, such as studying the effects of Covid-19 on the retina (see `https://www.thelancet.com/journals/eclinm/article/PIIS2589-5370(20)30294-7/fulltext` for details).

REMEMBER

The clock is ticking, and health organizations need every second they can get to chase down and kill off virulent diseases. Lest you think that modern medicine has been hugely successful in eliminating such diseases, think again. Only two serious pathogens have been eradicated to date: smallpox and rinderpest. (See the story at `https://asm.org/Articles/2020/March/Disease-Eradication-What-Does-It-Take-to-Wipe-out`.) We have a lot of work to do. Using products such as MATLAB makes researchers more efficient, creating the possibility of eradicating a virus or bacteria sooner — but only if the researcher actually knows how to use the software.

Becoming a Computer Chip Designer

Knowing how computers work at a detailed level — all the way down into the chip — opens an entirely new world of math calculations. A truly in-depth knowledge of computer chip technology involves not just electronics or chemistry, but a mix of both, with other technologies added in for good measure. (Strong math skills are a requirement because you can't actually see the interactions take place; you must know that they will occur based on the math involved.) The people who design chips today enter an alternative reality, because things really don't work the way you think they will when you're working at the level of individual atoms. Yet, jobs exist for designing a System on a Chip (SoC) or Application-Specific Integrated Circuit (ASIC). And given the advances in chip technology, you can count on lots of opportunities in this area.

Working with Robots

Many movies and television shows contain examples of robots interacting with humans in both the home and business environment. We're slowly getting there. For example, many people have a Roomba (see https://www.irobot.com/roomba for details) in their homes today to do the mundane task of vacuuming, although, help is sometimes required to make it happen. Oddly enough, you can use MATLAB to program a special programmable Roomba (see https://www.usna.edu/Users/weaprcon/esposito/roomba-matlab.php for the toolbox description). A Roomba is nothing like having the humanoid robots shown in *I, Robot* (https://www.amazon.com/exec/obidos/ASIN/B00005JN0T/datacservip0f-20/), but it's a practical example of what is available today.

Whether you're working with a reprogrammable Roomba or not, MathWorks has tools you can use to interact with robots (see https://www.mathworks.com/discovery/robot-programming.html, https://www.mathworks.com/help/robotics/ug/build-a-robot-step-by-step.html, and https://www.mathworks.com/help/robotics/getting-started-with-robotics-system-toolbox.html). These tools help you create robots for all sorts of environments, such as the industrial applications described at https://emerj.com/ai-sector-overviews/robots-in-retail-examples/. Because the use of robots is increasing dramatically (see the report at https://ifr.org/ifr-press-releases/news/record-2.7-million-robots-work-in-factories-around-the-globe for industrial use increases alone), you can be sure that plenty of jobs exist in this area.

Interestingly enough, the Boston Dynamics dancing robots described in the *Anderson Cooper Gets a Behind-the-Scenes Look at Dancing Robots on '60 Minutes'* article at https://www.yahoo.com/entertainment/anderson-cooper-gets-behind-scenes-072111140.html rely on MATLAB. You can learn more about this use in the "The First DRC Event: Guiding a Simulated Robot in a Virtual Environment" section of the *Designing a Nonlinear Feedback Controller for the DARPA Robotics Challenge* article at https://www.mathworks.com/company/newsletters/articles/designing-a-nonlinear-feedback-controller-for-the-darpa-robotics-challenge.html.

Keeping the Trucks Rolling

Designing better trucks may not seem like something you can do with MATLAB, but a modern truck consists of complicated machinery that has to operate safely on increasingly crowded roads. For example, just designing the air suspension systems used to couple the truck to the trailer and ensure that the two remain attached is a difficult task. You can read about this particular need at `https://www.mathworks.com/company/user_stories/continental-develops-electronically-controlled-air-suspension-for-heavy-duty-trucks.html`. The point is that designing trucks is a complex task.

REMEMBER

You can easily extend this particular need to other sorts of vehicles. The amount of engineering to design a modern car is daunting. Think about everything that a car needs to do now — everything from braking so that the car doesn't slide to ensuring that the people in the vehicle remain safe during an accident. Modern vehicles do all sorts of nonvehicle things, such as entertain the kids in the back seat so that you can drive in peace. All these capabilities require engineering that is better done using math software. Also consider self-driving cars. MATLAB can help you design a self-driving car using the Automated Driving Toolbox (`https://www.mathworks.com/products/automated-driving.html`).

Designing Equipment Used in the Field

Even though the story at `https://www.mathworks.com/company/user_stories/electrodynamics-associates-designs-high-performance-generator-controller-for-the-military.html` tells about a generator designed for use by the military, the story speaks to a much larger need. Every outdoor activity, such as a construction project, requires equipment of various sorts to make the task feasible. Generating power is a huge need. Trying to build something would be nearly impossible without the electricity required to run various kinds of tools.

Generators, tools, appliances, and all sorts of other devices require special design today. Not only does the equipment need to work well, but it has to do so at a low energy cost, reduced use of materials, nearly invisible maintenance costs, and with a small environmental impact. Trying to design such equipment without the proper math software would, again, be nearly impossible.

Reducing Risks Using Simulation

Many endeavors that businesses or people need to undertake entail the risk of failure. At issue isn't whether you have enough money to throw at any problems that may arise; rather, you can't know whether the project has a chance to succeed at all. Using simulation can greatly increase the likelihood of success, however. For example, consider the salvaging of the Russian submarine Kursk. (See the story at https://www.mathworks.com/company/user_stories/international-salvage-team-brings-home-the-kursk-submarine-using-a-simulation-developed-in-simulink.html for details.) It wouldn't be possible to know whether the salvage could ever succeed without simulating it first. Simulations require both math skills and plotting, both of which are found in MATLAB.

However, just knowing that the project can succeed isn't enough. Using a simulation lets you identify potential risks at the outset, before the project is under way. Risk identification and management are important parts of many endeavors today. Beyond having procedures in place that guarantee success, you need procedures for those times when things do go wrong. Creating a great simulation helps an engineer with the required knowledge to figure out the risky situations in advance and do everything needed to avoid them, and then also create procedures for when things go wrong anyway.

Creating Security Solutions

Feeling safe is an essential human need. Consequently, you find security solutions for a huge number of needs — everything from protecting a home to protecting a website. Security statistic sites such as Statista (https://www.statista.com/topics/2188/security-services-industry-in-the-us/) boggle the mind with the vast range of security solutions they list. It doesn't take long to figure out that there are a lot of security jobs available. Variety isn't a problem, either. Most MATLAB direct-security solutions involve website data (see https://www.mathworks.com/help/mps/security.html and https://www.mathworks.com/products/polyspace/application-security.html for details), which makes sense given that you use MATLAB to interact with data. However, when you consider how the various MATLAB toolboxes are designed, you could also create secure solutions for just about any industry today, such as the home security system at https://www.mathworks.com/help//stateflow/ug/modeling-a-security-system.html.

Appendix A

MATLAB Functions

This appendix provides you with an overview of the MATLAB functions. It would be impossible to provide a complete list of every function here, but these functions see common use for solving math problems. To see an exhaustive list of MATLAB functions, check out `https://www.mathworks.com/help/matlab/referencelist.html?type=function`.

Each function helps you perform a specific task within MATLAB. Tables A-1 through A-21 contain function names and a short description of the task that each function performs. If you need more information than this appendix gives you, you can obtain additional information by typing **help** *<function_name>* in the MATLAB command window and pressing Enter. Of course, you also have the information found in this book, and you can search in Help. See the "Getting additional help with MATLAB" section of Chapter 2 for some useful information on using the Help resources that MATLAB provides.

TABLE A-1 Arithmetic Functions

Function	Description
`ceil`	Rounds toward positive infinity
`diff`	Differences and approximate derivatives
`fix`	Rounds toward zero
`floor`	Rounds toward negative infinity
`idivide`	Integer division with rounding option
`ldivide`	Left-array division — comparable to .\
`minus`	Minus — comparable to - subtracting two objects
`mldivide`	Solves systems of linear equations Ax = B for x — comparable to \
mod	Modulus after division
mpower	Matrix power — comparable to ^
`mrdivide`	Solves systems of linear equations xA = B for x — comparable to /

(continued)

TABLE A-1 *(continued)*

Function	Description
`mtimes`	Matrix multiplication — comparable to *
`plus`	Plus — comparable to + adding two objects
`power`	Array power — comparable to .^
`prod`	Product of array elements
`rdivide`	Right-array division — comparable to ./
`rem`	Remainder after division
`round`	Rounds to nearest integer
`sum`	Sum of array elements
`times`	Array multiply — comparable to .*
`uminus`	Unary minus — comparable to – acting on one object
`uplus`	Unary plus — comparable to + acting on one object

TABLE A-2 **Trigonometric Functions**

Function	Description
`acos`	Inverse cosine; result in radians
`acosd`	Inverse cosine; result in degrees
`acosh`	Inverse hyperbolic cosine
`acot`	Inverse cotangent; result in radians
`acotd`	Inverse cotangent; result in degrees
`acoth`	Inverse hyperbolic cotangent
`acsc`	Inverse cosecant; result in radians
`acscd`	Inverse cosecant; result in degrees
`acsch`	Inverse hyperbolic cosecant
`asec`	Inverse secant; result in radians
`asecd`	Inverse secant; result in degrees
`asech`	Inverse hyperbolic secant

Function	Description
asin	Inverse sine; result in radians
asind	Inverse sine; result in degrees
asinh	Inverse hyperbolic sine
atan	Inverse tangent; result in radians
atan2	Four-quadrant inverse tangent in radians
atan2d	Four-quadrant inverse tangent; result in degrees
atand	Inverse tangent; result in degrees
atanh	Inverse hyperbolic tangent
cos	Cosine of argument in radians
cosd	Cosine of argument in degrees
cosh	Hyperbolic cosine
cot	Cotangent of argument in radians
cotd	Cotangent of argument in degrees
coth	Hyperbolic cotangent
csc	Cosecant of argument in radians
cscd	Cosecant of argument in degrees
csch	Hyperbolic cosecant
hypot	Square root of sum of squares
sec	Secant of argument in radians
secd	Secant of argument in degrees
sech	Hyperbolic secant
sin	Sine of argument in radians
sind	Sine of argument in degrees
sinh	Hyperbolic sine of argument in radians
tan	Tangent of argument in radians
tand	Tangent of argument in degrees
tanh	Hyperbolic tangent

TABLE A-3

Exponentials, Logarithms, Powers, and Roots

Function	Description
exp	Exponential
expm1	Computes exp(x)–1 accurately for small values of x
log	Natural logarithm
log10	Common (base 10) logarithm
log1p	Computes log(1+x) accurately for small values of x
log2	Determines base 2 logarithm and dissects floating-point numbers into exponent and mantissa
nextpow2	Computes the exponent of the next higher power of 2
nthroot	Real nth root of real numbers
pow2	Base 2 power and scale floating-point numbers
reallog	Natural logarithm for nonnegative real arrays
realpow	Array power for real-only output
realsqrt	Square root for nonnegative real arrays
sqrt	Square root

TABLE A-4

Complex Number Functions

Function	Description
abs	Absolute value and complex magnitude
angle	Phase angle
complex	Construct complex data from real and imaginary components
conj	Complex conjugate
cplxpair	Sorts complex numbers into complex conjugate pairs
i	Imaginary unit
imag	Imaginary part of complex number
isreal	Checks whether input is a real array
j	Imaginary unit
real	Real part of complex number

Function	Description
`sign`	The `sign()` (or `signum`) function returns `1` if the corresponding element is greater than 0, `0` if the corresponding element is zero, and `-1` if the corresponding element is less than 0
`unwrap`	Shifts the phase angles of vector P

TABLE A-5

Discrete Math Functions

Function	Description
`factor`	Prime factors
`factorial`	Factorial function
`gcd`	Greatest common divisor
`isprime`	Array elements that are prime numbers
`lcm`	Least common multiple
`matchpairs`	Solves a linear assignment problem
`nchoosek`	Binomial coefficient or all combinations
`perms`	All possible permutations
`primes`	Generates list of prime numbers
`rat`, `rats`	Rational fraction approximation

TABLE A-6

Polynomial Functions

Function	Description
`poly`	Polynomial with specified roots
`polyder`	Polynomial derivative
`polyeig`	Polynomial eigenvalue problem
`polyfit`	Polynomial curve fitting
`polyint`	Integrates the polynomial analytically
`polyval`, `polyvalm`	Polynomial evaluation
`roots`	Polynomial roots

TABLE A-7

Special Functions

Function	Description
erf	Error function
erfc	Complementary error function
erfcinv	Inverse complementary error function
erfcx	Scaled complementary error function
erfinv	Inverse error function

TABLE A-8

Cartesian, Polar, and Spherical Functions

Function	Description
cart2pol	Transforms Cartesian coordinates to polar or cylindrical coordinates
cart2sph	Transforms Cartesian coordinates to spherical coordinates
pol2cart	Transforms polar or cylindrical coordinates to Cartesian coordinates
sph2cart	Transforms spherical coordinates to Cartesian coordinates

TABLE A-9

Constants and Test Matrix Functions

Function	Description
compan	Returns the corresponding companion matrix
eps	Floating-point relative accuracy
i,j	Imaginary units
Inf	Infinity
pi	Ratio of circle's circumference to its diameter
NaN	Not-a-Number
isfinite	Array elements that are finite
isinf	Array elements that are infinite
isnan	Array elements that are NaN
gallery	Test matrices
magic	Magic square

TABLE A-10 **Matrix Operation Functions**

Function	Description
`cross`	Vectors cross product
`ctranpose`	Performs complex conjugate transpose
`dot`	Vectors dot product
`funm`	Evaluates a general matrix function
`kron`	Kronecker tensor product
`transpose`	Transposes

TABLE A-11 **Linear Equation Functions**

Function	Description
`inv`	Matrix inverse
`linsolve`	Solves linear system of equations

TABLE A-12 **Eigenvalue Functions**

Function	Description
`eig`	Eigenvalues and eigenvectors
`eigs`	Largest eigenvalues and eigenvectors of matrix

TABLE A-13 **Matrix Analysis Functions**

Function	Description
`det`	Matrix determinant
`norm`	Vector and matrix norms
`rank`	Rank of matrix
`rref`	Reduced row echelon form
`trace`	Sum of diagonal elements

TABLE A-14 Matrix Functions

Function	Description
arrayfun	Applies function to each element of array
expm	Matrix exponential
logm	Matrix logarithm
sqrtm	Matrix square root

TABLE A-15 Statistical Functions

Function	Description
bounds	Obtains the smallest and largest elements of an array
corrcoef	Correlation coefficients
cov	Covariance matrix
max, maxk	Largest elements in array
mean	Average or mean value of array
median	Median value of array
min, mink	Smallest elements in array
mode	Most frequent values in array
std	Standard deviation
var	Variance

TABLE A-16 Random Number Generator

Function	Description
rng	Controls random number generation
rand	Uniformly distributed pseudo-random numbers
randn	Normally distributed pseudo-random numbers
randi	Uniformly distributed pseudo-random integers

TABLE A-17

Interpolation

Function	Description
interp1	1-D data interpolation (table lookup)
interp2	2-D data interpolation (table lookup)
interp3	3-D data interpolation (table lookup)
interpn	*n*D data interpolation (table lookup)
griddedInterpolant	Interpolant for gridded data
griddata	Interpolates scattered data
ndgrid	Rectangular grid in *n*D space
meshgrid	Rectangular grid in 2D and 3D space
spline	Cubic spline data interpolation

TABLE A-18

Optimization Functions

Function	Description
fminbnd	Finds minimum of single-variable function on fixed intervals
fminsearch	Finds minimum of unconstrained multivariable function using derivative-free method
fzero	Finds root of continuous function of one variable
lsqnonneg	Solves a non-negative linear least-squares problem

TABLE A-19

Ordinary Differential Equation Functions

Function	Description
ode23, ode45, ode113	Solve nonstiff differential equations
ode15s, od23s, ode23t, od23tb	Solve stiff differential equations
ode15i, decic	Solve fully implicit differential equations

TABLE A-20

Sparse Matrix Manipulation Functions

Function	Description
spy	Visualizes a sparsity pattern
find	Finds indices and values of nonzero elements

TABLE A-21

Elementary Polygons

Function	Description
polyarea	Area of polygon
inpolygon	Determines whether points are inside or on the edges of a polygon
rectint	Rectangle intersection area

Appendix B

MATLAB's Plotting Routines

The tables in this appendix, B-1 through B-5, list each of the plotting routines in MATLAB with a brief description and an example. To save space, some entries show a single encompassing example for multiple commands. An example may also use an ellipsis (. . .) to show the continuation of a pattern. To assist you, the code includes several functions to generate test matrices, for example, `rand()`, `magic()`, `peaks()`, `cylinder()`, `ellipsoid()`, and `sphere()`. It's also important to note that MATLAB is always changing and that some functions, like `ezpolar()`, are eventually going to go away. After all, many of the other ez* functions have retired and are now sipping beverages on a sandy beach on a tropical island.

TABLE B-1 **Basic Plotting Routines**

Routine	Description	Example

`comet`	Just like `plot`, but `comet` animates the trajectory; it helps to have a larger vector to slow down the comet trace a tad	`x=[0:2*pi/100:2*pi];`
		`y=exp(-0.4*x).*sin(x);`
		`plot(x,y); figure(2)`
		`comet(x,y); figure(3)`
		`ribbon(x,y);figure(4)`
		`y2=100*exp(-0.4*x).*cos(x);`
		`plotyy(x,y,x,y2)`

`plot`	Plots data passed in by vectors	
`plotyy`	Plots data where y values may differ greatly — it makes two y-axes	
`ribbon`	Like `plot`, but it displays the data as 3D ribbons	
`fplot`, `fimplicit`	Plots the expression or function	`fplot(@(x) sin(x),[-2*pi 2*pi])`

`loglog`	Plots data on a log scale on both x- and y-axes — y proportional to a power of x is straight on this plot	`x=[0:2*pi/100:2*pi];`
		`y=10*x.^pi;`
		`loglog(x,y)`

(continued)

TABLE B-1 *(continued)*

Routine	Description	Example

`semilogx`	X-axis data is on a log scale and the y-axis is on a linear scale — y linearly related to `log(x)` is straight on this plot	`x=[0:2*pi/100:2*pi];`
		`y=10*log(x)+pi;`
		`semilogx(x,y);`

`semilogy`	Y-axis data is on a log scale and x-axis linear — y proportional to exponential of `x` is straight on this plot	`x=[0:2*pi/100:2*pi];`
		`y=10*exp(pi*x);`
		`semilogy(x,y);`

TABLE B-2 **Beyond Basics**

Routine	Description	Example

`area`	Acts just like `plot` except for the fact that it fills in the area for you	`x=[0:2*pi/100:2*pi];`
		`y=exp(-0.4*x).*sin(x);`
		`area(x,y)`

`bar`	Creates a standard bar chart that can handle both grouping and stacking	`x=[8,7,6;13,21,15;32,27,32];`
		`bar(x); figure(2)`
		`y=sum(x,2);`
		`bar(y); figure(3)`
		`bar(x,'stacked'); figure(4)`
		`barh(x); figure(5)`
		`bar3(x); figure(6)`
		`bar3h(x)`

`barh`	Just like `bar` except plot is horizontal	
`bar3`	Adds a little 3D pizazz to a standard bar chart	
`bar3h`	Creates a 3D horizontal bar chart	
`compass`	Like `polar`, but `compass` shows data as vectors from origin	`compass(rand(1,3)-0.5,...` `rand(1,3)-0.5)`
`ezpolar`	Creates a `polar` plot where the distance from origin vs. angle is plotted — argument is function expression	`ezpolar('cos(2*x)^2');`

`fill`	Fills polygons (vector inputs define vertices) with the specified color	`y=sin([0:2*pi/5:2*pi])`
		`x=cos([0:2*pi/5:2*pi])`
		`fill(x,y,'g')`

Routine	Description	Example

`pie`	Creates a standard pie chart	`x=[2,4,6,8];`
		`pie(x); figure(2)`
		`pie3(x)`

`pie3`	Adds some 3D pizazz to `pie`	

`polar`	Creates a polar plot like `ezpolar` but does so by accepting vector arguments — x value corresponds to angle and y value to distance from origin	`x=[0:2*pi/100:2*pi];`
		`y=(exp(-0.1*x).*sin(x)).^2;`
		`polar(x,y)`

TABLE B-3

Statistical Plotting Routines

Routine	Description	Example

`errorbar`	Like `scatter` but adds error bars	`x=[0:2*pi/100:2*pi];`
		`y=10*x+pi+10*randn(1,101);`
		`scatter(x,y); figure(2)`
		`stem(x,y); figure(3)`
		`errorbar(x,y,10*ones(1,101), ones(1,101))`
		`figure(4);`
		`hist(y); figure(5)`
		`histogram(y)`
		`histc(y,[-40:20:80])`
		`histcounts(y,5)`

`hist`, `histogram`	Creates a histogram — a bar chart showing the frequency of occurrence of data vs. value. These two routines handle the bins differently	
`scatter`	Plots `(x,y)` data points	
`stem`	Like `scatter`, but adds a line from the x axis to a data point	
`histc`	Related to `hist`, but rather than making a plot, it makes a vector of counts	`histc(y,[-40:20:80])`
`histcounts`	Like `histc` except it creates *n* bins	`histcounts(y,5)`

`stairs`	Like `scatter`, but makes stairsteps when y values change	`x=[0:2*pi/10:2*pi];`
		`y=10*x+pi+10*randn(1,11);`
		`stairs(x,y)`

`rose`	A cross between `polar` and `histogram`; it displays frequency vs. angle	`rose(randn(1,100),5)`
`pareto`	A bar chart arranged with the highest bars first	`histc(randn(1,100),[-4:1:4])` `pareto(ans)`

(continued)

TABLE B-3 *(continued)*

Routine	Description	Example

spy	A scatter plot of zeros in a matrix	`mymat=rand(5);`
		`mymat=(mymat>0.5).*mymat;`
		`spy(mymat)`

plotmatrix	A scatter plot of all permutations of columns of x and y	`plotmatrix(magic(3),magic(3))`

TABLE B-4 3D Graphics

Routine	Description	Example

contour	Creates a contour plot with matrix arguments	`x=[-2*pi:4*pi/100:2*pi];`
		`y=[-2*pi:4*pi/100:2*pi];`
		`z=cos(x)'*cos(y);`
		`contour(x,y,z);title ('contour');figure(2)`
		`contourf(x,y,z);title ('contourf');figure(3)`
		`contour3(x,y,z);title ('contour3');figure(4)`
		`surf(x,y,z);title ('surf');figure(5)`
		`surface(x,y,z);title ('surface');figure(6)`
		`mesh(x,y,z);title ('mesh');figure(7)`
		`waterfall(x,y,z);title ('waterfall');figure(8)`
		`surfc(x,y,z);title ('surfc');figure(9)`
		`meshc(x,y,z);title ('meshc');figure(10)`
		`meshz(x,y,z);title ('meshz');figure(11)`
		`surfl(x,y,z);title ('surfl');figure(12)`
		`pcolor(z);title('surfl')`

contourf	Same as `contour`, except `contourf` fills in the contours	
contour3	Same as `contour`, except `contour3` provides a 3D perspective	
mesh	Creates a wireframe mesh surface	
meshc	Wireframe mesh with contour	
meshz	Wireframe mesh with a curtain around the plot	
pcolor	Shows values of the matrix as colors	
surfc	Same as `surf`, but with contour plot added	
surf, surface	Creates a filled surface	
surfl	Same as `surf`, except `surfl` simulates light and shadow	
waterfall	Like mesh, but all column lines are omitted	

Routine	Description	Example

`fcontour`	Plots the function as a contour	`x = (-5:.1:5);`
		`y = (-5:.1:5);`
		`f = @(x,y) sin(x) + cos(y);`
		`fcontour(f)`

`fill3`	Like `fill` except in 3D; note that the fill area may not be coplanar	`fill3([0,1,1,0],[0,0,1,0],[0,1,0,1],'g')`

`fmesh`	Plots the function as a mesh	`x = (-5:.1:5);`
		`y = (-5:.1:5);`
		`f = @(x,y) sin(x) + cos(y);`
		`fmesh(f)`

`fplot3`	Plots a series of three functions expressing x-, y-, and z-axes as a parametric curve	`t = (-5:.1:5);`
		`x=sin(t);`
		`y=cos(t);`
		`z=t;`
		`xt = @(t) sin(t);`
		`yt = @(t) cos(t);`
		`zt = @(t) t;`
		`fplot3(xt,yt,zt)`

`fsurf`	Plots the function as a 3D surface	`x = (-5:.1:5);`
		`y = (-5:.1:5);`
		`f = @(x,y) sin(x) + cos(y);`
		`fsurf(f)`

`plot3`	Like `ezplot3`, except it takes vector arguments	`a=[-2*pi:4*pi/100:2*pi];`
		`x=sin(a); y=cos(a);`
		`z=sin(3*a/2);`
		`plot3(x,y,z);`
		`figure(2)`
		`scatter3(x,y,z)`

`scatter3`	Like `plot3`, but shows individual points	

(continued)

TABLE B-4 *(continued)*

Routine	Description	Example
`stem3`	Stem plot of 3D data	`stem3(rand(5))`
`surfnorm`	Creates surface plot with normal vectors	`[x,y,z]=peaks; % Test function` `surfnorm(x,y,z)`

TABLE B-5 **Vector Fields**

Routine	Description	Example

`coneplot`	Like `quiver3`, but `coneplot` shows velocity as cones	`[x,y,z]=meshgrid([-5:2:3],[-3:2:3],[-3:2:3]);`
		`r=sqrt(x.^2+y.^2+z.^2);`
		`u=ones(4,5,4)+(10./r.^2).*cos(atan2(y,x)).*sin(acos(z./r));`
		`v=(10./r.^2).*sin(atan2(y,x)).*sin(acos(z./r));`
		`w=10.*z./r; quiver3(x,y,z,u,v,w);hold on;`
		`streamribbon(x,y,z,u,v,w,-5,0,.1);figure(2)`
		`coneplot(x,y,z,u,v,w,x,y,z); figure(3)`
		`quiver3(x,y,z,u,v,w);hold on;`
		`streamtube(x,y,z,u,v,w,-5,0,.1);`

`quiver3`	Like `quiver` for 3D; the example adds a uniform velocity field to a `1/r2` velocity field	
`streamribbon`	Like `streamline`, but shows ribbons	
`streamtube`	Like streamline, but plots tubes (cylindrical 3D flow lines)	
`feather`	Similar to compass, except it steps once for each element in `x` and y	`feather(rand(1,3)-0.5, rand(1,3)-0.5)`

`quiver`	Works like `feather`, except `quiver` plots vectors in an x-y plane	`[x,y]=meshgrid([-5:5],[-5:5]);`
		`u=ones(11)+(4./(sqrt(x.^2+y.^2)).*cos(atan2(y,x)));`
		`v=(4./(sqrt(x.^2+y.^2)).*sin(atan2(y,x)));`
		`v(6,6)=0; u(6,6)=0;`
		`quiver(x,y,u,v); hold on`
		`streamline(x,y,u,v,[-5,-5,-5],[-1,.01,1]); hold off`

`streamline`	Plots line from the vector field; the example plots 2D `streamline` on the same plot as `quiver`	

Index

Symbols and Numerics

A

B

D

E

F

G

H

I

J

K

L

M

N

O

P

Q

R

S

T

U

V

W

X

Y

Z

About the Authors

John Mueller is a freelance author and technical editor. He has writing in his blood, having produced 118 books and more than 600 articles to date. The topics range from networking to artificial intelligence and from database management to heads-down programming. Some of his current books include discussions of data science, machine learning, and algorithms. His technical editing skills have helped more than 70 authors refine the content of their manuscripts. John has provided technical editing services to various magazines and performed various kinds of consulting. He also writes certification exams. Be sure to read John's blog at http://blog.johnmuellerbooks.com/. You can reach him on the Internet at John@JohnMuellerBooks.com. He also has a website at http://www.johnmuellerbooks.com/. Be sure to follow John on Amazon at https://www.amazon.com/John-Paul-Mueller/e/B004MOD0YS/.

Jim Sizemore is a physics and engineering professor earning an MS in physics from UC-San Diego and a PhD in Materials Science from Stanford University. He was employed many years in the semiconductor industry working on several projects, including diffusion and oxidation, radiation hardening, and optoelectronics. After his private sector career, he turned to teaching, and is currently a physics and engineering professor at Tyler Junior College in Tyler, Texas. There, he started a popular science club where students were able to design and build several projects, including a 2 meter trebuchet, just in case they needed to attack any castles in the area. He currently teaches programming for engineers, with MATLAB being the primary language taught. He has enthusiasm for teaching and learning, but also enjoys photography and bicycle riding in his spare time.

Dedication

John Paul Mueller: Dedicated to Amaya Funmaker on the occasion of her graduation. I wish you all the success you've earned through your diligent work in school and hope you have an amazing life!

Jim Sizemore: I wish to dedicate this book to many mentors including Bob Abel, Greg Sherman, and Gene Branum. Also my thoughts and dedication go to my brother, Bill, who is afflicted with Parkinson's disease, and to my son, Daniel, because the future belongs to his generation.

Authors' Acknowledgments

John Paul Mueller: Thanks to my wife, Rebecca. Even though she is gone now, her spirit is in every book I write, in every word that appears on the page. She believed in me when no one else would.

Rod Stephens has been a friend and colleague for many years now. His technical edit of this book was quite thorough and I greatly appreciate his efforts. Rod took the time to talk with me about book topics through email as I wrote about them. He offered insights on how to write better examples, more easily understood text, and supplied access to resources that might not otherwise appear here.

Matt Wagner, my agent, deserves credit for helping me get the contract in the first place and taking care of all the details that most authors don't really consider. I always appreciate his assistance. It's good to know that someone wants to help.

A number of people read all or part of this book to help me refine the approach, test application code, verify the extensive text, and generally provide input that all readers wish they could have. These unpaid volunteers helped in ways too numerous to mention here. I especially appreciate the efforts of Eva Beattie who provided general input, read the entire book, and selflessly devoted herself to this project.

Finally, I would like to thank Kelsey Baird, Susan Christophersen, and the rest of the editorial and production staff.

Jim Sizemore: Thanks to Sara McCaslin who freely shared in order to start our MATLAB programming course.

Thanks also go to Gene Branum, Doug Parsons, and other Tyler Junior College colleagues who supported me in order to devote time to this project.

Finally, I wish to thank my coauthor John Mueller, whose writing skill and experience were essential to taking my original vision and finely polishing it.

Publisher's Acknowledgments

Acquisitions Editor: Kelsey Baird

Project Manager and Copy Editor: Susan Christophersen

Technical Editor: Rod Stephens

Production Editor: Vivek Lakshmikanth

Cover Image: © Cover image by John Mueller recreated with permission of The Mathworks, Inc.

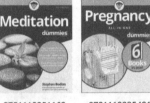

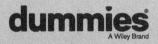

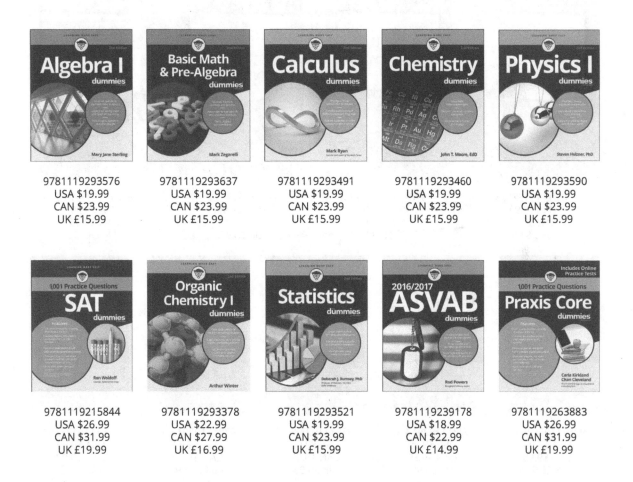

Small books for big imaginations

Unleash Their Creativity

dummies.com